Geomechanics of Failures. Advanced Topics

E.E. Alonso • N.M. Pinyol • A.M. Puzrin

Geomechanics of Failures. Advanced Topics

Springer

Eduardo E. Alonso
Department of Geotechnical Engineering
Civil Engineering School
Calle Gran Capitan s/n
Edificio D-2
08034 Barcelona
Spain
eduardo.alonso@upc.edu

Núria M. Pinyol
International Center for Numerical Methods in Engineering (CIMNE)
Campus Nord UPC
Edifici C-1
08034 Barcelona
Spain
nuria.pinyol@upc.edu

Alexander M. Puzrin
Institut für Geotechnik
ETH Zürich
Wolfgang-Pauli-Str. 15
8093 Zürich
Switzerland
alexander.puzrin@igt.baug.ethz.ch

ISBN 978-94-007-9227-2 ISBN 978-90-481-3538-7 (eBook)
DOI 10.1007/978-90-481-3538-7
Springer Dordrecht Heidelberg London New york

Softcover re-print of the Hardcover 1st edition 2010

Printed on acid-free paper

Springer is Part of Springer Science+Business Media (www.springer.com)

Table of Contents

Preface

Geotechnical failures, specially the catastrophic ones, are an excellent experience and a source of inspiration to improve our current understanding of phenomena and our procedures and tools for analysis and prediction.

This unconventional manner to learn Geomechanics is the essence of this book. In general, Soil Mechanics and Geotechnical text books describe first the concepts and the theoretical developments and then apply them to interpret or solve a particular application. This book has a different approach. The case (a failure) is first described and then an explanation is sought. This approach is developed through a set of steps which can be summarized as follows.

1. Identify the nature of the problem
2. Develop a dedicated and specific formulation of the case, based on established basic concepts. In general, no single existing theory or procedure is available to solve the case at hand.
3. Provide a solution within an acceptable degree of complexity.
4. Extract the fundamental aspects of the problem and highlight its relevance.

The cases selected have been grouped into three main topics: Landslides, Embankments and Dams and Dynamics of Failures. No attempt to provide a comprehensive account of known catastrophic failures has been done. But the cases selected (Vaiont, Aznalcóllar, Brattas-St. Moritz) are rather unique and illustrate a number of relevant and to some extent controversial issues which are of wide interest.

Finite element methods have not been used. In the landslides analysed (Vaiont and Brattas-St. Moritz) currently available commercial programs are of limited utility. In the remaining cases the analysis performed provides a sufficient insight and interpretation of the field behaviour.

The book teaches how to build the necessary models to understand the failures. Balance and equilibrium equations are formulated at different scales which are selected having in mind the abstract representation of the key concepts of each case. In some of the Chapters calculation tools, included in well known and widely available programs (Excel, Maple, etc.) have been used. Some details of the "ad hoc" programs developed have also been included in Appendices to help the readers to follow the details of the calculation.

Chapters include also a short description of the changes in the original design and the mitigation measures which could have prevented the failure. Also a summary section of lessons learned is provided. Finally, selected topics and more advanced reading are suggested.

The book is associated with a Master/Doctorate courses being offered at the Department of Geotechnical Engineering and Geosciences of UPC, Barcelona and at the Institute of Geotechnical Engineering, ETH Zurich. Potential readers for the book include graduate students, faculty and professionals in the fields of Civil and Geotechnical Engineering.

The authors acknowledge with thanks the contribution of some of their colleagues and fellow researchers to the Chapters of this book. Chapter 1 is based on the research performed in collaboration with Ivo Sterba (ETH), Andreas Schmid (ETH), Michael Iten (ETH), Markus Schwager (ETH) and Dr. Sophie Messerklinger (ETH). Professor Antonio Lloret (UPC) and Dr. Enrique Romero (UPC) helped in the interpretation and the experimental work reported in Chapter 2. Discussions held with Dr. Gabriel Fernández (University of Illinois) about the Vaiont slide were very helpful. He provided also first-hand information on the slide. Prof. Sebastià Olivella (UPC) contributed to the formulation of the thermo-hydro-mechanical phenomena invoked to explain the dynamics of the slide motion (Chapter 5). Jubert Pineda (UPC) performed the laboratory heating tests described in Chapter 5. Professor Antonio Gens (UPC) participated actively in the analysis of Aznalcóllar dam failure (Chapters 4 and 6). Dr. José Moya (UPC) investigated the geological aspects of Aznalcóllar and helped with the interpretation of field observations (Chapter 5). Laboratory tests on the foundation clay of Aznalcóllar dam were conducted by Professor Antonio Lloret, Dr. Enrique Romero, Dr. Roberto Rodríguez (University of Girona) and Joan Rius (UPC).

The authors express also their gratitude to Raúl Giménez (UPC) and María del Mar Obrador (UPC) for their dedication to the drafting of figures and the formatting of texts.

E.E. Alonso, N.M. Pinyol, A.M. Puzrin
Barcelona, Zürich. December 2009

PART I

LANDSLIDES

Chapter 1

A Constrained Creeping Landslide: Brattas-St. Moritz Landslide, Switzerland

TABLE OF CONTENTS

E.E. Alonso et al., *Geomechanics of Failures. Advanced Topics*,
DOI 10.1007/978-90-481-3538-7_1, © Springer Science+Business Media B.V. 2010

Chapter 1

A Constrained Creeping Landslide: Brattas, St. Moritz Landslide, Switzerland

1.1 Case Description

Landslides are one of the major geotechnical hazards affecting economy and life in the subjected areas. Though most of the landslides take place in rural areas, they cause significant damage to the infrastructure such as roads, railways, etc. Of major concern to the community are, however, the landslides active in urban areas. Among these, the Brattas Landslide of St. Moritz (Fig. 1.1a), with its most prominent landmark, the Leaning Tower (Fig. 1.1b), is of special interest, because:

- remarkably, the landslide stops in the middle of the town;
- in spite of the landslide, the town has enormous real estate prices;
- the town had to adopt special construction laws for the affected areas;
- there is an extensive displacement monitoring program in progress;
- there is an interesting experience related to the behaviour of existing structures;
- original engineering solutions have been put forward for the construction of new structures.

The Institute of Geotechnical Engineering at the ETH Zurich has been actively involved over the last 30 years in all geotechnical aspects of the problem, both in research and in providing expert service to the community.

1.1.1 Geometry, geology and displacements

The unstable northern slope above the village of St. Moritz (Fig. 1.1a) may be divided into two zones (Fig. 1.2a, after Müller and Messina, 1992). The upper zone is the Gianda Laret rockfall, which extends from the detachment zone at an altitude of 2400 m above sea level down to the local rock outcrop at an altitude of about 2100 m. The lower zone is the Brattas landslide composed of a thick soil mass, which is moving downhill but is blocked at its foot by the Via Maistra rock ridge, after which the movement stops (Fig. 1.2b). It stretches from an altitude of 2100 m down to 1800 m over a horizontal distance of 800 m. The landslide is 600 m wide and is bounded on both sides by parallel shear surfaces. The slope has an average inclination of 20°. The main sliding surface revealed in one boring reaches a depth of about 50 m. The sliding mass is built of soil layers of great heterogeneity, both in terms of stratigraphy and material properties.

In the portion where it is approached by the landslide, Via Maistra is getting narrower by about 0.5 cm per year. Uphill from Via Maistra, the displacement rate increases (Fig. 1.2b). The movement has only been measured in the developed areas, which is not sufficient to conclude if there is any interaction between the Gianda Laret rockfall and the Brattas landslide.

(a)

(b)

Figure 1.1 The St. Moritz Landslide: (a) an aerial photo; (b) the Leaning Tower.

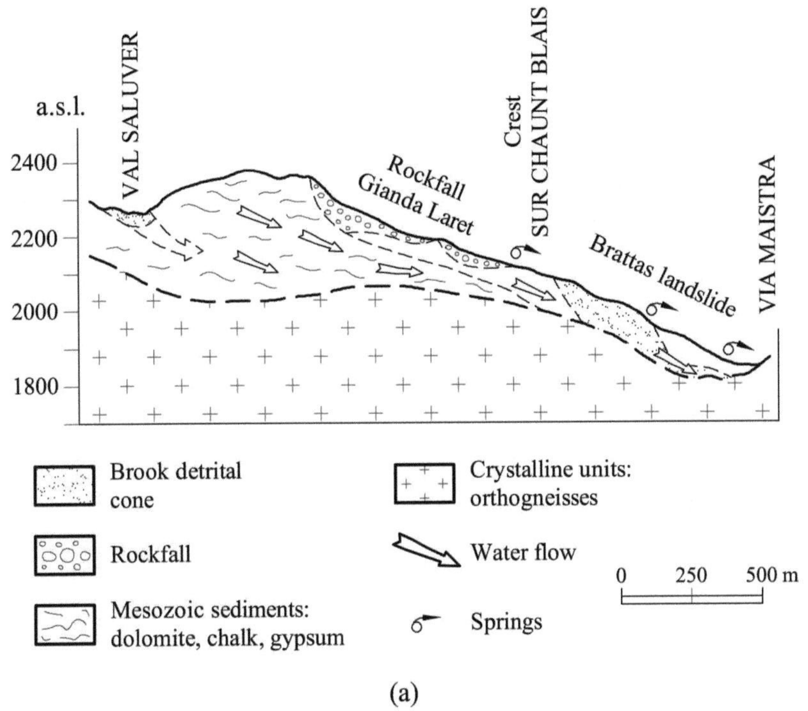

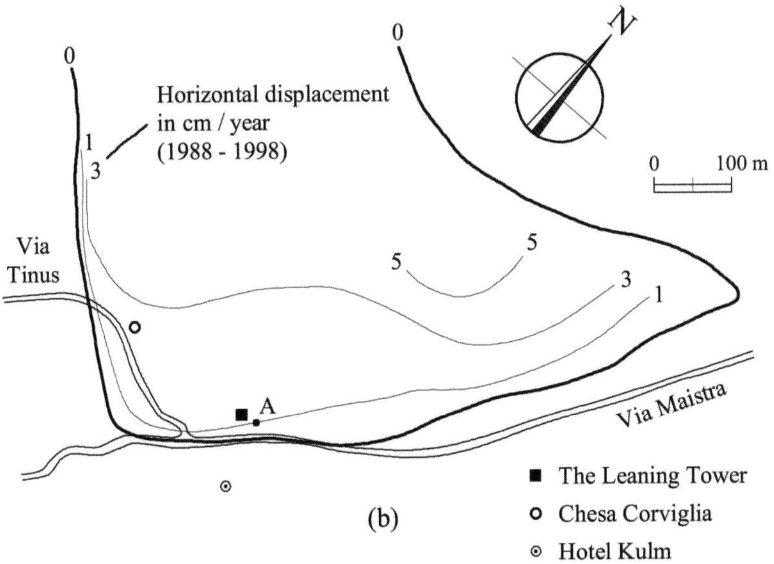

Figure 1.2 The Brattas Landslide: (a) the geology (Müller and Messina, 1992); (b) yearly horizontal displacements (after Schlüchter, 1988, with the data from Tschudi and Angst, 1998).

1.1.2 The leaning tower of St Moritz

The effects of the landslide are most evident in the behaviour of the oldest structures. The Leaning Tower of St Moritz Church, which was built in the 12th century at the foot of the landslide (Fig. 1.3a, Schlüchter, 1988), exhibits an alarming tilt of 5°. The church itself had to be demolished in 1893 because of the excessive differential settlements. Since 1908, regular tilt and displacement measurements have been carried out (Fig. 1.3b). Stabilization efforts in 1928 and 1968 were not successful in the long-term, and an alarming reaction of the tower to the earthquake in Friaul on May 6, 1976 was detected. As a result, it was decided to undertake an additional stabilization attempt, which was accomplished in 1983.

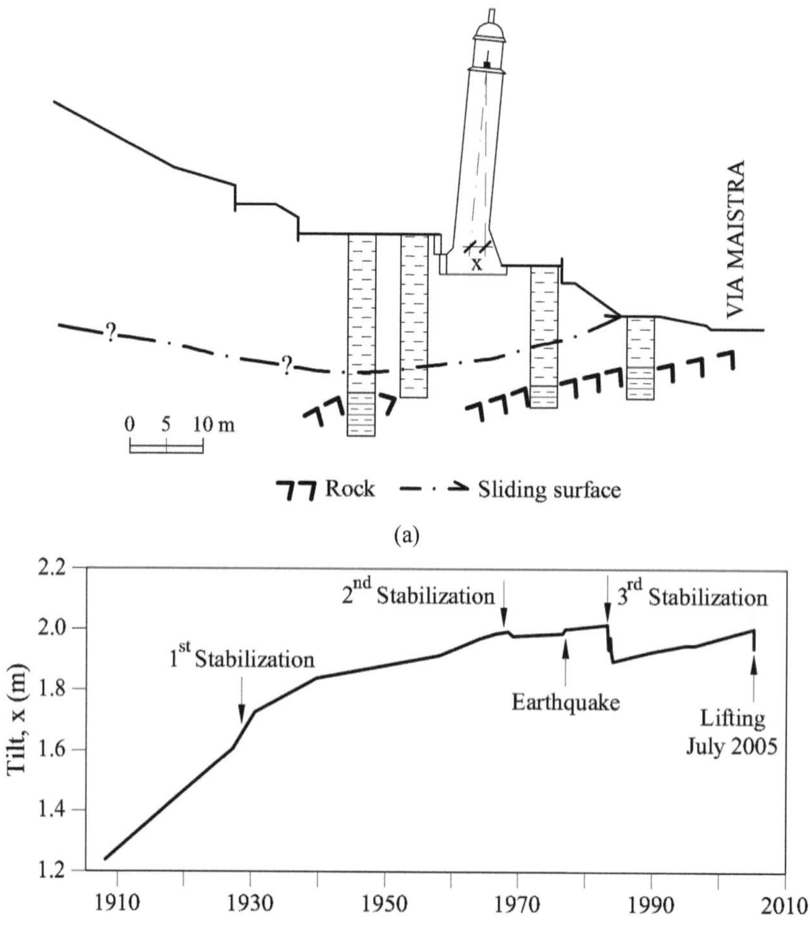

Figure 1.3 The Leaning Tower: (a) geological profile; (b) inclination in time.

1.1.3 Chesa Corviglia

Many modern buildings have also experienced the negative effects of the landslide. Chesa Corviglia is a terrace-type structure (Fig. 1.4a) on the south-western edge of the Brattas slope (Fig. 1.2b). The original building concept included a 20 m high anchored concrete pile wall (Fig. 1.4b) completely and permanently separated from the structure. Both the piles and the anchors were supposed to penetrate into the stable rock. Unfortunately, the prognoses concerning the location of the rock turned out to be false. As a result, both the wall and the house are moving downhill at different speeds, causing the gap between them to close (Fig. 1.4c), and threatening the stability of the house.

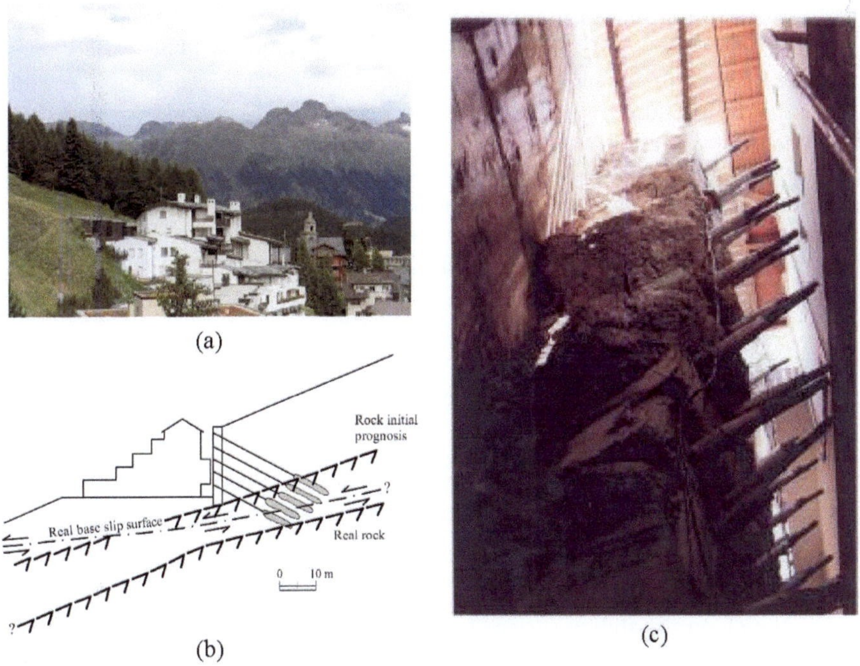

Figure 1.4 Chesa Corviglia: (a) the building; (b) the retaining structure (Gysi, 1999); (c) the closing gap between the retaining wall and the building.

1.1.4 The problem

The large-scale geologic situation of the area is seen as the primary cause of the landslide: the Mesozoic sediments of the Bernina nappe were pushed over the crystalline rock of the Err nappe. (Nappe is a large sheet of rock that has been moved a considerable distance from its original position). The hydrological conditions constitute further causes of instability. Various deep aquifers were observed in the landslide, which create independent water tables. An increase in the pore water pressure due to the snow melt causes shear strength degradation, which leads to the yearly movements, but sometimes also to recurring large-scale landslide events. According to Schlüchter (1988), there is geological evidence of a number of these landslide events which have occurred in the last 5000 years –the last one approximately 700 years ago.

Since the only global stabilization possibility –dewatering of the entire landslide– is extremely expensive, it may be a better solution just to adjust the existing and new structures to the yearly displacements, provided they do not exceed a reasonable level. In this context, it becomes of crucial importance to be able to assess the long-term stability and displacements of the landslide.

1.1.5 Long-term stability and displacements

The unusual feature of this landslide is that it has "nowhere" to go and its downhill movement is slowing down in time, which intuitively implies the landslide stability. However, in spite of this landslide slowing, and in certain scenarios exactly because of it, the shear strength of the sliding surface may decrease, leading to an increase in compressive stresses at the landslide foot and, ultimately, to a failure. A simple model of a constrained landslide developed by Puzrin and Sterba (2006) to provide an assessment for the long-term stability and displacements of the St. Moritz landslide is briefly described below. It is based on an inverse analysis, which allows for the safety factor to be determined solely by curve fitting the observed displacement data. For a safety factor lower than one, the time of failure can be predicted using additional earth pressure measurements in the sliding layer.

1.2 The Theory

The material presented in this section has been previously published in Puzrin and Sterba (2006) and is reproduced here with kind permission of Thomas Telford Limited.

1.2.1 Model assumptions

The schematic layout of the boundary-value problem of a slowing constrained landslide is given in Figure 1.5. Equilibrium of the sliding layer relates the shear stress $\tau(x,t)$ on the sliding surface to the average effective normal stress in the layer $p'(x,t)$ and the effective active earth pressure p'_a acting at the top of the layer:

$$p'(x,t)H + \int_x^L \tau(x,t)\,\mathrm{d}x = \gamma_t H(L-x)\sin\alpha + p'_a H. \tag{1.1}$$

Here γ_t is the total unit weight of soil; α is the slope inclination; L and H are the landslide length and thickness, respectively. In (1.1) we use effective earth pressures assuming that the average pore water pressure is constant along the slope: $u(x,t) = u(t)$, i.e. there is a flow parallel to the slope surface.

In order to solve this boundary-value problem, we would supplement Equation (1.1) with constitutive equations, relating stresses τ and p' to displacements δ, strains $\varepsilon = \partial\delta/\partial x$ and their rates. Solving this system of equations, together with the boundary and initial conditions, would allow for displacements $\delta(x,t)$, strains $\varepsilon = \partial\delta/\partial x$, and earth pressures $p'(x,t)$ to be calculated in space and time and the landslide behaviour to be predicted. This conventional approach to the boundary-value problem is called *forward analysis*.

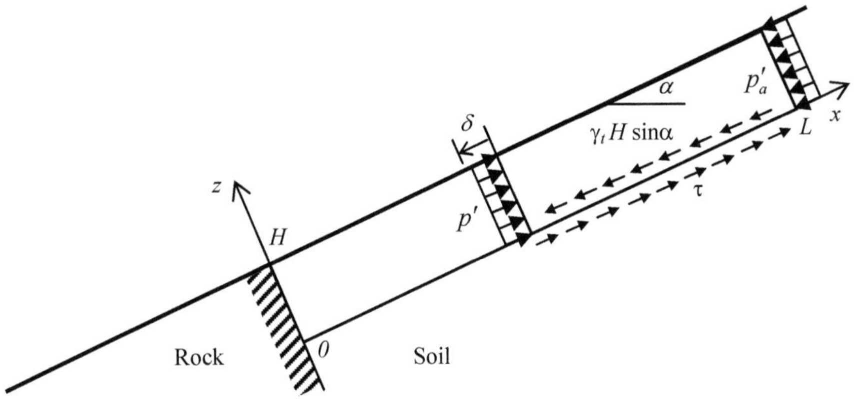

Figure 1.5 Schematic layout of the constrained landslide model.

Because the processes in a constrained landslide are slower than in the one which is free to slide, we assume that the excess pore water pressure caused by the shearing has enough time to dissipate. Therefore, the time-dependency of displacements is solely due to rate effects. These rate effects are known to exhibit themselves in the secondary compression (see, e.g. Chapter 2 in "Geomechanics of Failures" by the same authors), but also, as demonstrated by Skempton (1985), in the rate dependency of the residual strength (Fig. 1.6a).

As seen from Equation (1.1) and Figure 1.5, the weight and the active force in the layer are resisted by the earth pressure in the layer and the shear stress on the sliding surface. These are schematically represented in Figure 1.6b by the elastic spring (with elastic modulus *E*) and the slip element (with the slip stress τ_r), respectively. In order to introduce time-dependency, we include two dashpot elements: one (with viscosity η_p) to describe rate-dependent processes within the soil layer, the other (with viscosity η_τ), on the sliding surface, resulting in the following constitutive equations:

$$p' = E\varepsilon + \eta_p(\varepsilon,\dot{\varepsilon})\dot{\varepsilon}, \qquad \tau = \tau_r + \eta_\tau\dot{\delta}, \qquad \varepsilon = \partial\delta/\partial x. \tag{1.2}$$

(In order to facilitate the future analysis we assume here that η_p is not necessarily a constant parameter but can be a function of strain and strain rate).

In fact, the residual strength is also a function of displacement, continuously decreasing as displacement grows: $\tau = \tau_r + \eta_\tau\dot{\delta} - \chi\delta$. As the landslide slows down, the displacements are still growing and both effects contribute to the shear strength degradation. In this simplified analysis, however, we consider only rate effects (i.e., $\chi = 0$), which is not conservative.

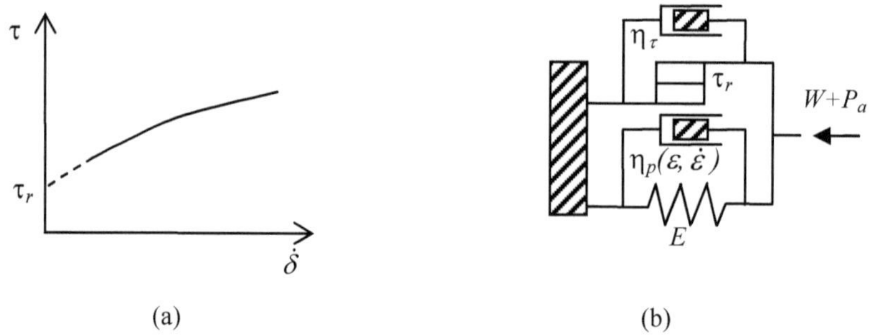

Figure 1.6 (a) Viscous nature of residual strength; (b) Schematic layout of the constitutive model.

1.2.2 Curve fitting of slope displacements

A disadvantage of the *forward analysis* approach described above is that it does not take into account the observed slope displacements. Spatial variability of the soil properties results in high levels of indeterminacy in constitutive models and their parameters obtained in laboratory tests. This often causes large discrepancies between the calculated and observed behaviour. In contrast to the forward approach, the *inverse analysis* of the problem would allow for the material properties to be back-calculated directly from the observed displacements. This would account for the global slope behaviour, as opposed to the behaviour of the locally extracted soil samples, and would provide a more reliable basis for the future predictions.

For the St. Moritz landslide, the yearly displacements are plotted in Figure 1.7a against the distance from the rock outcrop. A typical development of the displacements in time is plotted in Figure 1.7b (Lang and Sterba, 2002). The displacements were monitored between 1979 and 1999 at the point A on the slope located 15 m east from the Leaning Tower (Fig. 1.2b).

These displacements, as a function of time and space, can be directly used in the inverse analysis, but this requires a numerical solution of the differential equations. In an attempt to obtain an analytical solution, Puzrin and Sterba (2006) proposed fitting the observed normalized displacements data $\bar{\delta}(\bar{x},t) = \delta(x/L,t)/L$ in Figure 1.7 using the following analytical function:

$$\bar{\delta}(\bar{x},t) = \bar{\delta}_x(\bar{x})\bar{\delta}_t(t) = \bar{x}(a - b\bar{x})(1 - \exp(-ct + d)), \quad (1.3)$$

where

$$\bar{x} = x/L, \quad 0 \le b/a < 0.5, \quad c > 0. \quad (1.4)$$

This function describes displacements that are zero at the landslide foot and increase monotonically (when $0 \le b/a < 0.5$) along the slope towards its crest (Fig. 1.8a), while slowing with time and approaching an asymptotic value (Fig.

1.8b). The function (1.3) is simple and yet provides sufficient flexibility to fit the observation data both in the space (parameters a and b) and time domains (parameters c and d). As seen in Figure 1.7, this function (solid line) for $b/a = 0.15$ and $c = 0.045$ provides a reasonably good fit to the observed displacement data of St. Moritz landslide both in space and time.

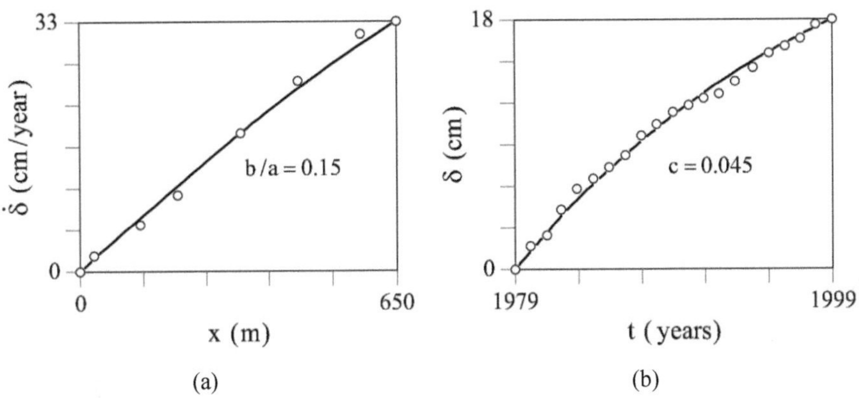

Figure 1.7 Displacement data monitored for the St. Moritz landslide: (a) distribution of average yearly displacements in space in 2006-2008; (b) development in time at the point A on the slope located 15 m east from the Leaning Tower (Fig. 1.2b).

In fact, as shown by Puzrin and Sterba (2006), this function also has a certain theoretical background: in represents the closed form solution of the boundary-value problem (1.1) – (1.2) for a particular case of $\eta_\tau = 0$ and $\eta_p = \text{const}$.

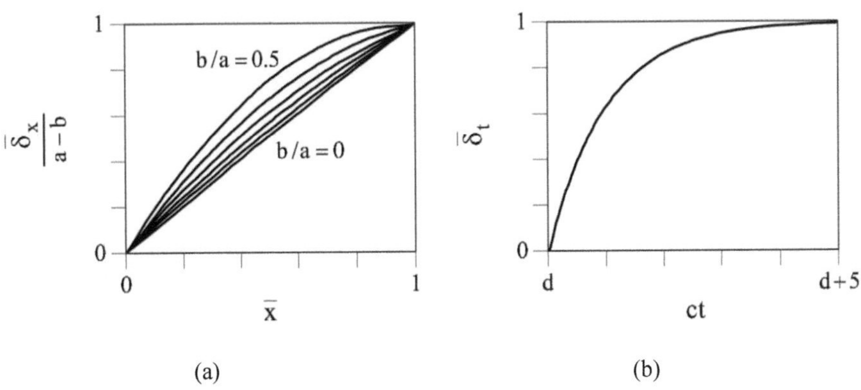

Figure 1.8 Normalized functions for curve-fitting of slope displacements: (a) in space; (b) in time.

1.2.3 The inverse analysis procedure

In the inverse analysis procedure the parameters of the constitutive model (1.2) are derived from the observed displacements (1.3). First, we substitute the constitutive relations (1.2), into the equation of equilibrium (1.1):

$$p'(x,t)+\frac{\eta_\tau}{H}\int_x^L \dot{\delta}(x,t)\,\mathrm{d}x=(\gamma_t\sin\alpha-\tau_r/H)(L-x)+p'_a. \tag{1.5}$$

Next, by substitution of the observation data expressed via the analytical function (1.3) into (1.5), differentiation and integration we obtain

$$\begin{aligned}p'(\bar{x},t)&=p'_a+\frac{L}{H}(\gamma_t H\sin\alpha-\tau_r)(1-\bar{x})-\\&\frac{L^2}{H}c\eta_\tau e^{-ct+d}\left[\frac{a}{2}\left(1-\bar{x}^2\right)-\frac{b}{3}\left(1-\bar{x}^3\right)\right].\end{aligned} \tag{1.6}$$

Expressions for the linear strain and its rate are derived from Equation (1.3):

$$\varepsilon(\bar{x},t)=\frac{\partial\bar{\delta}}{\partial\bar{x}}=(a-2b\bar{x})(1-\exp(-ct+d)),$$

$$\dot{\varepsilon}(\bar{x},t)=(a-2b\bar{x})c\exp(-ct+d), \tag{1.7}$$

which can then be resolved with respect to t and x:

$$\exp(-ct+d)=\frac{\dot{\varepsilon}/c}{\varepsilon+\dot{\varepsilon}/c},\qquad \bar{x}=\frac{a-\varepsilon-\dot{\varepsilon}/c}{2b} \tag{1.8}$$

and substituted into Equation (1.6):

$$\begin{aligned}p'&=p'_a+\frac{L}{H}(\gamma_t H\sin\alpha-\tau_r)\left(1-\frac{a-\varepsilon-\dot{\varepsilon}/c}{2b}\right)-\\&\frac{L^2}{H}\eta_\tau\frac{\dot{\varepsilon}}{\varepsilon+\dot{\varepsilon}/c}\left[\frac{a}{2}\left(1-\left(\frac{a-\varepsilon-\dot{\varepsilon}/c}{2b}\right)^2\right)-\frac{b}{3}\left(1-\left(\frac{a-\varepsilon-\dot{\varepsilon}/c}{2b}\right)^3\right)\right].\end{aligned} \tag{1.9}$$

It can be easily shown (by substitution) that Equation (1.9) becomes identical to the first constitutive Equation (1.2), when the following relationship exists between the parameters of the constitutive model (Eq. (1.2)) and the coefficients of the curve fitting Equation (1.3):

$$E=\frac{p'_a}{a-2b}\quad \tau_r=\gamma_t H\sin\alpha-\frac{2bp'_a}{a-2b}\frac{H}{L}, \tag{1.10}$$

$$\eta_p(\varepsilon,\dot{\varepsilon}) = \frac{p'_a}{ac-2bc} - \frac{L^2}{H}\frac{\eta_\tau}{\varepsilon+\dot{\varepsilon}/c}\left[\frac{a}{2}\left(1-\left(\frac{a-\varepsilon-\dot{\varepsilon}/c}{2b}\right)^2\right)-\frac{b}{3}\left(1-\left(\frac{a-\varepsilon-\dot{\varepsilon}/c}{2b}\right)^3\right)\right]. \tag{1.11}$$

Unfortunately, the derived expression, via inverse analysis, for the non-constant term of the viscosity coefficient (1.11) is rather complex. The good news, however, is that, unlike the viscosity coefficient η_τ, it does not enter Equation (1.6) for the earth-pressure evolution, which at the landslide bottom ($x = 0$) can be expressed as

$$p'(0,t) = \frac{p'_a}{1-2b/a} - \frac{L^2}{H}c\eta_\tau \exp(-ct+d)\left[\frac{a}{2}-\frac{b}{3}\right]. \tag{1.12}$$

1.2.4 The safety factor

The safety factor for the slope stability can be defined as the ratio between the soil resistance (passive earth pressure p'_p) and the maximum earth pressure that can develop at the foot of the landslide in time, calculated from (1.12):

$$F_s = \frac{p'_p}{p'(0,\infty)} = \frac{1-2b/a}{p'_a/p'_p}. \tag{1.13}$$

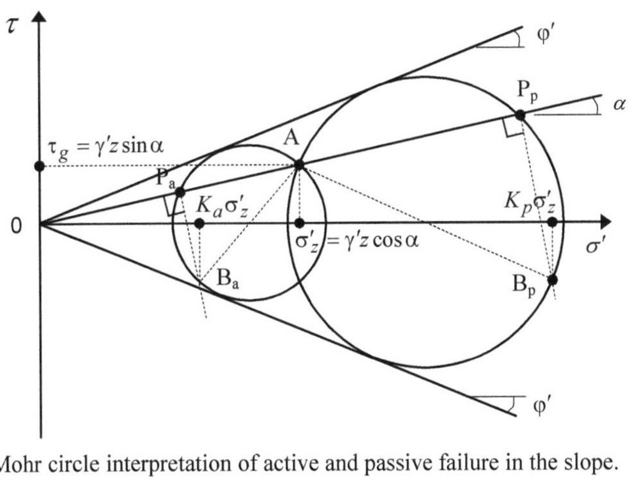

Figure 1.9 Mohr circle interpretation of active and passive failure in the slope.

Note, that the entire stability analysis can be performed using only the observed displacement data and the values of the effective active and passive earth pressures p'_a and p'_p (acting parallel to the slope). These pressures can be found from trigonometric analysis of Figure 1.9 (Chu, 1991). The stress state on a plane parallel to the slope α at the depth z is represented by point A. The failure takes place when a Mohr circle passing through point A touches the failure envelope $\tau = \sigma' \tan\varphi'$, which produces two circles for the passive and active failure, respectively. Their poles P_p and P_a are found by drawing a straight line inclined by angle α through point A up to the intersection with the corresponding circle. Drawing a straight line inclined by angle $90^\circ - \alpha$ through pole P up to the intersection with the corresponding circle produces the state of stress at failure B at the plane perpendicular to the slope. The normal stresses at B_a and B_p are the effective active and passive earth pressures, respectively, acting parallel to the slope at the depth z. The respective earth pressure coefficients are given by

$$\begin{Bmatrix} K_a \\ K_p \end{Bmatrix} = 1 + 2\tan^2\varphi' \mp 2\sqrt{\left(1+\tan^2\varphi'\right)\left(\tan^2\varphi' - \tan^2\alpha\right)} . \tag{1.14}$$

The average values of the effective active and passive pressures over the thickness of the sliding layer H are then given by the following formula:

$$\begin{Bmatrix} p'_a \\ p'_p \end{Bmatrix} = \frac{1}{2}\gamma' H \cos\alpha \left[1 + 2\tan^2\varphi' \mp 2\sqrt{\left(1+\tan^2\varphi'\right)\left(\tan^2\varphi' - \tan^2\alpha\right)}\right] . \tag{1.15}$$

where φ' and γ' are the effective angle of internal friction and effective unit weight of the soil in the sliding layer, respectively.

1.2.5 The long-term displacements

The safety factor defined by Equation (1.13) allows for distinguishing between the *safe* and *failure scenarios* of the landslide evolution. In the *safe scenario* of $F_s > 1$, the slope will eventually stop sliding and the final displacement increment for the point x on the slope are defined by

$$\delta_\infty(x) = \frac{\delta_M(x)}{1 - \exp\left(-c\left(t_M - t_1\right)\right)}, \tag{1.16}$$

where $\delta_M(x) = \delta(x, t_M)$ is the displacement of the point x at the time of measurement t_M; t_1 is the time of initial measurements, so that $\delta(x, t_1) = 0$.

1.2.6 The time to failure

In the *failure scenario* of $F_S < 1$, the earth pressure at the bottom of the slide will eventually reach the passive pressure p'_p. The time before the slope failure t_f can be back-calculated from Equation (1.12), but this requires an additional soil parameter η_τ. In order to determine η_τ though, the existing displacement observation data is not sufficient and has to be supplemented by some other kind of measurements. The time changes in the earth pressure p' at the slope bottom

$(x=0)$ are of the utmost importance in this analysis. If it were possible to measure the increase in the earth pressure Δp over a period of time Δt :

$$\Delta p = p'(0,t_0+\Delta t) - p'(0,t_0), \tag{1.17}$$

where t_0 is the time of the initial pressure measurement, substitution of (1.12) into (1.17) would produce the desired expression for η_τ :

$$\eta_\tau = \frac{\Delta p\, H \exp(c t_0 - d)}{L^2 c(a/2 - b/3)(1-\exp(-c\,\Delta t))}, \tag{1.18}$$

so that the time before the slope failure would be given by

$$t_f = t_0 + \frac{1}{c}\ln\frac{\Delta p' / p'_p}{(1-\exp(-c\,\Delta t))(1/F_s - 1)}. \tag{1.19}$$

1.2.7 Summary

Based on the above derivations, the following procedure can be used for the stability analysis of a constrained landslide. We assume that the following displacement monitoring data is available: $\delta(x_i, t_j)$, where $i = 1,...,N$; $j = 1,...,M$; N and M are the number of measurement points in space and time, respectively; x_N is the highest point on the slope measured; and t_1 is the time of the first (zero) measurement in time, so that $\delta(x_i, t_1) = 0$ for all $i = 1,\ldots,N$.

1. Plot the normalized observation data $y_{i,j} = \delta(x_i, t_j)/\delta(x_N, t_j)$, for $i = 1,\ldots,N$; $j = 1,\ldots,M$ versus the normalized coordinate $x'_i = x_i / x_N$ (i.e., in space).
2. Determine the coefficient $k = b/a$ by fitting the following function to this data:

$$y = x'\frac{L' - k x'}{L' - k}, \tag{1.20}$$

   where $L' = L/x_N$ is the normalized length of the landslide.
3. Calculate effective active and passive earth pressures from Equations (1.15) and their ratio p'_a / p'_p :

$$\frac{p'_a}{p'_p} = \frac{1 + 2\tan^2\varphi' - 2\sqrt{(1+\tan^2\varphi')(\tan^2\varphi' - \tan^2\alpha)}}{1 + 2\tan^2\varphi' + 2\sqrt{(1+\tan^2\varphi')(\tan^2\varphi' - \tan^2\alpha)}}. \tag{1.21}$$

4. Substitute $k = b/a$ and p'_a/p'_p into (1.13) to calculate the safety factor. If $F_s > 1$, the slope is stable in the long-term, if $F_s < 1$, it is not.

Note that the only parameters required for this stability analysis (in addition to the observed displacements) are α, L and φ'. Not even the thickness of the sliding layer H or the unit weight of soil γ' is required. Also, it has not yet been necessary to define the time-related parameters c and d. These will be defined in the following steps and used to calculate the final slope displacements δ_∞ for the *safe scenario* ($F_s > 1$) or the time to failure t_f for the *failure scenario* ($F_s < 1$).

5. Plot the normalized observation data $w_{i,j} = \delta(x_i, t_j)/\delta(x_i, t_M)$, for $i = 1,\ldots,N$; $j = 1,\ldots,M$ versus t_j (i.e., in time).
6. Determine the coefficient c by fitting the following function to this data:

$$w = \frac{1 - \exp(-c(t - t_1))}{1 - \exp(-c(t_M - t_1))}. \tag{1.22}$$

7. If $F_s > 1$, the final displacement increments for each $i = 1,\ldots,N$ are:

$$\delta_\infty(x_i) = \frac{\delta(x_i, t_M)}{1 - \exp(-c(t_M - t_1))}. \tag{1.23}$$

8. If $F_s < 1$, starting from time t_0, measure the increase in the earth pressure Δp over a period of time Δt. Calculate the time of the future failure from formula (1.19).

1.3 Analysis of the Landslide

The theory presented above provides tools for a simplified inverse stability analysis of the Brattas, St. Moritz Landslide. In 2005, it was applied by Puzrin and Sterba (2006) to the monitored displacement data available at that time (in the lower 200 m of the landslide). Using the assumption that the two parts of the landslide in Figure 1.2a are connected, i.e. the total length of the landslide is $L = 1500$ m, the curve-fitting of Equation (1.20) produced a ratio $b/a = 0.39$ (Fig. 1.10). The effective angle of internal friction was assumed (after Vermeer, 1997) to be within the range of $\varphi' = 28° - 35°$, so that for the average slope of $\alpha = 20°$ from Equation (1.21) $p'_a/p'_p = 0.28 - 0.15$ was obtained and the corresponding safety factor followed from Equation (1.13): $F_s = 0.78 - 1.46$.

This preliminary result did not allow for the possibility of the failure scenario to be excluded. In order to improve the accuracy of prediction, additional observation data from the middle and upper parts of the landslide had to be collected. This could provide more data for the curve-fitting, but even more importantly, an uncertainty with respect to the interaction between the Gianda Laret rockfall and the Brattas landslide had to be resolved. This could help to define more accurately the upper boundary of the Brattas landslide and the static boundary conditions on it. To collect these and other data, an extensive field investigation program was carried out in the year 2006.

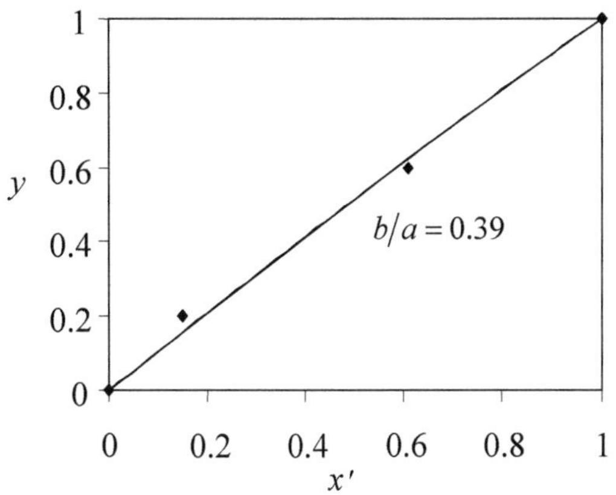

Figure 1.10 Curve fitting of the normalized yearly displacement data in space monitored for the St. Moritz landslide before 2005.

1.3.1 Geodetic measurements

The main part of the field campaign was geodetic measurements. A grid of measurement points was established covering the Gianda Laret rockfall and the Brattas landslide (Fig. 1.11a) with a denser grid at the boundary between them. The coordinates of the points were measured with a precision of up to 1 cm. The coordinates were then taken again in 2007 and 2008 with the corresponding yearly displacements plotted in Figure 1.11b against the distance from the rock outcrop at Via Maistra.

The plot in Figure 1.11b exhibits a number of interesting features. First of all, the yearly displacements 2007–2008 are almost twice larger than the yearly displacements 2006–2007. This can be explained by significantly larger precipitation in the 2007–2008 period and represents a typical fluctuation of the yearly displacements due to changes in climatic conditions.

Second, in spite of this fluctuation, the shape of the displacement distribution along the landslide remains remarkably stable and, when normalized by the maximum displacement, it produces almost the same curve (Fig. 1.12a), which is a normalized version of the plot in Fig. 1.7a. The yearly displacement is normalized by 33 cm, the distance – by 650 m.

Finally, and very importantly, it was possible to establish that the movement rate at the top of the Brattas landslide (600–700 m away from Via Maistra)

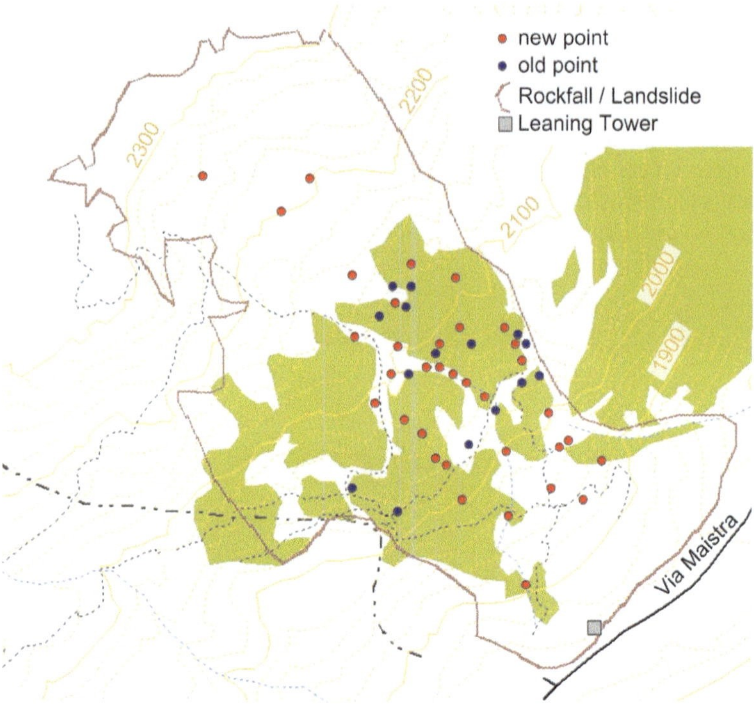

(a)

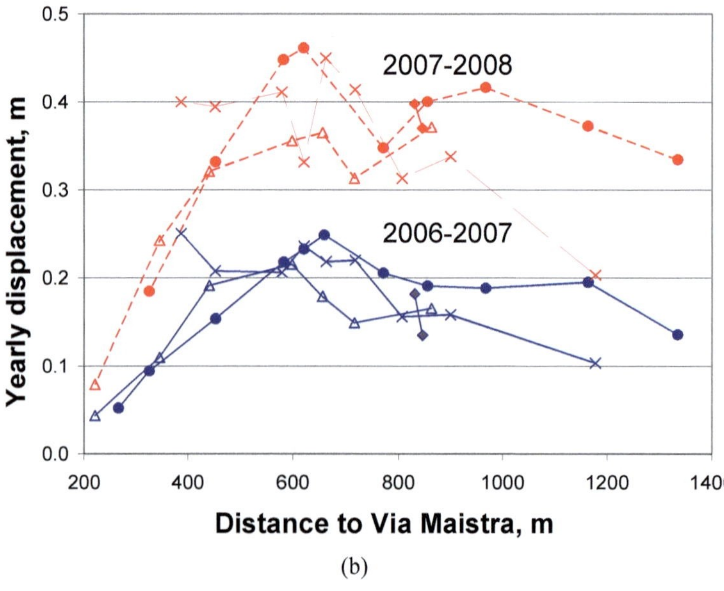

(b)

Figure 1.11 Geodetic measurement: (a) the points; (b) yearly displacements.

reaches 45 cm/year, while at the bottom part of the Gianda Laret rockfall (700–800 m away from Via Maistra), only 30 cm/year. It is, therefore, likely that the rockfall movement is stopped by a rock ridge at an altitude of about 2,100 m and does not fully transfer earth pressures to the Brattas landslide. The upper boundary of the Brattas landslide, however, is apparently slowly shifting upwards with blocks from the rock ridge gradually collapsing into the sliding mass.

1.3.2 Simplified model

Based on the latest observation data, we assume that there is no interaction between the Gianda Laret rockfall and the Brattas landslide, i.e., the following landslide parameters can be adopted: $L = 700$ m and $\alpha = 20^\circ$. The effective angle of internal friction is assumed (after Vermeer, 1997) to be within the range of $\varphi' = 28^\circ - 35^\circ$, so that from (1.20): $p'_a/p'_p = 0.28 - 0.15$.

The displacement data in Figure 1.11b is averaged over the two years, combined with the existing data from the lower 200 m and normalized as suggested in Section 1.2.7 (Table 1.1).

Table 1.1 Normalized yearly horizontal displacements along the landslide.

x, m	$x' = x/650$	$\dot{\delta}$, cm/year	$y = \dot{\delta}/33$	y_{fitted}
0	0	0	0	0
30	0.046	2	0.061	0.053
122	0.188	6	0.182	0.212
200	0.308	10	0.303	0.342
330	0.508	18	0.545	0.548
450	0.692	25	0.758	0.727
580	0.892	31	0.939	0.908
650	1	33	1	1

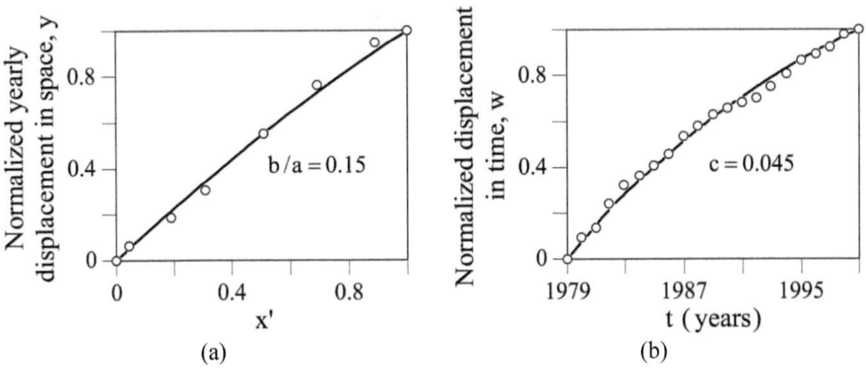

Figure 1.12 Curve fitting of the normalized displacement data monitored for the St. Moritz landslide: (a) in space; (b) in time.

1.3.3 The safety factor

The normalized displacement data from Table 1.1 is plotted in Figure 1.12a. The best fit by the analytical curve (1.20) to this normalized displacement data is achieved at $b/a = 0.15$ (Fig. 1.12a and Table 1.1). By substituting these parameters into Equation (1.13), we obtain the range for the safety factor: $F_s = 2.49 - 4.63$. As is seen, analysis excludes the possibility of a future failure. It has to be mentioned, however, that this conclusion has a preliminary nature, because the observations over the 2 years are not sufficient for long-term analysis.

1.3.4 The long-term displacements

Since we find ourselves within the *safe scenario*, we should be able to predict the long-term displacements of the landslide. For example, the normalized displacement data monitored between 1979 and 1999 at point A on the slope located 15 m east from the Tower (Fig. 1.2b) is plotted in time in Figure 1.12b (Lang and Sterba, 2002). The best fit to this data is achieved using the analytical curve (1.22) with $c = 0.045$ (Fig. 1.12b). The total downhill displacement of point A between 1979 and 1999 was 177 mm, therefore in the *safe scenario* case, according to formula (1.23), the final displacement of this point will be 298 mm.

1.3.5 Discussion

The above stability analysis is based on fitting the observed displacements using a simple and yet sufficiently versatile curve fitting function. Inverse analysis of a visco-elastic visco-plastic model of the landslide allows for both the *safe* and *failure scenarios* of the landslide evolution to be identified and explored within the same unified framework. Preliminary analysis shows that the Brattas landslide is stable.

Unfortunately, as mentioned above, the data available for this analysis are not sufficiently reliable to entirely exclude the possibility of the *failure scenario* for the Brattas landslide. The definite time related predictions can be only achieved by monitoring the long-term displacements along the entire landslide length over a long period of time and by measuring the earth pressure changes in the area of high compression at the landslide bottom, which are planned in the future. Effects of the climatic changes and groundwater conditions should be also studied and incorporated in the more detailed analysis.

1.4 Analysis of the Leaning Tower

In addition to the long-term stability and displacement predictions presented above, we shall attempt to answer another important question: is the tower inclination indeed due to the landslide movements or could it be caused by a leaning instability or a bearing capacity failure? To answer this question, we could apply the theory from Chapter 3 in "Geomechanics of Failures" by the same authors (referred as Chapter 3* below), but this requires parameters of the soil strength and stiffness in the vicinity of the tower. As a part of the field investigation program carried out in 2006, the soil strength and stiffness were measured directly in the field, using two different types of dilatometer tests.

1.4.1 Dilatometer tests

The field investigation program included installation of an inclinometer and a piezometer in the vicinity of the Leaning Tower. In the same borehole, two types of dilatometer tests were performed to determine the soil stiffness at different depths (Puzrin *et al.*, 2008). The first type of test involved a Cambridge dilatometer (Fig. 1.13a): a cylindrical probe with an inflatable rubber membrane inserted into a predrilled borehole. The second type of test involved a Marchetti dilatometer (Fig. 1.13b): a spade-like probe with a round flat inflatable steel membrane pushed into soil up to a depth of 1 m from the bottom of the borehole. In spite of difficult soil conditions (gravelly clays), both tests produced meaningful results (Fig. 1.14).

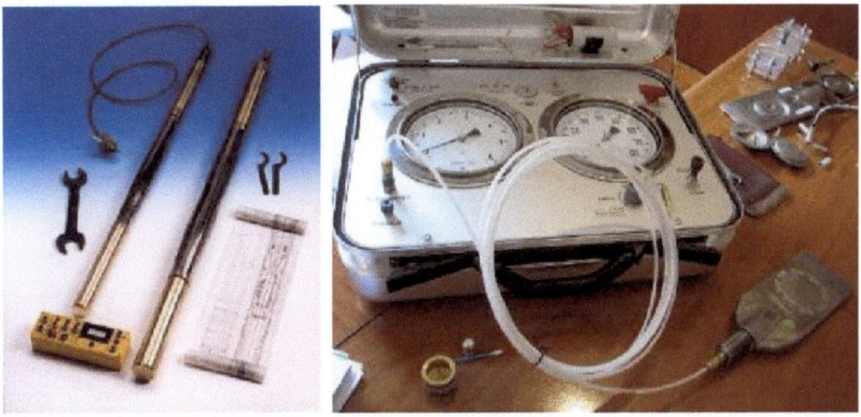

(a) (b)

Figure 1.13 Dilatometers: (a) Cambridge in-situ probe; (c) Marchetti probe.

1.4.2 Leaning instability

Before its stabilization, the tower was based on a square foundation (Sterba *et al.*, 2002) and had the following geometry:

- the height of the centre of gravity: $H_c \approx 13.0$ m;
- the (half)width of the square foundation: $b = 3.0$ m;
- the average radius of the square ring foundation: $r = 1.5$ m. (see Chapter 3*)

The tower high slenderness ratio of $H_c/r = 8.7$ requires checking against the leaning instability. The tower is built on a 15 m thick layer of gravely clay. The properties of this clay were derived from the dilatometer tests (Fig.1.14):

- the compression index: $C_c = 0.12$;
- the swelling index: $C_e = 0.02$;
- the in-situ void ratio: $e_0 = 0.5$.

Inequality (3*.26) gives the following lower estimate for the critical ratio (note, that $\rho = 0.5$, so that $4\rho = 2.0$ and $f = 1.3$ from the plot in Fig. 3*.9a):

$$\frac{H_c}{r} \leq \frac{4\rho^2 + 1/2}{3f(4\rho)\rho} \cdot \frac{2.3 \cdot (1 + e_0)}{C_c + C_e} = 19.0 \tag{1.24}$$

The tower slenderness ratio $H_c/r = 8.7$ appears to be much smaller than the critical value. We have established that the excessive inclination of the tower is not due to the leaning instability. Can it be then a bearing capacity failure?

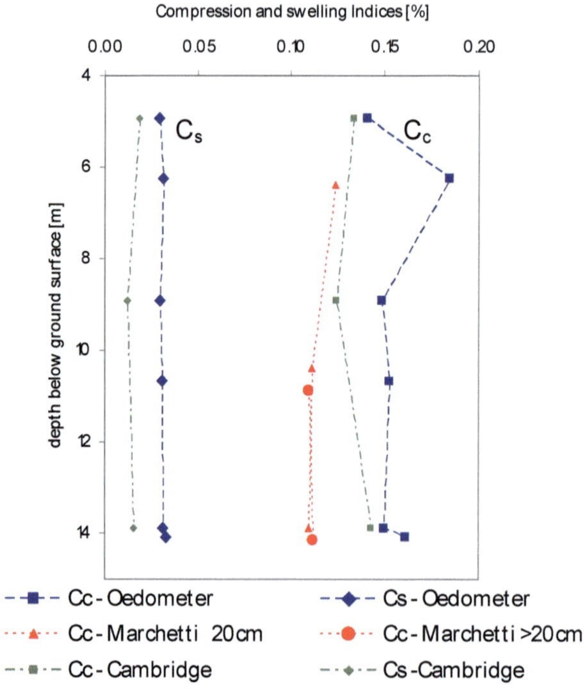

Figure 1.14 Compression and swelling indices from dilatometer and oedometer tests.

1.4.3 Bearing capacity

Let us first calculate the bearing capacity of the tower close to the end of its construction when it was still standing straight. This requires the length of the equivalent footing, which is calculated using the first Eq. (3*.6) from Chapter 3*:

$$L = 4r = 4 \times 1.5 = 6.0 \text{ m}$$

The width of the equivalent footing is $b = 3.0$ m, its depth $t = 2$ m. Substitution of the above parameters into the formulas (3*.35) – (3*.37) gives:

$$N_q = 14.72, \quad s_q = 1.27, \quad d_q = 1.18,$$

$$N_\gamma = 13.13, \quad s_\gamma = 0.80, \quad d_\gamma = 1.00.$$

A conservative estimate of the depth of the groundwater level is 2 m below the ground surface, of the angle of internal friction $\varphi' = 28°$, and of the total unit weight of soil $\gamma = 20\ \text{kN/m}^3$. The bearing capacity (3.34) can be calculated as

$$\begin{aligned}\sigma_f &= 20\times 2.0\times 14.72\times 1.27\times 1.18 + \\ &\frac{1}{2}\times 3.0\times 10\times 13.13\times 0.80\times 1.00 = 1.04\ \text{MPa}\end{aligned} \tag{1.25}$$

The weight of the tower is G = 1264 tons, which results in an average contact stress of σ = 344 kPa. Then the safety factor against the bearing capacity failure for the not inclined tower was

$$F_s = \frac{\sigma_f}{\sigma} = \frac{1{,}040}{344} = 3.0. \tag{1.26}$$

This is sufficient to ensure that the bearing capacity failure was not possible. We also need to keep in mind that this calculation is extremely conservative.

Before its stabilization, when the tower was inclined, the pressure under one of the footings was much higher than the average. Calculating from the second Equation (3*.6): $B = 3.0$ m, from the second Equation (3*.33), we obtain

$$\sigma_L(\alpha) = \frac{1{,}264\times 9.81}{3.0\times 6.0}\left(\frac{1}{2} + \frac{13.0}{3.0}\tan 5°\right) = 606\ \text{kPa}. \tag{1.27}$$

Assuming that the bearing capacity of this foundation did not change (which is conservative, because its depth increased due to the settlement), the safety factor is still significantly larger than unity:

$$F_s = \frac{\sigma_f}{\sigma_L} = \frac{1{,}040}{606} = 1.72. \tag{1.28}$$

Thus, the bearing capacity failure can also be excluded as the source of the tower inclination.

1.4.4 Discussion

We have established that the excessive inclination of the tower is neither due to the leaning instability nor to the bearing capacity. Furthermore, other structures in the landslide area, with much lower height to width ratios and contact pressures, also appear to be inclined. All this confirms that the most likely reason for the tower inclination is the landslide displacement.

1.5 Mitigation Measures

The above analysis shows that, although the landslide will probably not fail catastrophically, its movements keep damaging the structures. There are two ways to mitigate this damage: to stabilize the structures locally and let them "swim" with the landslide, or to stabilize the landslide globally by drainage.

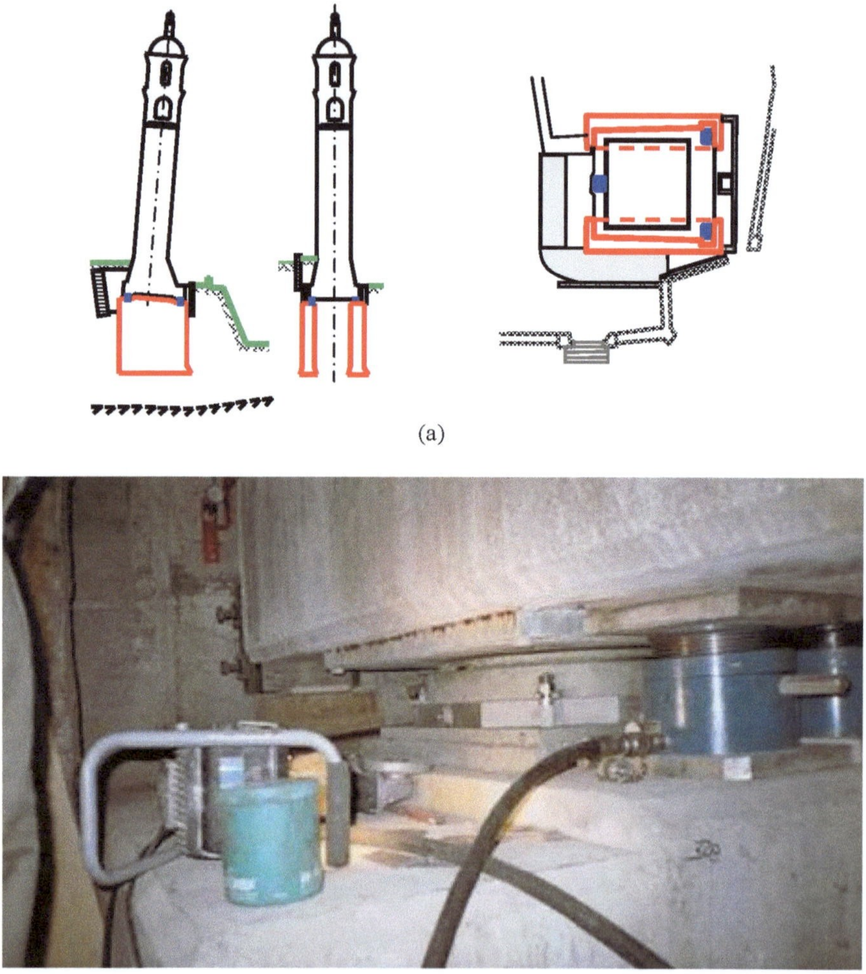

(a)

(b)

Figure 1.15 The Leaning Tower: (a) new foundations; (b) the jacking-up procedure.

1.5.1 Stabilization of the leaning tower of St. Moritz

This stabilization procedure consisted of placing pre-stressed reinforced concrete collars in the area at the foot of the tower and sinking two reinforced concrete barrettes to a depth of 10 m below the original foundation level (Fig. 1.15a). The tower was then lifted by hydraulic jack-ups from its original foundation and its weight (1,264 tons) was placed on the foundation barrettes via three Teflon bearing pads (Fig. 1.15b), and its tilt was decreased. The masonry of the tower was reinforced using vertical internal pre-stressing. Thus, the tower is "swimming" in the creeping mass, allowing for its tilt being periodically corrected by lifting and introducing additional plates into the bearing pads (last correction carried out in 2005).

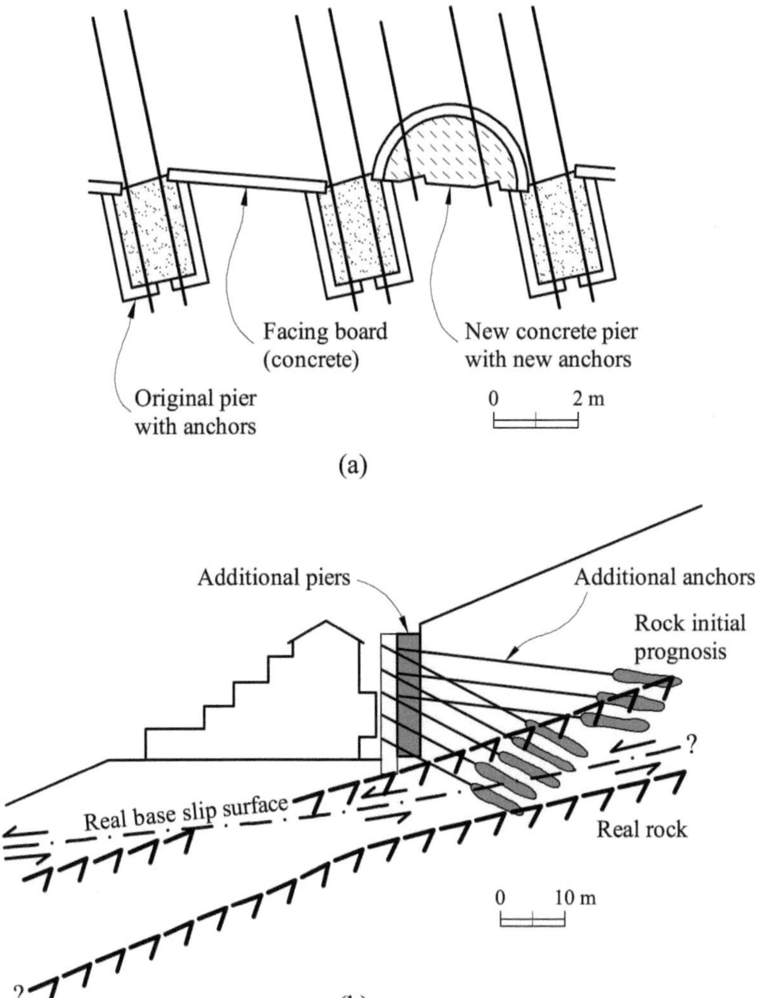

Figure 1.16 Chesa Corviglia (after Gysi, 1999): (a) a horizontal cross-section with the additional piers; (b) a vertical cross-section with the additional anchors.

1.5.2 Stabilization of Chesa Corviglia

Stabilization of Chesa Corviglia required ingenious and expensive geotechnical measures designed by the Gysi Leoni Mader AG engineering company. The wall was reinforced by additional piers (Fig. 1.16a), and additional anchors were placed above the original ones (Fig. 1.16b). The new anchors do not penetrate below the sliding surface (which could cause a dangerous increase in the anchor forces in time due to the landslide movements). However, they go sufficiently far away from the house into the sliding layer in the hope of bringing about identical displacements of the wall and the house, rather than hindering the creep movement of the entire landslide, which is probably an impossible task.

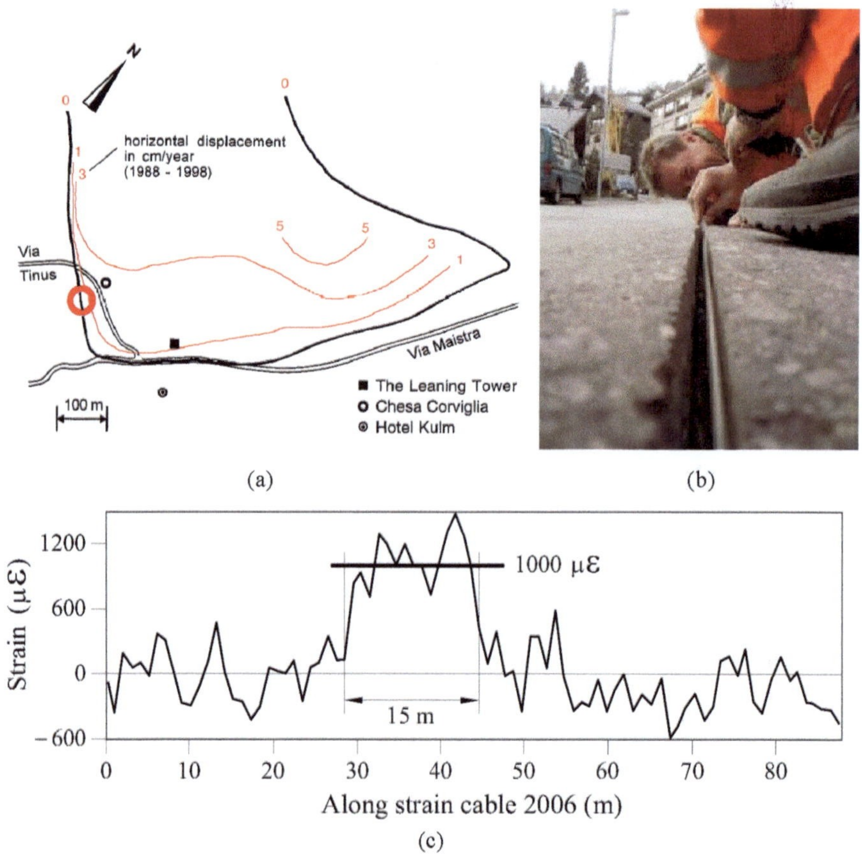

Figure 1.17 Fiberoptics: (a) the location; (b) the cable; (c) the measured strain.

1.5.3 Special regulations for new construction

As a consequence of the negative experiences with the Chesa Corviglia, the municipality of St. Moritz had to react. Since the only global stabilization possibility –dewatering of the entire landslide– was found to be not feasible, new permit specifications that regulate construction activities in the Brattas landslide were established. In addition to the legal problems (limited property rights, responsibility and economical considerations, rapid ageing of the construction, monitoring costs, etc.), these regulations address the planning, construction, and monitoring of the structures. For example, during the project phase of new constructions, weight compensation and equilibrium is a priority. The base of the structure should be made as rigid as possible, and the use of the permanent ground anchors is not permitted.

1.5.4 Defining landslide boundaries using fiberoptics

Because of the municipality regulations described above, the new construction within the landslide area is significantly more complex and expensive than outside this zone. Unfortunately, the western boundary of the landslide crosses the town and cannot be clearly identified from the geodetic measurements. This creates a legal and technical uncertainty when the owners of the boundary properties initiate a new construction.

To resolve this problem, it was decided to use the Via Tinus road, which crosses the boundary (Fig. 1.17a) as a gigantic strain gauge, by equipping it with a 90 m long fiberoptics cable, glued within a 7 cm deep trench in the asphalt (Fig. 1.17b). This was a novel application of the fiberoptic strain-monitoring technology (Iten *et al.*, 2008). By propagating light waves in the cable and using BOTDR (Brillouin Optical Time-Domain Reflectometry) for the signal processing, it was possible after one year of landslide movement to detect the relative elongation in the cable in the 15 m long zone of the shear boundary of the landslide (Fig. 1.17c).

1.5.5 Monitoring of the earth pressure at the landslide bottom

Information about the earth pressure changes in a sliding layer of a creeping landslide is a crucial component for understanding, analysis, and stabilization of creeping landslides. This information is especially important for constrained landslides where the pressures in the compression zone could reach the passive pressure and lead to a catastrophic failure. For a *failure scenario*, combining the measured increase in pressure with geodetic measurements allows for back-calculating parameter η_τ from Equation (1.18) and predicting when the constrained landslide will fail (Eq. (1.19)).

Unfortunately, measuring the earth pressures is one of the most challenging problems in the geotechnical monitoring. Conventional methods, such as pressure cells, require additional boreholes and produce unreliable results. In order to overcome these problems, a novel device – inclinodeformometer (IDM) was developed at the Institute of Geotechnical Engineering, ETH Zurich, to measure changes of earth pressure in a sliding layer of a creeping landslide (Schwager *et al.*, 2010). The device makes use of the existing and widely used technology of the inclinometer measurements. The change of earth pressures in the sliding layer leads to the changes in the inclinometer pipe shape and dimensions. The IDM probe is being lowered down the depth of the pipe on three wheels guided along the channels of the inclinometer pipe (Fig. 1.18a).

The upper and lower wheels are rolling in the same channel. These wheels are fixed to the probe. The middle wheel is connected via a lever with two springs so that it can be pressed against the opposite channel. A change in the diameter of the pipe leads to a change of the position of the middle wheel in respect to the probe. There are three tilt sensors detecting the relative inclination between the probe and the lever of the middle wheel (Fig. 1.18a). Continuous diameter measurements in two perpendicular directions can be taken. Diameter changes measured over a period of time allow for the pressure change to be back-calculated from a solution of a boundary value problem with properly described constitutive behaviours of the pipes and the surrounding soil.

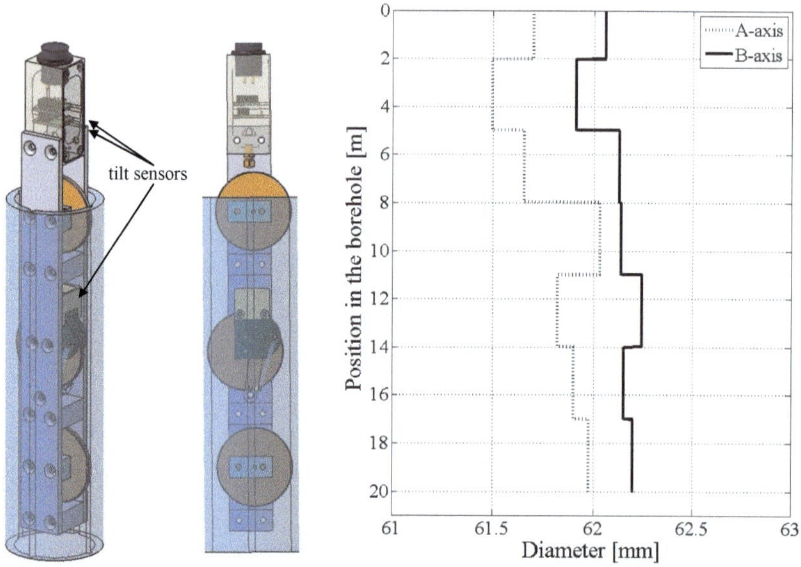

(a) (b)

Figure 1.18 Inclinodeformometer (IDM): a) device in the pipe; b) the measured diameters of an inclinometer pipe in the compression zone of the St Moritz landslide.

An advantage of the inclinodeformometer is that it does not require any additional infrastructure than standard inclinometer pipes, which are being installed anyway for landslide monitoring. Furthermore, these pipes can be used for pressure change measurements in the sliding layer long after they were sheared and became unsuitable for inclinometer measurements. Full-scale laboratory tests performed in a 2 m high calibration chamber demonstrated that the pressure measurement accuracy can be as high as 5 kPa.

Initial field measurements performed on the St. Moritz landslide confirmed significant stress anisotropy in the compression zone of this constrained creeping landslide (Fig. 1.18b). The A-axis in Figure 1.18b coincides with the direction of the landslide velocity, the B-axis is perpendicular to it. The pipe diameters are averaged every 3 meters – e.g., within each continuous pipe section between the installation joints. The measurements consistently demonstrate an elliptical pipe shape with a smaller diameter parallel to the landslide velocity. The difference between the pipe sections is most likely due to the variation in the initial pipe diameters. The first measurements of the diameter changes in three different inclinometer pipes were taken in 2009 and the back-calculated yearly earth pressure changes in the compression zone ranged between 5 and 15 kPa (Schwager *et al.*, 2010).

1.6 Lessons Learned

1.6.1 Stability of constrained creeping landslides

Creeping landslides, which are constrained by natural or man-made obstacles (e.g., a retaining wall) have "nowhere" to go and their downhill movements are slowing in time. This intuitively implies that the landslide stability is not an issue, which in some cases can be wrong, in particular when residual strength at the slip surfaces increases with shearing strain rate. In such a case, because the landslide is slowing, the shear strength on the sliding surface will decrease, leading to an increase in compressive stresses at the landslide foot and, possibly, to a catastrophic failure.

1.6.2 Inverse analysis

Spatial variability of soil properties results in high levels of indeterminacy in constitutive models and their parameters obtained in laboratory tests. This causes large discrepancies between the predicted and observed landslide behaviour. In contrast to the forward approach, inverse analysis allows for the material properties to be back-calculated directly from the observed displacements. This accounts for the global slope behaviour and provides a more reliable basis for the future predictions.

1.6.3 Landslide monitoring

The inverse analysis can produce reliable predictions only if the long-term displacements are monitored along the entire landslide length and the earth pressure changes are measured in the area of high compression at the landslide bottom. In addition, fiberoptic strain-monitoring technology can be used to better define landslide boundaries, in particular, in urban areas. A novel device – inclinodeformometer (IDM) – can be used to back-calculate changes in earth pressure in a sliding layer of a creeping landslide.

1.6.4 Stabilization of structures

An attempt should be made to separate a structure from a landslide, e.g. the foundation may be allowed to "swim" in the creeping mass with the structure tilt being periodically corrected. In any case, all parts of the structure should be ensured to move with the same velocity. The weight of the structure should compensate for that of the excavated ground. The base of the structure should be made as rigid as possible.

1.6.5 Environmental factors and landslide stabilization

Sharp changes in the yearly displacements are normally caused by the changes in environmental factors, such as precipitations. These can represent a danger to the short-term stability of the landslide, even if its long-term stability is ensured. This long-term stability can, in turn, be negatively affected by global climatic changes. Stabilization of the landslide using a drainage system can successfully mitigate these hazards.

References

Chu, S. (1991) Rankine analysis of active and passive pressures in dry sands. *Soils and Foundations* 31 (4), 115 – 120.

Gysi, H.J. (1999) Geschichte und Sanierung eines permanenten Einschnittes in einem Kriechhang. Gemeinschaftsveranstaltung Hochschule Rapperswil HRS, Rapperswil, Schulungszentrum TFB, Wildegg, Veranstaltung 4401.

Iten, M., Puzrin, A.M. and Schmid, A. (2008) Landslide monitoring using a road-embedded optical fiber sensor. *Proceedings of SPIE 6933*, 693315 – 9.

Lang, H.J. and Sterba, I. (2002). Schiefer Turm, St. Moritz – Stellungnahme zu den Messungen. *Bericht 3676/10, Institut für Geotechnik, ETH Zürich*, 10 p.

Müller, E.R. and Messina, G. (1992) Geotechnisches Gutachten, Bericht Nr. 2570-1, Rutschhang Sass Runzöl-Brattas, St. Moritz. Büchi und Müller AG, Chur (unpublished).

Puzrin, A.M., and Sterba, I. (2006) Inverse long-term stability analysis of a constrained landslide. *Géotechnique* 56 (7), 483 – 489.

Puzrin, A.M., Messerklinger, S. and Schmid, A. (2008) The in-situ stiffness of the sliding layer in a creeping landslide. *Proceeding of the Fourth International Symposium on Deformation Characteristics of Geomaterials*. Atlanta, Georgia, USA, 1, 407 – 412.

Schlüchter, C. (1988) Instabilities in the area of St. Moritz, Switzerland – Geology, chronology, geotechnology. *Proceedings of the Fifth International Symposium on Landslides*, Lausanne, Switzerland. Ed. C. Bonnard, 1, 1375–1380.

Schwager, M.V., Schmid, A.M. and Puzrin, A.M. (2010) Inclinodeformometer for earth pressure measurements in creeping landslides (accepted to *Soils and Foundations*).

Skempton, A.W. (1985). Residual strength of clays in landslides, folded strata and the laboratory. *Géotechnique* 35 (1), 3 – 18.

Sterba, I., Lang, H.J. and Amann, P. (2000) The Brattas Landslide in St. Moritz. In: *Proceedings of the International Conference on Geotechnical and Geological Engineering: GeoEng 2000*, Melbourne, Australia, 2, 144–149.

Tschudi, D. and Angst, R. (1999) Diplomvermessungskurs Samedan 1998, Rutschhang Brattas, St. Moritz. Bericht 283, *Institut für Geodäsie und Photogrammetrie, ETH Zürich,* 31 p.

Vermeer, P.A. (1997) PLAXIS Practice I: The leaning Tower of St. Moritz. *PLAXIS Bulletin*, No. 4, 4 – 7.

Chapter 2

Catastrophic Slide: Vaiont Landslide, Italy

TABLE OF CONTENTS

E.E. Alonso et al., *Geomechanics of Failures. Advanced Topics*,
DOI 10.1007/978-90-481-3538-7_2,

Chapter 2

Catastrophic Slide: Vaiont Landslide, Italy

2.1 The Landslide

An impressive double curvature arch dam, 276 m high, was built in the years 1957 – 1960 to store the waters of the Vaiont River, located in the Italian Alps, approximately 80 km north of the city of Venice. The dam was built in a narrow canyon, cut by the river in massive Jurassic limestone (Fig. 2.1a). The photograph shows, in the foreground, the limestone abutments of the dam and, in the background, the steep slope of the left bank of the river, which was actually the toe of an ancient landslide. The ancient slide became unstable in October 1963, when the level of the reservoir was close to its maximum, and invaded the reservoir at great speed. The displaced water generated a gigantic wave, 220 m high, which flew over the dam (which stood without bursting) and destroyed several villages downstream, causing more than 2,000 casualties. The failure sent seismic waves, recorded in seismographs across Europe.

Figure 2.1 View of Vaiont Dam from downstream: (a) before the catastrophic landslide; (b) after the slide (Valdés Díaz-Caneja, 1964).

Figure 2.1b is a view of the left bank of the river after the slide. The dam, in the lower part of the photograph, was not directly hit by the slide. A small lake remains between the dam and the toe of the slide. The bridge topping the dam has been destroyed. The slide scarp and a newly created lake may be seen in the background of the photograph. This catastrophe caused a great impact, which was deeply felt by dam and geotechnical engineers around the world.

A brief account of the events leading to the landslide of the left bank of the reservoir is given in the following paragraphs.

Dedicated geological surveys of the left margin of the reservoir started in 1958 under the supervision of L. Müller-Salzburg, an expert in rock mechanics. It was soon realized that a large proportion of the left bank of the reservoir was in fact a very large prehistoric landslide which, sometime in the past, filled the Vaiont valley. The valley had been excavated by the river at the end of the last glacial period (Würm) (Semenza and Ghirotti, 2000). After this prehistoric landslide, the river excavated again a deep valley through the slipped mass. The geological history of the landslide, an aspect which is always of interest in stability problems, is reviewed later.

At the end of 1960, once the dam was built and the reservoir partially impounded, a long, continuous peripheral crack, 1 m wide and 2.5 km in length, marked the contour of a huge mass, creeping towards the reservoir in the northern direction (Fig. 2.2).

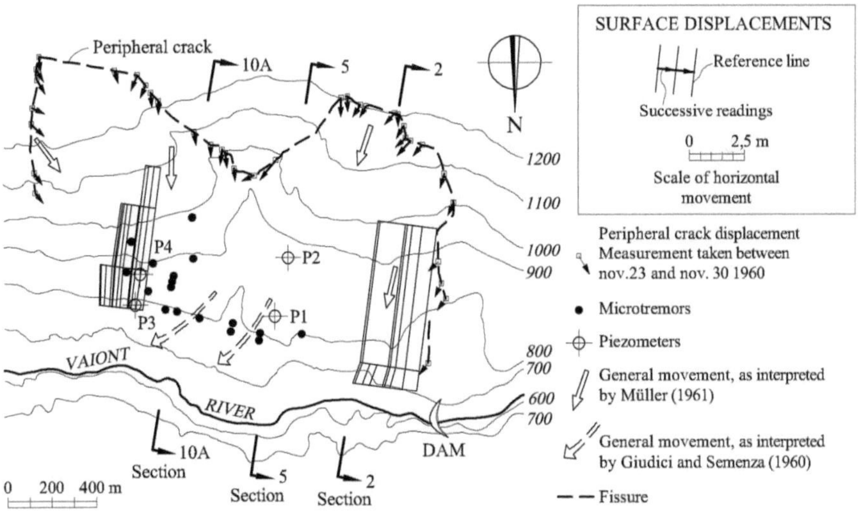

Figure 2.2 Map of the Vaiont sliding area. Note the position (and comparative size!) of the arch dam on the lower right-hand corner of the figure. (Simplified from Belloni and Stefani (1987) (© 1987 with permission from Elsevier) with additional information from several authors.)

In the following three years, the downward motion of the slide was monitored by means of surface markers. Some of the data provided by them are also plotted in Figure 2.2. In addition, water pressures in perforated pipes, located in four boreholes (location shown in Fig. 2.2), were monitored, starting in July 1961. The history of rainfall, reservoir level, rate of surface displacement, and water levels in piezometers in the four years preceding the failure is shown in Figure 2.3. Geophysical campaigns were also performed in December 1959 and 1960. Notice also, in Figure 2.3, that two slides of limited size took place during the first partial filling of the reservoir in 1960. Project engineers were by that time convinced that a large landslide could partially fill the reservoir, isolating the dam from the upstream part of the reservoir, and a by-pass tunnel was built in 1961 as a precautionary measure.

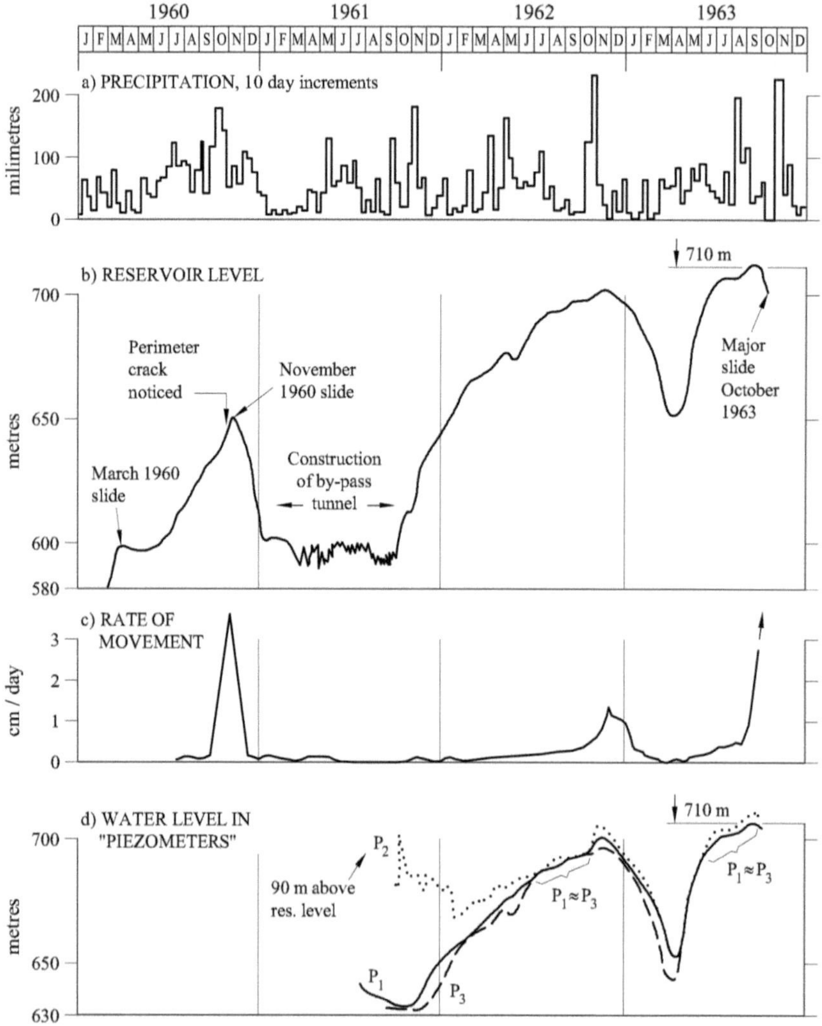

Figure 2.3 Relationship between precipitation, reservoir elevation, maximum velocity of horizontal surface displacements, and water level in piezometers. (After Hendron and Patton (1985), based on a figure by Müller (1964).)

However, all the investigation efforts provided limited information on some key aspects of the landslide such as the position and shape of the sliding surface and the pore water pressures acting on it. The measured rate of displacements of surface markers could be roughly correlated with the water level of the reservoir (Fig. 2.3). After two cycles of reservoir elevation, which partially filled and emptied the reservoir in the period 1960–1962, the water level reached a maximum (absolute) elevation of 710 m, at the end of September 1963. At that time, the accumulated displacements of surface markers had reached values in excess of 2.50–3 m (Fig. 2.4). The figure shows a good correlation between the increase in water level in the reservoir and the acceleration of the landslide.

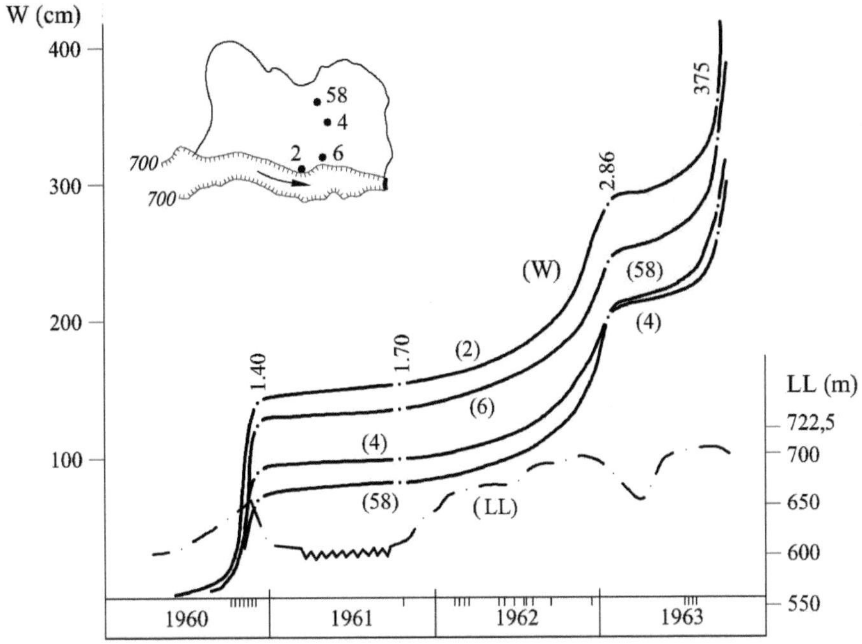

Figure 2.4 Accumulated displacements of surface markers (W) in the period 1960–1963 and its correlation with reservoir elevation (LL). Seismic events are marked in the time scale (reprinted from Nonveiller, 1987, © 1987, with permission form Elsevier).

Surface velocities of 20–30 cm per day were registered in the days preceding the final rapid motion that took place in October 9, 1963. An estimated total volume of rock of $280 \times 10^6\ m^3$ became unstable, accelerated, and invaded the reservoir at an estimated speed of 30 m/s (around 110 km/hour).

Figure 2.5 is a photograph of the slide taken in 1979. The landslide has filled the valley of the Vaiont River, which can be seen in the background. A residual lake can be seen in the lower left part of the image. The upper planar sliding plane (clear colours) is now exposed. The simplified map in Figure 2.5b, taken from Broili (1967), shows the position of the dam (not seen in the photograph), which maintains a small reservoir, the residual lake within the sliding area and the contours of the landslide before and after the failure.

(a)

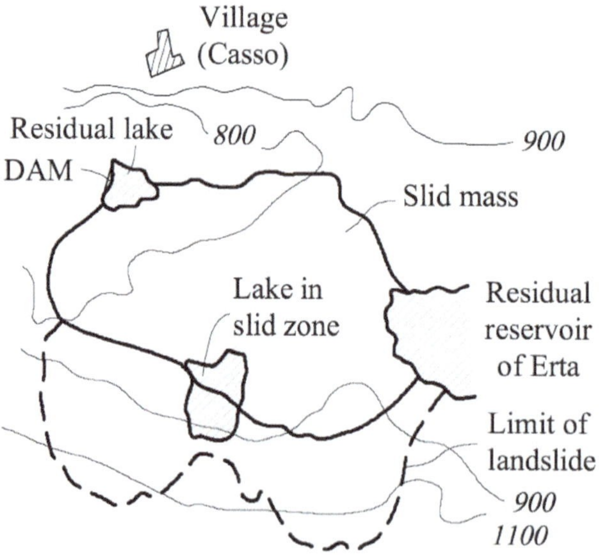

(b)

Figure 2.5 (a) Photo of the slide area, taken in 1979 (courtesy of G. Fernández); (b) plan view of the area after the slide (Broili, 1967).

The Vaiont landslide has attracted world wide attention concerning the causes and processes involved in the failure. Interest in Vaiont has never decreased within the technical community despite the 47 years that have elapsed since the accident. Papers analyzing the failure have been published at a maintained rate in journals and conferences. The landslide is one of the largest (in terms of volume of mobilized mass) in history. As stated by Hendron and Patton (1987) *"It is likely that more information has been published and more analyses have been made of the Vaiont data than for any other slide in the world"*. This chapter and Chapter 5 are additional contributions to this long list, with the aim of maintaining simplicity, but at the same time with the hope of capturing some fundamental aspects of the failure. Vaiont has been analyzed by researchers in rock and soil mechanics and some specific views of the mechanisms involved in the failure can sometimes be traced to the background of the people conducting the analysis.

One of the main reasons of this interest is the difficulty in explaining the extremely high velocity of the moving mass. The implication of this lack of understanding is that the risk associated with other landslide occurrences of a similar nature (natural slides affected in its toe by increasing water levels, a common situation in dam engineering) cannot be properly evaluated.

The issue of the velocity of the Vaiont landslide will be discussed in Chapter 5. But before this, the conditions for static equilibrium should be understood. Static models, even if they are simple, require an understanding of the main geological, geometrical, hydraulic, and geotechnical features of the slide. In the case of Vaiont, this information should ideally be extended to the old prehistoric landslide, which was reactivated by the reservoir impounding.

2.2 Geological Setting

The Vaiont River, which flows from east to west, cuts a large syncline structure which folds Jurassic and Cretaceous strata (Fig. 2.6). The syncline created the "open chair" shape of the Jurassic strata of the left margin of the river, which can also be seen in the figure. The axis of the syncline plunges a few degrees towards the east (normal to the plane of the figure). The syncline shape eventually defined the geometry of the failure surface, which is always important information for understanding the subsequent behaviour of the slide.

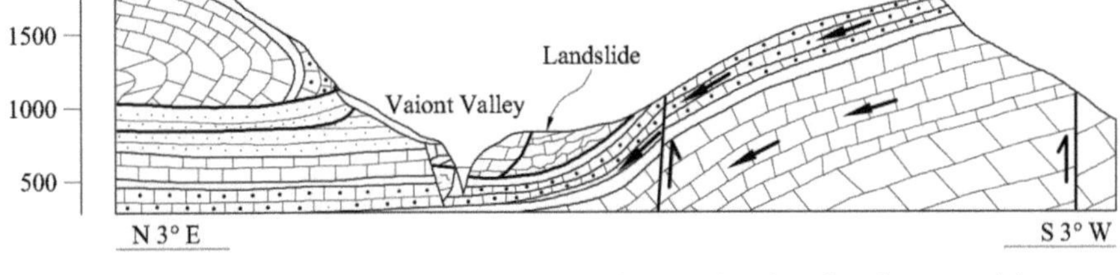

Figure 2.6 North (Monte Toc) to south (Monte Salta) section showing the general layout of the syncline, theVaiont gorge and the position of the ancient landslide (after Semenza and Ghirotti, 2000).

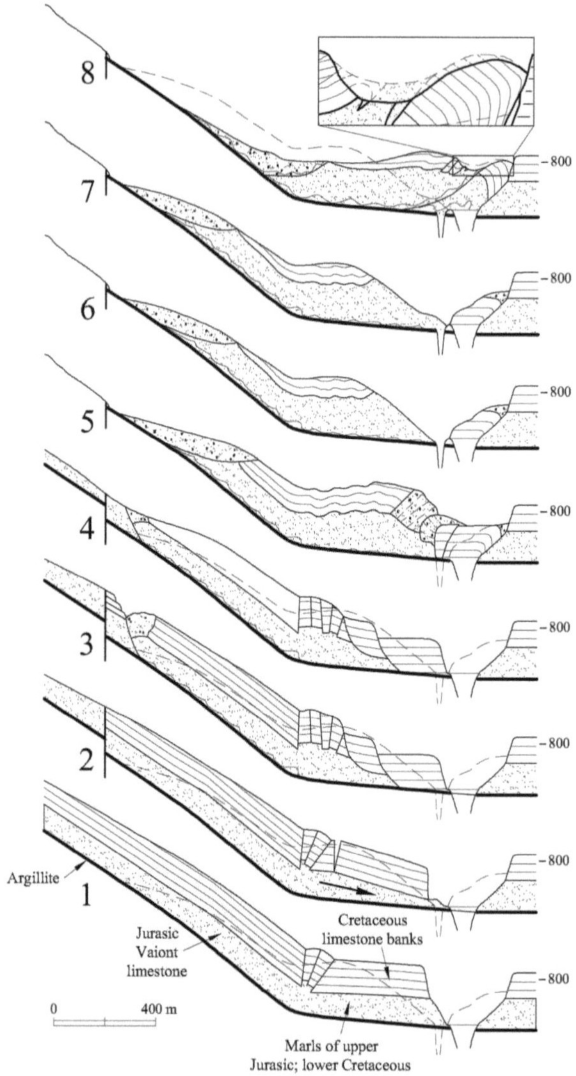

Figure 2.7 Tentative reconstruction of the paleo-slide of Vaiont. 1: Situation before the first motion (end of last glaciation?); 2: First motion of the slope; 3: Process of progressive sliding (undulated continuous line) and rotational slides at the toe; 4: Successive erosion phenomena on the upper parts; 5: Ancient landslide and intense fracturing of strata. The valley is invaded by the gigantic slide; 6: The slide before November 4, 1960, after thousands of years of erosion. The river has cut a new, narrow gorge; 7: The profile after a "small" landslide on November 4, 1960; 8: The final shape of the cross-section after the slide of October 8, 1963 (present situation). The inset shows an eroded part of the slide surface by the rapidly moving waters displaced by the slide (simplified and adapted from Semenza, 2001).

E. Semenza, an engineering geologist son of the dam designer, made important geological contributions to understand the geology of the site. In his book "La storia del Vaiont raccontata del geologo che ha scoperto la frana" ("The story of Vaiont told by the geologist who discovered the slide", Semenza, 2001), he includes a tentative reconstruction of the past history of the slide in a series of representative cross-sections, which are reproduced in Figure 2.7.

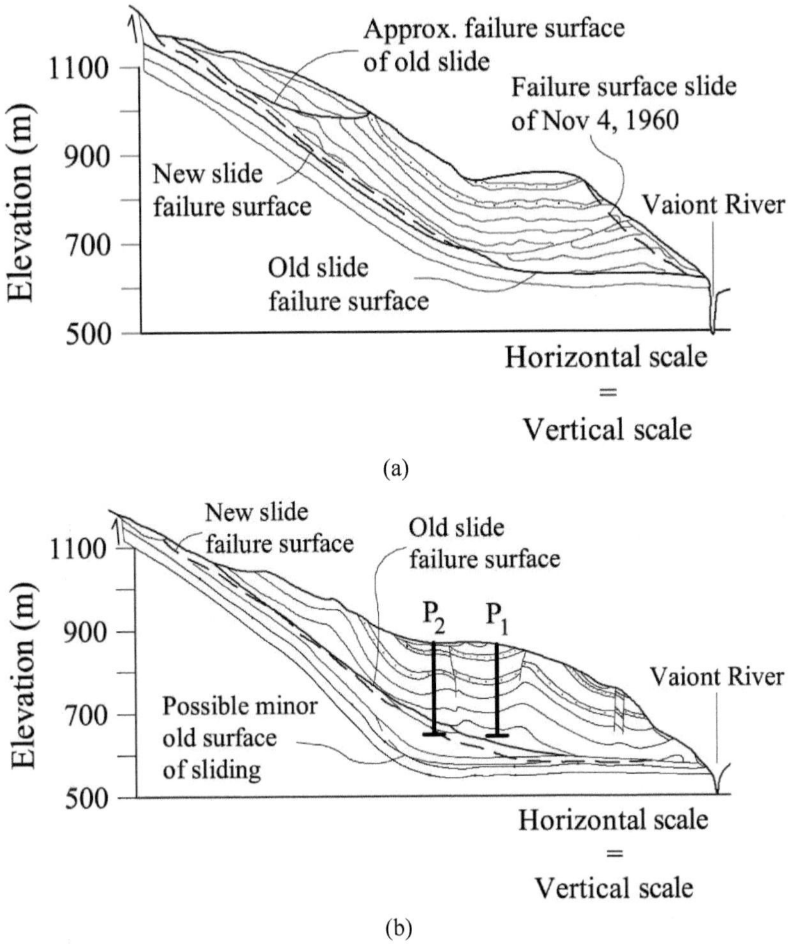

Figure 2.8 Two representative cross-sections of the landslide: (a) Section 2; (b) Section 5 (see the location in Fig. 2.2). After Hendron and Patton, 1985. The position and length of piezometers P_1 and P_2 are shown on Cross-section 5.

This reconstruction conveys a clear message from a geomechanical point of view: the failure surface, which was probably initiated several tens of thousands of years ago, has been subjected to an ever-increasing story of accumulated relative displacements. The second important point is that the rock mass affected by the

1963 landslide had suffered a history of cracking and "damage" during recent geological times. The sliding surface is located in strata of the upper Mälm period (upper Jurassic). Clays and marls were found in these layers (see the description of the failure surface below). Above the sliding surface, finely stratified layers of marl and limestone from the Mälm period were identified. Below the sliding surface, the Jurassic limestone banks of the Dogger period remained unaffected. In the upper part, limestone strata from the lower Cretaceous crowned the moving mass. In general, the folded layers of limestone and marl were strongly fractured (drilling water was often lost in the exploratory borings performed in 1960).

Two representative cross-sections of the slide, located upstream of the dam's position at distances of 400 and 600 m, respectively, are reproduced in Figure 2.8 (Sections 2 and 5; Hendron and Patton, 1985). The two cross-sections will be used later to analyze the stability conditions of the landslide.

2.3 The Sliding Surface

In their comprehensive report of 1985, Hendron and Patton (1985) describe the detailed investigation performed to identify the nature of the sliding surface. The conclusion is that thin (a few centimetres thick) continuous layers of high plasticity clay were consistently found in the position of the failure surface. A photograph of the surface is shown in Figure 2.9.

Figure 2.9 A striated continuous clay layer belonging to the sliding surface (courtesy of G. Fernández).

Samples from these clay layers were tested by different laboratories and the results are described in Hendron and Patton (1985). The clays were found to be highly plastic (a plasticity chart is given in Fig. 2.10), a result explained by their significant Ca-montmorillonite content. Liquid limits well in excess of 50% were often found. More recently, Tika and Hutchinson (1999) reported the values w_L = 50% and PI = 22%.

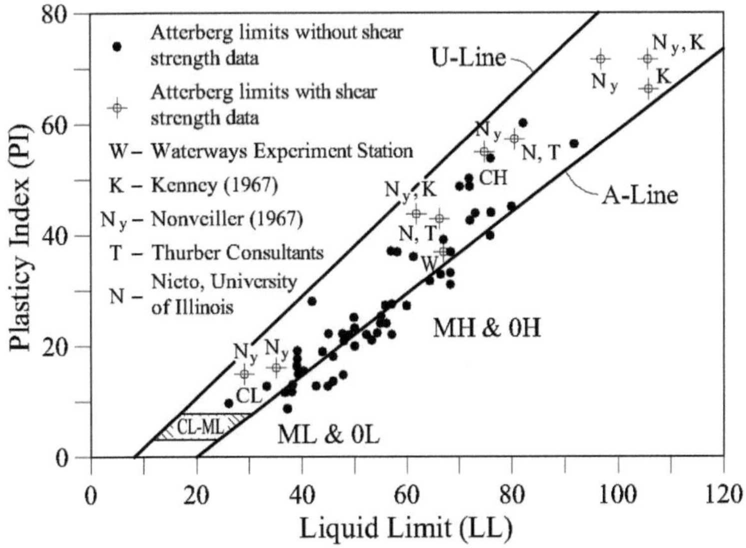

Figure 2.10 Plasticity of clay samples from the Vaiont sliding surface (Hendron and Patton, 1985).

Direct shear tests on remoulded specimens were also reported by Hendron and Patton (1985). In some cases, stress reversals were applied in order to find residual conditions. In fact, the past history of the landslide indicates that the residual friction angle was the relevant strength parameter along the failure surface. Measured average values of residual friction angle ranged between 8 and 10 degrees. These values are consistent with existing correlations between residual friction angles and clay plasticity (Lupini *et al.,* 1981). Tika and Hutchinson (1999) used the ring shear apparatus to find the residual strength. This test, conducted on remoulded specimens, approximates better the large relative shear displacements experienced in nature by the actual sliding surface. They also measured a residual friction angle of 10 degrees for a relative shear displacement in excess of 200 mm (Fig. 2.11a).

Tika and Hutchinson (1999) also examined the effect of the shearing rate. They found (Fig. 2.11b) a further reduction in residual friction which reached low values (5°) for shearing rates of 0.1 m/s, a velocity which is still far lower than the estimated sliding velocities of the real failure. However, it is a common experience that increasing strain rate leads to an increase in the strength of soils. More data on the effect of the shearing rate on residual strength is probably needed before reaching definite conclusions on this issue.

Hendron and Patton (1985) estimate that some factors (areas of the sliding surface without clay, some localized shearing across strata, irregularities in the geometry of the sliding surface) could increase the average residual friction angle operating in the field and they estimate that $\varphi'_{res} = 12°$ is a good approximation for static conditions.

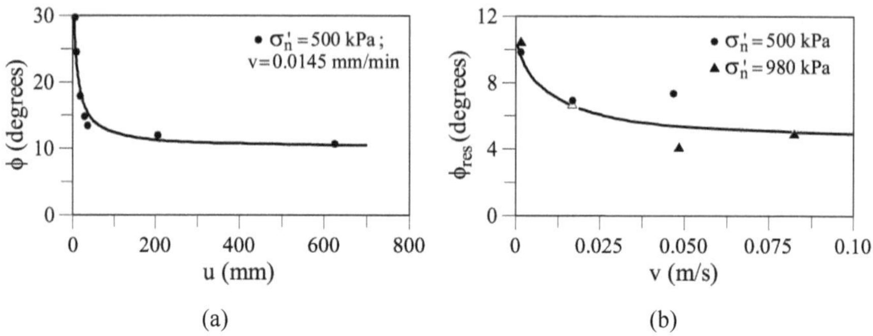

Figure 2.11 Ring shear tests on a clay specimen from the vicinity of the Vaiont sliding surface: (a) static residual friction determined at a shearing rate of 0.0145 mm/min; (b) effect of shearing rate (Tika and Hutchinson, 1999).

2.4 Monitoring Data before the Slide

Significant monitoring data taken during the three years preceding the failure were given in Figures 2.3 and 2.4. The main purpose behind the limited instrumentation available was to relate the level of the reservoir with the measured vertical and horizontal displacements of a number of topographic marks distributed on the slide surface. Data on horizontal displacements, plotted as a function of position and time in several profiles following the south-north direction in Figure 2.2, suggest that the slide was essentially moving as a rigid body. The direction of the slide is also indicated in the figure by several arrows. Some of them (small arrows along the peripheral crack) indicate that the moving mass was actually detaching from the stable rock, implying no friction resistance along the eastern and western boundaries of the slide.

Seismic (volumetric P-wave) velocities were measured in central parts of the slide in December 1959 and again in December 1960. A drop in velocity from v_p = 5–6 km/s in 1959 to v_p = 2.5–3 km/s was recorded. This information may be interpreted as an indication of the progressive weakening of the rock mass due to the distortion induced by the creeping motion of the slide. The velocities initially recorded at the end of 1959 are very high and they correspond to a rock of good quality (Barton, 2007). This is perhaps surprising in view of the prehistoric landslide motions described above. The strong drop in seismic velocity in just one year, which is a tiny fraction of time within the complex life of the landslide, seems exaggerated but it is pointing towards significant shear distortions within the rock mass, motivated by the first impoundment of the reservoir which implied a raise of the water level of 200 m (see the history of events in Fig. 2.3). The associated increase in pore water pressures on the sliding surface is very large and it is unlikely that past rainfall events could have produced such a strong drop in effective stress, especially in the lower part of the slide.

It should be emphasized that these P-wave velocities are much higher than the velocities measured in soils, even if they are dense and compact. In other words, the strength that may be associated with the shearing of the rock mass above the sliding surface is orders of magnitude larger than the strength available at the clay-dominated thin layers at the base of the slide, being sheared along sedimentation planes of very high continuity.

2.5 Water Pressures and Rainfall

The position of piezometers (they were open perforated pipes) was indicated, in plan view, in Figure 2.2 and in cross-section in Figure 2.8. A perforated pipe only provides information on the average water pressures crossed by the tube. Note too that the pipes did not reach the position of the sliding surface. Therefore, they did not provide direct information on the water pressures actually existing in the vicinity of the sliding surface, which is fundamental information to perform a drained stability analysis of the landslide.

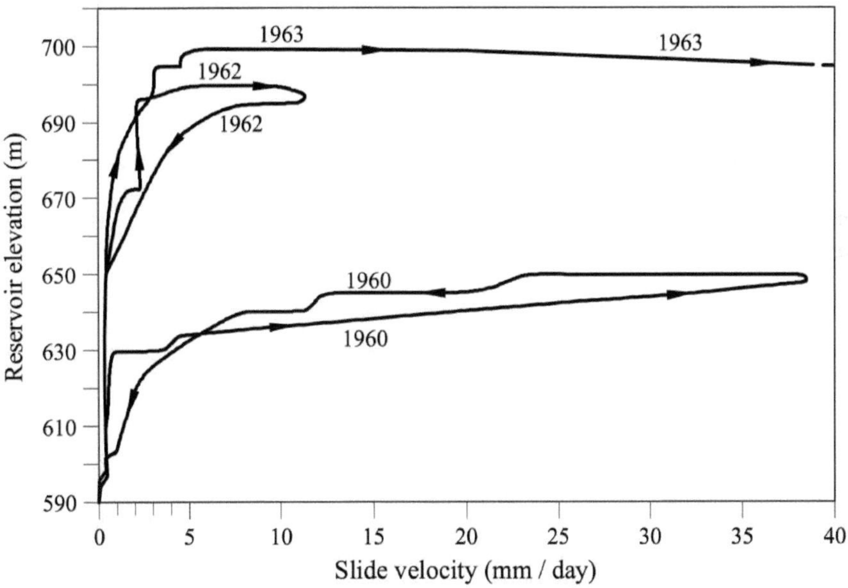

Figure 2.12 Relationship between water level in the reservoir and sliding velocity (courtesy of G. Fernández).

In general, the water levels recorded by the piezometers follow closely the changing levels of the reservoir (compare Figs. 2.3b and 2.3d). The exception is Piezometer 2, at least during the initial part of the recording period. The initial readings in this piezometer indicated water pressures significantly above (90 m of water column) the reservoir surface. This information has been interpreted as an indication of additional factors, other than the level in the reservoir, which may control the water pressure at the sliding surface. Since the cretaceous limestone affected by karstic phenomena is a rather pervious mass, rainfall water infiltrating at high elevations may result in artesian pore pressures against the impervious Mälm formations located at the base of the landslide. Arrows showing the circulation of water in Figure 2.6 illustrate this possibility. However, no further and direct evidence of this possibility was recorded. On the other hand, the simultaneous variation of piezometer and reservoir levels is a good indication of the high permeability of the rock mass above the sliding surface.

When the water level in the reservoir is plotted against the recorded slide velocity (Fig. 2.12), an interesting result is obtained. An increasing water level leads to an increase in sliding velocity. The relationship is highly nonlinear and it tends towards an asymptotic limit, which is an indication of failure. The problem with Figure 2.12 is that this relationship is not unique, a result which is not expected if the slide motion is thought to be governed by the effective normal stresses acting on the sliding surface, which, in turn, are controlled by the reservoir level. In fact, the second reservoir filling led to a second asymptotic value for the water level in the reservoir.

This result was probably the main reason behind the decision to increase the water level for the third time in search of a higher (but safe) level in the reservoir, which would allow the normal operation of the dam. The idea behind this decision, apparently put forward by L. Müller, is that the rock reacts in a different way when it is wetted for the first time, compared with its reaction when it has previously been wetted. There is no fundamental mechanical basis for this proposition, however. The fact is that, during the third attempt to raise the water level, displacement velocities increased continuously and the final attempts to reduce the velocity of the slide, by lowering the level of the reservoir (Fig. 2.3b), did not work.

An explanation for the apparent inconsistency of results in Figure 2.12 could be found if the reservoir water level and rainfall are combined in the spirit that the prevailing water pressures on the sliding surface, irrespective of their origin, should control the stability.

Hendron and Patton (1985) found a reasonably good explanation if rainfall, averaged over the preceding 30 days, and water level are jointly considered to explain the landslide velocity (Fig. 2.13). The boundary line between "stable" and "unstable" situations, plotted in Figure 2.13, could even provide the equivalent reservoir elevation for a given rainfall intensity.

The actual failure occurred for a 30-day precipitation of 240 mm, when the reservoir was at an elevation of 700 m. Leonards (1987) analyzed further the rainfall records and the history of reservoir elevation and could not find a satisfactory explanation, free of inconsistencies, for the relationship between velocities of the slide, reservoir elevation, and previous rainfall. The pore pressure regime prevailing at the sliding surface remains rather uncertain in the Vaiont landslide.

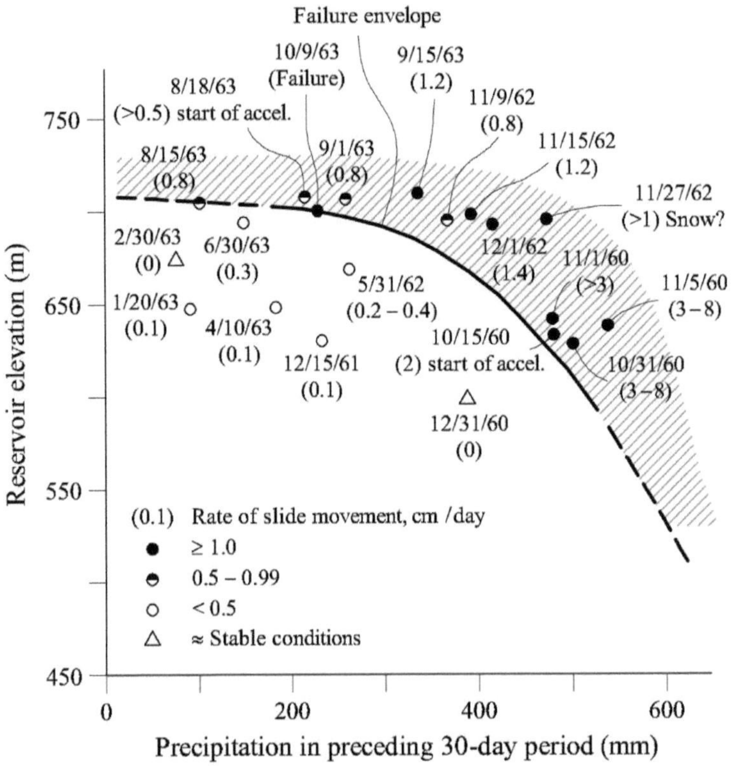

Figure 2.13 Sliding rate related with precipitation (averaged over the preceding 30-day period) and reservoir elevation (Hendron and Patton, 1985).

2.6 A Simple Stability Model

The two representative cross-sections, 2 and 5 in Figure 2.8, are represented in Figure 2.14 in a simplified version, which is, however, close to the original drawings. The two plots highlight that the failure surface could be described by two planes: a lower horizontal plane daylighting at the river canyon wall and an inclined planar surface. A rock wedge whose thickness decreases upwards rests on the inclined plane. The rock mass reaches its maximum thickness, 270 m, in the central lower part of the slide, above the horizontal sliding plane.

A good proportion of reported stability analyses of Vaiont, especially in the years following the failure, have concentrated on the determination of the friction angle necessary for stability (Jaeger, 1965; Nonveiller, 1967; Mencl, 1966; Skempton, 1966; Kenney, 1967). Classic procedures for stability analysis in soil mechanics using limit equilibrium methods were used. These methods explain the instability for friction angles in the range 18 – 28°. The preceding account of the relevant information on Vaiont, namely the data presented by Hendron and Patton (1985) indicates, however, that the friction angle at the failure surface could hardly be larger than 12 degrees.

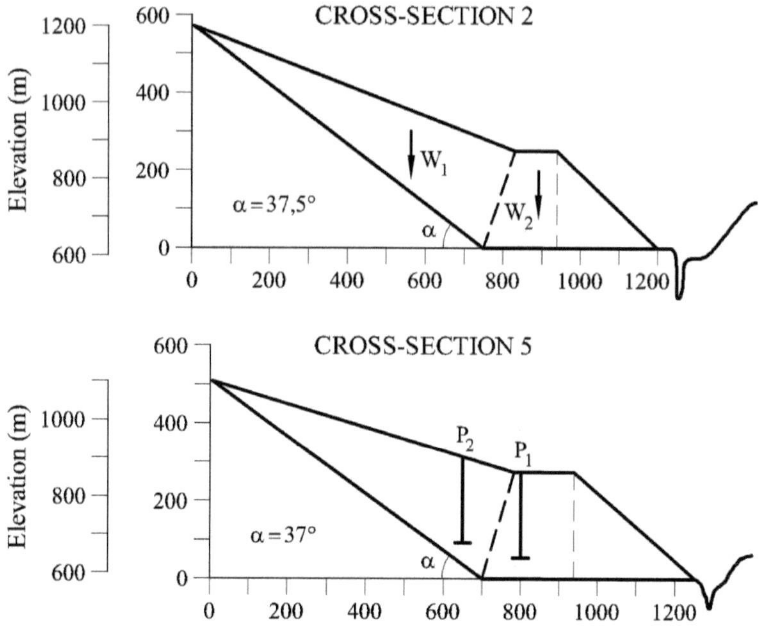

Figure 2.14 Cross-sections 2 and 5 of the Vaiont landslide. Initial geometry.

Two main reasons support this statement: the fact that Vaiont was a case of landslide reactivation (which implies large previous shearing displacements at the sliding surface and, hence, a clear situation of residual strength conditions) and the small residual friction angles (8–10°) measured in the highly plastic clays (Ca-montmorillonite rich) associated with the sliding surface. Therefore, a relevant question is: are the representative cross-sections in Figure 2.14 stable, given the value of the basal friction angle and the estimated conditions of pore water pressure, when the reservoir reached elevations in the range of 650 to 700 m?

The cross-sections plotted in Figure 2.14 suggest that the slide may be defined as two interacting wedges: an upper one (Wedge 1) sliding on a plane having a dip of 36–37° and a lower one (Wedge 2) sliding on a horizontal plane. Since a (common) friction angle of 12 degrees is acting at the basal sliding surfaces, the upper wedge is intrinsically unstable and will push the lower resisting wedge. The weights of the two wedges and the distribution of pore water pressures prevailing on the sliding plane will, as a first approximation, dictate the stability conditions. However, the interaction between the two wedges also plays a relevant role in explaining the stability, as discussed below.

2.6.1 Kinematics of the slide

It is worth at this point to examine the kinematics of the slide. If the motion starts, one may imagine the slide as a train sliding downwards, an image which is brought to justify that the absolute velocity in the upper and lower parts of the slide are essentially the same. Surveying data plotted in Figure 2.2 support this simple hypothesis, which is to be expected in the reactivation of an old landslide. The difference in velocity (or displacement) when comparing the upper and lower parts of the slide obviously lies in the direction of these vectors: they will be parallel to the underlying failure surface. A conflict arises, however, at the kink or junction between the two sliding planes. Within the train analogy, if the wagon passing over this kink is to maintain contact with the kinked rail, it will be bent and sheared. It is hard to imagine that voids will develop in the layered sequence of marl and limestone at 270 m depth. The alternative is the bending and shearing of strata. In fact, a single shearing plane may be invoked to accommodate the sudden change in the direction of velocity at the kink. This is indicated in Figure 2.15, where sliding velocity vectors $\mathbf{v}_1$ (in the direction of the upper inclined surface) and $\mathbf{v}_2$ (horizontal, parallel to the basal plane) are plotted with a common origin. This velocity diagram represents the conditions at the kink (point A), where the rock approaches A with velocity $\mathbf{v}_1$ and leaves it with velocity $\mathbf{v}_2$. Since the absolute velocity of the two wedges is the same, the relative motion of the two wedges (vector $\mathbf{v}_{12}$) is directed in the direction of the bisector of the angle between the upper and lower sliding surfaces. Therefore, a change in the direction of the velocities of the two wedges may be accommodated by a relative shear in the direction of the bisector plane, plotted in Figure 2.14.

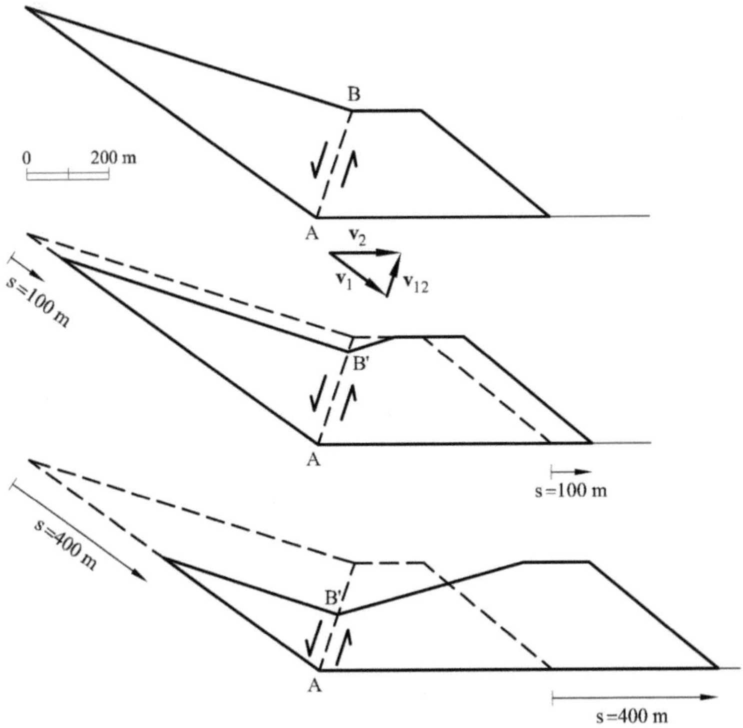

Figure 2.15 Kinematics of sliding. Section 5.

The motion of the slide implies that the (unstable) mass from the upper wedge becomes the (stable) mass of the lower wedge. In this process, the sliding resistance along the common plane separating the two wedges has to be overcome. If it is accepted, because of the preceding discussion, that the common plane of intense shear bounding the two wedges is the bisector plane (Fig. 2.15), the evolution of the geometry of the sliding mass may be approximated by the successive cross-sections shown in Figure 2.15 for total slide displacements $s = 0$ m, $s = 100$ m and $s = 400$ m. Figure 2.15 is a graphic expression of the condition of mass conservation during landslide motion. It will be used later to perform a dynamic analysis of the failure.

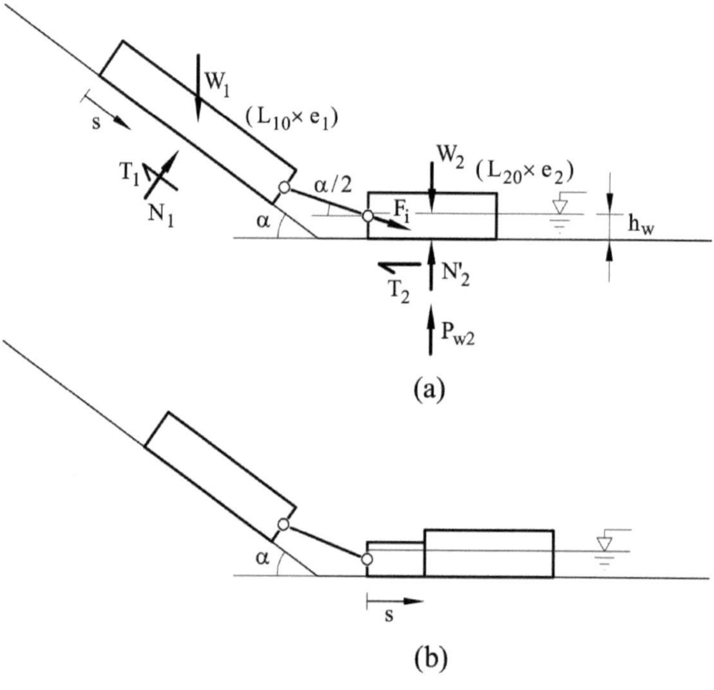

Figure 2.16 Two-block model of the Vaiont slide: (a) definition of geometry and forces (initial stage); (b) the slide after a displacement s.

2.6.2 Two-block model

Consider in Figure 2.16, the "unstable" and "stable" blocks mentioned before in a very simple representation: two solid blocks connected by double hinged bar normal to the bisector plane. The interaction between the two blocks is simply given by a force, F_i. Note that this force introduces normal and shear forces on the common plane between the two blocks. The lower block is partially submerged and the level of water has a height h_w with respect to the lower horizontal sliding plane. The upper block is not affected by water.

The sketch in Figure 2.16a provides a definition of forces acting on each block. A simple problem is defined as follows: find the angle of basal shearing resistance for equilibrium. This is an elementary problem in mechanics which is solved by expressing equilibrium of forces for each block and then forcing a common value for the interaction between the two blocks. Static equilibrium expressions (normal and parallel to the direction of sliding) are written as follows, in terms of effective stresses:

- Upper block 1:

$$W_1 \cos\alpha + F_i \sin(\alpha/2) = N_1, \tag{2.1a}$$

$$W_1 \sin\alpha = T_1 + F_i \cos(\alpha/2), \tag{2.1b}$$

$$T_1 = N_1 \tan\varphi_b', \tag{2.1c}$$

since no water is acting on the upper sliding block, $N_1 = N_1'$.

- Lower block 2:

$$W_2 + F_i \sin(\alpha/2) = N_2' + P_{w2}, \tag{2.2a}$$

$$F_i \cos(\alpha/2) = T_2, \tag{2.2b}$$

$$T_2 = N_2' \tan\varphi_b', \tag{2.2c}$$

where $\tan\varphi_b'$ is the effective friction coefficient on the sliding planes.
Isolating F_i in (2.1) and (2.2), respectively, and making them equal, results in

$$\frac{W_1(\sin\alpha - \cos\alpha\tan\varphi_b')}{\sin(\alpha/2)\tan\varphi_b' + \cos(\alpha/2)} = \frac{(W_2 - P_{w2})\tan\varphi_b'}{\cos(\alpha/2) - \tan\varphi_b'\sin(\alpha/2)}, \tag{2.3}$$

which is a second-order algebraic equation for $\tan\varphi_b'$. The volumes of blocks 1 and 2 are estimated as follows for Section 5: V_{10} = 112,590 m^3/m and V_{20} = 93,000 m^3/m, where the subscript $_0$ indicates initial value (no displacement of the slide). The indicated volumes correspond to a landslide "slice", one meter thick.
The value of P_{w2} may be calculated as $P_{w2} = L_{20}h_w$ if a length for Block 2 is estimated. The length of the basal horizontal plane in Figures 2.14 or 2.15 is L_{20} = 560 m. Finally, a specific weight, γ_r = 23.5 kN/m^3 was taken for the rock in order to compute the weights of the blocks. Accepting these values, the following friction angles are derived for Cross-section 5 (α = 37°):

$$\varphi_b' = 21.1° \text{ for } h_w = 120 \text{ m},$$

$$\varphi_b' = 19.4° \text{ for } h_w = 60 \text{ m}.$$

The lower horizontal plane in Section 5 is approximately at elevation 590 m (Fig. 2.13) and the maximum reservoir level attained was 710 m (Fig. 2.3).

Therefore, the first case defines the maximum water pressure experienced by the lower block before the failure. Slide displacements (which, in practice, are interpreted as a condition of strict static equilibrium) were also recorded at lower water elevations (h_w = 60 m, which corresponds to the situation in November 1960, see Fig. 2.3). However, the actual pore water pressure is also controlled by the rainfall regime, as previously discussed, and uncertainties remain on the actual value of the operating pore water pressures against the sliding surface.

Despite its simplicity, the block model provides some hints on the effect of water level and slide displacement on safety factor. If $\varphi_b' = 21.1°$ is taken as the real effective friction angle along the failure surface, the safety factor, F, is defined as

$$F = \frac{\tan(21.1°)}{\tan(\varphi'_{\text{mob}})},$$

where φ'_{mob} is the "mobilized" friction angle, i.e. the friction angle that ensures strict equilibrium for another situation of the slide and, in particular, for changing water levels in the reservoir. Values of φ'_{mob} were calculated through Equation (2.3) for different values of h_w and the calculated safety factor is plotted in Figure 2.17a. The explanation of this figure is straightforward: as water level increases, it reduces the effective weight of the lower block, (W_2 – P_{w2}), and the friction required for equilibrium has to increase. Note, however, that the upper block is not affected by the water level in this simplified model, a situation that may change in other cases. In Vaiont, as shown later, the maximum reservoir level introduces pore water pressures in the lower part of the upper wedge. It should be added that the trend shown in Figure 2.17a (decreasing safety factor as the water level increases) is not a general result for other slide geometries and stronger changes in water elevation.

The effect of changing geometry as the slide is set in motion, may be also analyzed. Figure 2.16b includes a proposal to transfer mass from the upper block to the lower one. It is a rough approximation to the more refined model sketched in Figure 2.15. It simply states that the current weights of the two blocks, for a slide displacement s is given by

$$W_1 = W_{10} - e_1 s \gamma_r, \tag{2.4a}$$

$$W_2 = W_{20} + e_1 s \gamma_r, \tag{2.4b}$$

where e_1 is the thickness of the upper block ($V_{10} = L_{10}e_1$; for Section 5, L_{10} = 700 m and the volume of the upper block is V_{10} = 112,590 m^3/m; therefore, e_1 = 160.8 m). In addition, the water uplift under block 2 is calculated as $P_{w2} = (L_{20} + s)h_w$.

Equation (2.3) provides again the value of φ'_{mob} for the current weights, and therefore safety factors may be found for increasing slide displacements. They are plotted in Figure 2.17b, for Cross-section 5. The result is to be expected: the moving slide becomes progressively more stable because the lower stabilizing weight increases at the expense of the upper unstable block whose mass is continuously decreasing.

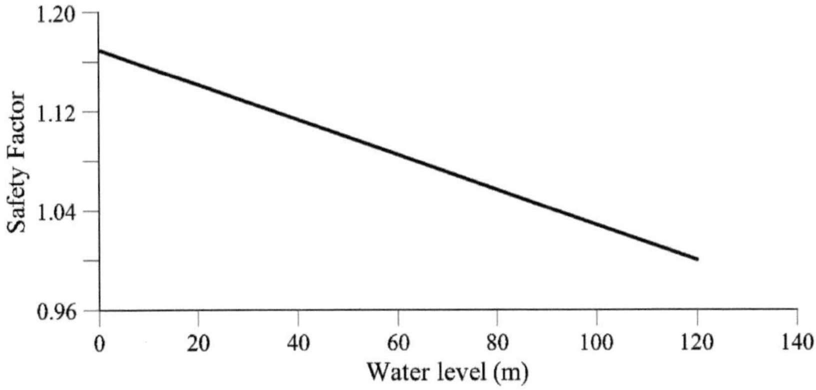

(a)

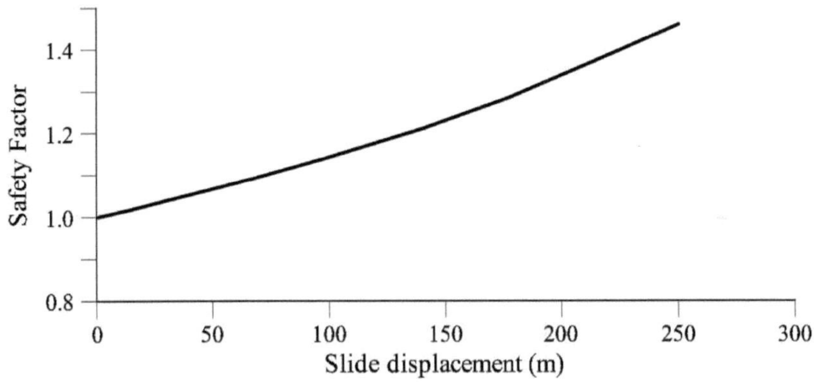

(b)

Figure 2.17 Two-block model. Effect of (a) water level – for zero displacement – and (b) slide displacement – for h_w = 120 m – on safety factor. Section 5.

Unfortunately, the real behaviour of Vaiont was totally different: it accelerated downwards despite the prediction of the simple two-block model. Somehow, the resisting forces had to decrease substantially in order to transform a self-stabilizing mechanism (the two-block model) into an increasingly unstable mass, able to accelerate.

The two-block model has a further limitation: the effective friction angle for equilibrium (φ'_b = 21.1° for h_w = 120 m or φ'_b = 19.4° for h_w = 60 m, both in Cross-section 5; the "small" difference is non-relevant here) is far higher than the residual friction angle, φ'_{res} = 12°, which is the most likely value as justified above. This is an inconsistent result which indicates that the simple two-block model is too crude to represent the actual conditions of the Vaiont slide (equally inconsistent results are obtained for Cross-section 2).

The next step will be to remove some of the limitations of the simple two-block model in order to approximate more realistically the sliding conditions summarized in Figure 2.15.

2.6.3 Two interacting wedges

Shearing across the common plane AB between the upper and lower wedges (Fig. 2.15) has a direction approximately perpendicular to the sedimentation planes of marls and limestones of the Mälm period overlying the failure surface. The shear resistance offered by plane AB is difficult to estimate because of the intricate geometry involved at several scales and the limited continuity of joints. Some researchers in rock mechanics, notably E. Hoek, have made efforts to provide an answer to this difficult problem from a practical perspective. An account of Hoek's work may be found in the rock mechanics textbook (Hoek, 2007).

Following Hoek, the strength of rock masses may be approximated if some basic characteristics are determined (rock matrix unconfined strength; degree of jointing and state of the surfaces, lithology, etc.). As an example, Figure 2.18 shows the strength envelope in a Mohr stress plane for a rock mass that may approximate the Mälm layers above the sliding surface of Vaiont. The envelope was defined using the free access "virtual laboratory" found on the preceding web page. Details of the defined rock mass are given in the caption of Figure 2.18. It may correspond to the Vaiont rock mass, which was described as follows by Müller (1987), after the failure:

"The part of the stratigraphic column exposed in the slide mass consists of beds of partially crystalline limestones, limestones with hard siliceous inclusions, marly limestones, and marls. Many beds are strongly folded and show indications of slope tectonics. Its geological structure and also its geological sequence has remained essentially unchanged. The entire rock mass remained intact and the sediment facies is nearly unchanged. Apart from some newly formed faults, the only other effects of the slide were the opening of existing joints and the development of new joints, resulting in an overall volume increase of 4 – 6% and an associated reduction of the mechanical coherence of the rock mass."

The strength envelope is nonlinear but a Mohr – Coulomb approximation is also shown in Figure 2.18 for a range of normal stresses centered at $\sigma'_n = 2$ MPa, a stress which may represent average conditions on the bisector plane AB (Fig. 2.15). The Mohr – Coulomb strength parameters ($c'_r = 0.787$ MPa; $\varphi'_r = 38.5°$) define the linear Mohr – Coulomb approximation in Figure 2.18.

The relevant point is that the shear plane AB may offer a substantial resistance to be sheared and this resistance probably has a significant role in stability. Shearing across a rock mass is typically associated with the release of energy. In fact, in the years preceding the failure, when three attempts to fill the reservoir were made, seismic events were recorded on the slide surface. Their location is plotted in Figure 2.2. They approximately span, in plan view, the position of the shear plane AB plotted in Figure 2.15. Nonveiller (1987), quoting a report on these shocks mentions that "[...] *the shocks generated in the zone of the slide signify dilation of the material in a zone of sagging of the rock*".

These events had an increasing frequency in periods of slide acceleration, when the reservoir level increased. This is shown in Figure 2.4, where seismic events are plotted as small marks on the time axis (lower part of the figure).

It was also reported that the rock experienced a global degradation, reflected in a substantial drop of P-wave velocities, as a result of the slide motion during the period December 1959–December 1960. All this evidence supports the conclusion that a rock mass around the position of the ideal shear plane AB was subjected to intense shearing during the cycles of filling and emptying the reservoir in the years previous to the failure.

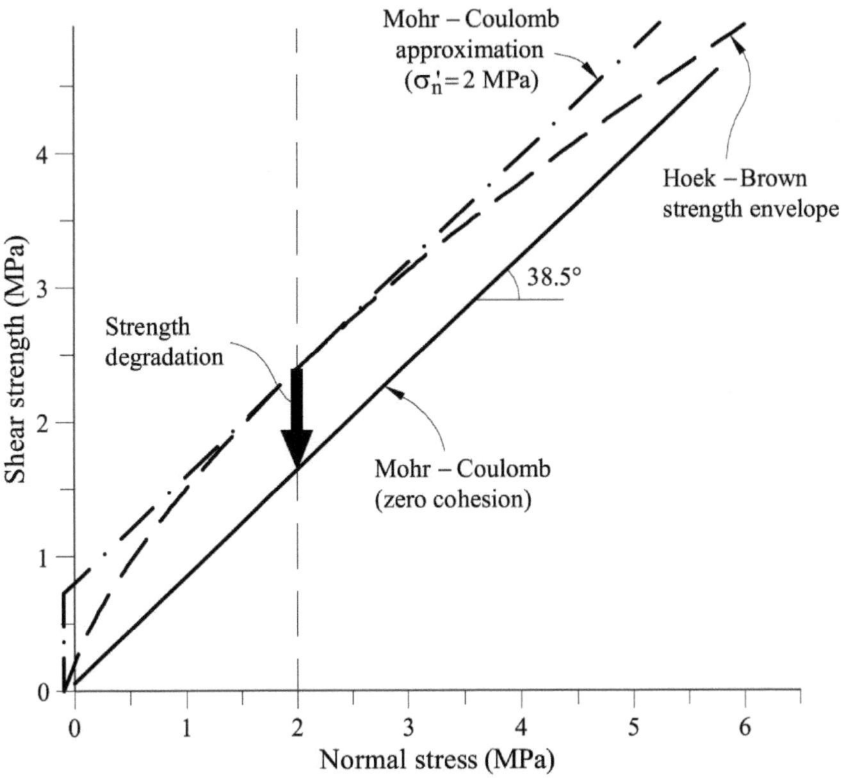

Figure 2.18 Strength envelope of a rock mass described as: strength of intact material: 50 MPa (limestone-claystone); Hoek Geological Strength Index (GSI = 50) (very blocky, interlocked, and partially disturbed, with multifaceted angular blocks formed by four or more joint sets), Hoek m_i parameter $m_i = 9$ (marls, soft limestones); degradation parameter $D = 0.5$ (in a scale 0 to 1) (according to the Hoek – Brown classification of rock masses; see www.rocscience.com). Also shown is the Mohr – Coulomb approximation for a normal stress of 2 MPa ($c_r' = 0.787$ MPa, $\varphi_r' = 38.5°$) and an arrow showing the degradation of cohesive intercept at constant φ_r' value.

A loss of strength (*reduction of mechanical coherence* in Müller's words) was certainly a consequence of this straining. Typically, cohesion is first lost but friction tends to remain without much change. This drop of cohesion as a result of straining along plane AB was shown in Figure 2.18. In the model described below, the apparent cohesion in the shear plane AB will be reduced as the slide moves forward.

Going back again to Figure 2.15, as slide displacement increases, "new" planes of rock cross the shearing position AB that remains fixed at the position of the bisector plane, which is independent of the slide motion. The consequence is that the shear strength along this plane will not decrease in a sudden and intense manner. Certainly, the motion of the slide will have some weakening effect, which is difficult to quantify. Finally, to complicate matters, progressive failure mechanisms along AB are to be expected in view of the brittle nature of rock strength, a phenomenon which will not be considered here but is mentioned because it will tend to partially destroy the strength available along shear plane AB.

A model based on the interaction of two wedges will now be developed. The main assumptions are:

- The upper and lower wedges change their geometry during sliding, as shown in Figure 2.15. The upper wedge looses mass which is added to the lower one.
- During the movement, the common plane AB reduces in length. Shearing across this plane (or, more generally, AB') is described by a Mohr-Coulomb strength criterion ($\tau = c_r' + \sigma' \tan\varphi_r'$). In addition, the cohesive intercept, c_r', is made dependent on slide displacement, s. This is a simplified procedure to introduce strength degradation of the rock mass during the slide motion. The friction angle is maintained constant.
- The lower sliding surface is assumed to be in residual conditions with strength parameters ($c' = 0$; $\varphi_b' = 12°$).
- Pore water pressures are given by a horizontal phreatic level.
- Equilibrium conditions are formulated in dynamic terms. In this way, it will be possible to analyze the effect of strength degradation of shearing plane AB' on slide motion. Static conditions of equilibrium are a particular case of the dynamic case. Only inertia terms are considered. No viscous effects are introduced.

The analysis follows the general procedure advanced before when considering the two hinged blocks but now dynamic equilibrium is fomulated: Newton's Second Law will be written for the upper and lower wedge, and a common interaction force across plane AB will be enforced. Newton's second law for a solid body motion states that the derivative of the solid momentum (mass times velocity) is balanced by the sum of forces acting on the body. Note that the mass of each wedge depends on displacement and therefore the term of time variation of mass can not be simplified when the time derivative of momentum is developed.

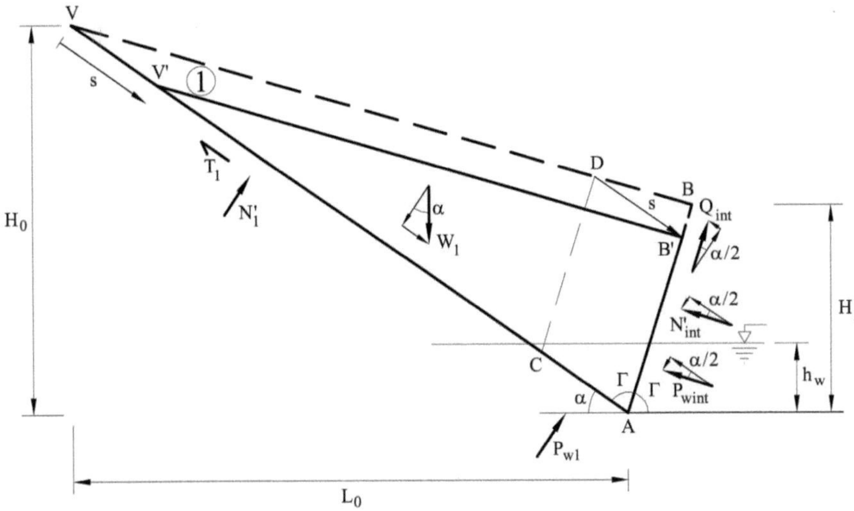

Figure 2.19 Geometry and forces on the upper wedge (1).

Upper Wedge (1)

Consider the wedge geometry and external forces in Figure 2.19. Dynamic equilibrium parallel to the motion (displacement s; velocity v = ds/dt) reads

$$W_1 \sin\alpha - T_1 - N'_{\text{int}} \cos(\alpha/2) - Q_{\text{int}} \sin(\alpha/2) - P_{w\text{int}} \cos(\alpha/2) = \frac{\mathrm{d}(M_1 v)}{\mathrm{d}t}, \tag{2.5}$$

where M_1 is the mass of Wedge 1, ($W_1 = M_1 g$; g: gravity acceleration). The time derivative of the right-hand side of Equation (2.5) can be developed as

$$\frac{\mathrm{d}(M_1 v)}{\mathrm{d}t} = M_1 \frac{\mathrm{d}v}{\mathrm{d}t} + \frac{\mathrm{d}M_1}{\mathrm{d}t} v \tag{2.6}$$

Equilibrium in normal direction to the basal sliding plane:

$$W_1 \cos\alpha - N'_1 + N'_{\text{int}} \sin(\alpha/2) - Q_{\text{int}} \cos(\alpha/2) - P_{w1} + P_{w\text{int}} \sin(\alpha/2) = 0 \tag{2.7}$$

where the interaction forces Q_{int} and N'_{int} are related through

$$Q_{\text{int}} = c'_r AB' + N'_{\text{int}} \tan\varphi'_r. \tag{2.8}$$

In addition, the shear resistance on the base of the wedge is given by

$$T_1 = N'_1 \tan\varphi'_b. \tag{2.9}$$

The motion Equation (2.5), in view of (2.7), (2.8), and (2.9), becomes

$$W_1 s_1 - N'_{\text{int}} s_2 + c'_r AB' s_3 - P_{w\text{int}} s_4 + P_{w1} \tan\varphi'_b = \frac{\mathrm{d}(M_1 v)}{\mathrm{d}t}, \tag{2.10}$$

where s_i are trigonometric constants, given by

$$s_1 = \sin\alpha - \tan\varphi_b' \cos\alpha, \tag{2.11a}$$

$$s_2 = \tan\varphi_b' \sin(\alpha/2) - \cos(\alpha/2)\tan\varphi_r' \tan\varphi_b' + \cos(\alpha/2) + \sin(\alpha/2)\tan\varphi_r', \tag{2.11b}$$

$$s_3 = \tan\varphi_b' \cos(\alpha/2) - \sin(\alpha/2), \tag{2.11c}$$

$$s_4 = \tan\varphi_b' \sin(\alpha/2) + \cos(\alpha/2). \tag{2.11d}$$

The effective interaction normal force, at this stage unknown, can be isolated from Equation (2.10):

$$N'_{\text{int}} = \frac{1}{s_2}\left(W_1 s_1 + c_r' AB' s_3 - P_{w\text{int}} s_4 + P_{w1} \tan\varphi_b' - \frac{\mathrm{d}(M_1 v)}{\mathrm{d}t}\right). \tag{2.12}$$

When the wedge slides a distance s along the basal plane, the length of the shear plane reduces from AB to AB′ (Fig. 2.19). Since triangles AVB and AV′B′ are similar, it is easy to find

$$AB' = \frac{L_0/\cos\alpha - s}{L_0/\cos\alpha}\frac{H_1}{\cos(\alpha/2)}, \tag{2.13}$$

where H_1 is the initial thickness of the lower wedge over the sliding plane (Fig. 2.19).

The volume of Wedge 1 can be expressed as a function of the initial geometric parameters and the displacement s as

$$V_{\text{Wedge 1}} = \frac{1}{2}\left(\frac{L_0}{\cos\alpha} - s\right)^2 \frac{H_1}{L_0}\frac{\cos\alpha}{\cos(\alpha/2)} \tag{2.14}$$

The mass and weight of the wedge can be now easily calculated by multiplying the volume of Equation 2.14 by the density (δ_r) and unit weight (γ_r) of the rock, respectively.
Time variation of mass can be obtained as follow:

$$\frac{\mathrm{d}M_1}{\mathrm{d}t} = \delta_r \frac{\mathrm{d}V_{\text{Wedge 1}}}{\mathrm{d}t} = -\delta_r\left(\frac{L_0}{\cos\alpha} - s\right)\frac{H_1}{L_0}\frac{\cos\alpha}{\cos(\alpha/2)}\frac{\mathrm{d}s}{\mathrm{d}t}, \tag{2.15}$$

where the time variation of the displacement ($\frac{\mathrm{d}s}{\mathrm{d}t}$) is equal to the velocity v.

Lower wedge (2)

The wedge geometry and external forces are given in Figure 2.20. The wedge is shown displaced forward at distance *s*.

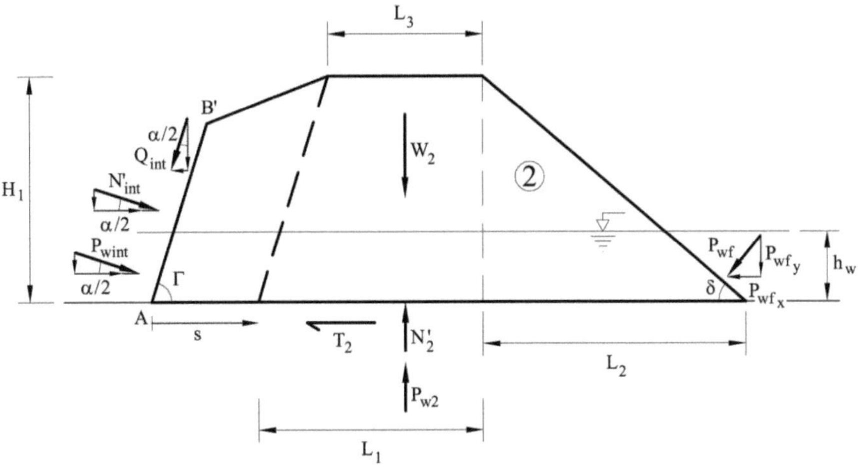

Figure 2.20 Geometry and forces on the lower wedge (2).

Dynamic equilibrium parallel to the direction of motion at a velocity $v = \mathrm{d}s/\mathrm{d}t$ reads

$$N'_{\text{int}}\cos(\alpha/2) - Q_{\text{int}}\sin(\alpha/2) - T_2 = \frac{\mathrm{d}(M_2 v)}{\mathrm{d}t}, \tag{2.16}$$

where M_2 is the mass of Wedge 2 ($W_2 = M_2\, g$; g: gravity acceleration). Note that the horizontal components of the water pressure forces $P_{w\text{int}}$ and P_{wf} acting on the slope surface are equal and opposite in sign. The terms on the right-hand side of the Equation (2.16) can be developed following Equation (2.6) and, since the total mass of the slide is constant, the time variation of M_2 will be equal to the time variation of M_1 indicated in Equation (2.6) but with an opposite sign.

The base resistance is given by

$$T_2 = N'_2 \tan\varphi'_b\,. \tag{2.17}$$

Taking Equation (2.8) into account, Equation (2.16) becomes

$$N'_{\text{int}}\cos(\alpha/2) - N'_2\tan\varphi'_b - (c'_r AB' + N'_{\text{int}}\tan\varphi'_r)\sin(\alpha/2) = \frac{\mathrm{d}(M_2 v)}{\mathrm{d}t}. \tag{2.18}$$

Equilibrium in a normal direction to the horizontal sliding plane reads:

$$\begin{aligned} W_2 - N'_2 + N'_{\text{int}}\sin(\alpha/2) + (c'_r AB' + N'_{\text{int}}\tan\varphi'_r)\cos(\alpha/2) + \\ P_{w\text{int}}\sin(\alpha/2) + P_{wf_y} - P_{w2} = 0. \end{aligned} \tag{2.19}$$

Equation (2.19) provides an expression for N'_2 which is introduced in Equation (2.18). The following expression is then found for the equation of motion in the direction of sliding:

$$N'_{\text{int}} s_5 - c'_r AB' s_6 - P_{w\text{int}} s_7 + (P_{w2} - P_{wf_y} - W_2) \tan \varphi'_b = \frac{\mathrm{d}(M_2 v)}{\mathrm{d}t}, \tag{2.20}$$

where s_i are trigonometric constants given by

$$s_5 = \cos(\alpha/2) - \tan \varphi'_b \sin(\alpha/2) - \cos(\alpha/2) \tan \varphi'_r \tan \varphi'_b - \sin(\alpha/2) \tan \varphi'_r, \tag{2.21a}$$

$$s_6 = \tan \varphi'_b \cos(\alpha/2) + \sin(\alpha/2), \tag{2.21b}$$

$$s_7 = \tan \varphi'_b \sin(\alpha/2). \tag{2.21c}$$

The effective interaction force between the two wedges is now found from Equation (2.20):

$$N'_{\text{int}} = \left(\frac{1}{s_5}\right)\left(c'_r AB' s_6 + P_{w\text{int}} s_7 + (P_{w2} - P_{wf_y} - W_2) \tan \varphi'_b + \frac{\mathrm{d}(M_2 v)}{\mathrm{d}t}\right). \tag{2.22}$$

A single motion equation may be found now if the expressions of N'_{int} from Equations (2.12) and (2.22) are made equal. Rearranging terms, the following equation of motion is derived:

$$W_1 s_1 s_5 + \left(W_2 - P_{w2} + P_{wf_y}\right) \tan \varphi'_b s_2 + c'_r AB'(s_3 s_5 - s_2 s_6) - P_{w\text{int}}(s_4 s_5 + s_7 s_2) + P_{w1} \tan \varphi'_b s_5 = s_5 \frac{\mathrm{d}(M_1 v)}{\mathrm{d}t} + s_2 \frac{\mathrm{d}(M_2 v)}{\mathrm{d}t}. \tag{2.23}$$

In order to simply the notation, Equation (2.23) can be rewritten introducing new trigonometric coefficients t_i:

$$W_1 t_1 + \left(W_2 - P_{w2} + P_{wf_y}\right) t_2 + c'_r AB' t_3 - P_{w\text{int}} t_4 + P_{w1} t_5 = s_5 \frac{\mathrm{d}(M_1 v)}{\mathrm{d}t} + s_2 \frac{\mathrm{d}(M_2 v)}{\mathrm{d}t} \tag{2.24}$$

where

$$t_1 = s_1 s_5, \tag{2.25a}$$

$$t_2 = \tan \varphi'_b s_2, \tag{2.25b}$$

$$t_3 = s_3 s_5 - s_2 s_6, \tag{2.25c}$$

$$t_4 = s_4 s_5 + s_7 s_2, \tag{2.25d}$$

$$t_5 = \tan \varphi'_b s_5. \tag{2.25e}$$

Under strict static equilibrium conditions, $(\mathrm{d}(M_1 v)/\mathrm{d}t = \mathrm{d}(M_2 v)/\mathrm{d}t = 0)$, Equation (2.24) could provide, for instance, the value of the apparent effective cohesion along shearing plane AB, in terms of the friction angle on AB, φ_r', wedge weights, pore pressure forces on their boundaries, and geometrical factors:

$$c_r' = \frac{-W_1 t_1 - \left(W_2 - P_{w2} + P_{wf_y}\right) t_2 + P_{w\mathrm{int}} t_4 - P_{w1} t_5}{AB' t_3}. \tag{2.26}$$

The water pressure forces entering the above equations are easily found as follows

$$P_{wf_y} = \frac{h_w^2 \gamma_w}{2 \tan \delta}, \tag{2.27a}$$

$$P_{w2} = (L_1 + L_2 + s) h_w \gamma_w, \tag{2.27b}$$

$$P_{w1} = \frac{h_w^2 \gamma_w}{2 \sin \alpha}, \tag{2.27c}$$

$$P_{w\mathrm{int}} = \frac{h_w^2 \gamma_w}{2 \cos(\alpha / 2)}. \tag{2.27d}$$

Initial ($s = 0$) wedge volumes, in view of Figures 2.19 and 2.20, are given by

$$V_{10} = \frac{L_0 H_1}{2 \cos \alpha}, \tag{2.28a}$$

$$V_{20} = \frac{L_1 + L_2 + L_3}{2} H_1, \tag{2.28b}$$

which allows the calculation of wedge weights.

2.6.4 Static equilibrium at failure

Cross-sections 2 and 5 (Fig. 2.14) are characterized by the geometrical parameters given in Table 2.1. The upper wedges of Sections 2 and 5 have similar volumes. However, the lower wedge of Section 2 has a significantly lower volume than Section 5. Therefore, Section 5 is more stable than Section 2, for a common set of strength parameters. Conditions for static equilibrium of these two sections will be first examined with the help of the set of relationships derived in the previous section. Since it has been argued that the residual friction at the basal sliding surface is a parameter known with sufficient certainty, the condition of stability may be used only to determine the strength parameters on shear plane AB. In fact, only combinations of the pair (c_r'; φ_r') may be found, since only one condition is available: the condition of static equilibrium at the initiation of failure (Eq.

(2.26)).

This is a nonlinear equation relating c'_r and φ'_r, which has been plotted in Figure 2.21 for Sections 2 and 5, assuming φ'_b equal to 12° and a rock specific weight of 23.5 kN/m^3.

Table 2.1 Geometrical parameters of Cross-sections 2 and 5.

	H_0 (m)	H_1 (m)	L_0 (m)	L_1 (m)	L_2 (m)	α (°)	δ (°)	V_1 (m^3/m)	V_2 (m^3/m)
Section 2	580	245	750	190	260	37.7	43.3	116142	68149
Section 5	510	260	700	240	320	36	39.1	112590	93000

Forces P_w (Eq. (2.27)), which provide the effect of water pressures on both wedges, should correspond to failure conditions. Since a horizontal water level has been assumed and the preceding rain was shown to have a non-negligible effect (see Fig. 2.13), all water pressure influence will be associated with the water level height above the lower horizontal sliding surface, h_w. The plot in Figure 2.13 provides the estimation of the equivalent value of h_w, i.e.: the reservoir water level, in the absence of rain in the preceding 30-day period, which explains the failure. This height corresponds approximately to the elevation 710 m and, therefore, in Section 5 (see Figs. 2.8 or 2.14) it implies a value h_w = 120 m. This reservoir elevation corresponds, in Section 2, to a water height of h_w = 90 m (the failure surface daylights at a higher elevation at Section 2; see Figs. 2.8 and 2.14). The (c'_r; φ'_r) values plotted in Figure 2.21 correspond to these two water elevations over the lower horizontal sliding plane.

Section 2 is "more demanding" in terms of required rock strength simply because of the relative weight of upper and lower wedges. This situation is reflected in the higher strength values required for the equilibrium calculated for Section 2 (Fig. 2.21). It is interesting to check that the (c'_r; φ'_r) combinations in Figure 2.21 are in fairly good agreement with the strength expected in rock sheared across bedding planes, discussed in 2.6.3. Since the variability of φ'_r values is small compared with the expected variation of cohesive intercepts (c'_r), a band of expected (c'_r; φ'_r) pairs, centered around φ'_r = 38°–40° has been plotted in Figure 2.21 as a reasonable estimation of the rock strength along shear plane AB.

If Section 5 is taken as a representative cross-section of the slide, the following combinations lead to strict equilibrium of Vaiont slide: (c'_r = 762.3 kPa; φ'_r = 38°); (c'_r = 564.0 kPa; φ'_r = 40°).

It is also interesting to examine the interaction forces between the two blocks and how they change as a function of the available friction on the basal sliding plane. Equations (2.12) and (2.22), for zero acceleration, provide this force for the two wedges. If Section 5 is selected for the analysis, the variation of N'_{int} with the base friction angle for two pairs of values (c'_r; φ'_r) is given in Figure 2.22. It was already stated that equilibrium is achieved if the interaction force N'_{int} between the upper and lower wedges is forced to have a common value. This condition also implies that the shear force, Q_{int}, and therefore the total interaction force are equal.

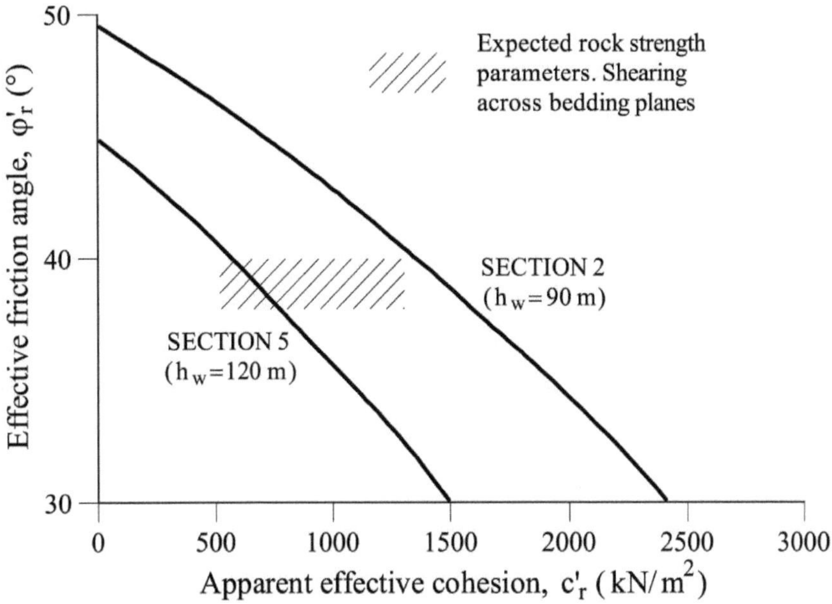

Figure 2.21 Strength parameters across shearing plane AB for equilibrium. Sections 2 and 5. Basal friction: $\varphi_b' = 12°$.

Figure 2.22 shows how the stabilizing N_{int}' force offered by the lower wedge increases fast as the friction at the sliding surface, φ_b', increases. On the other hand, the unbalanced N_{int}' force required for the equilibrium of the upper wedge decreases as φ_b' increases, but at a slower rate. Overall equilibrium is achieved when both forces are equal. For strength parameters $c_r' = 564.0$ kN/m^2 and $\varphi_r' = 40°$ equilibrium is achieved for $\varphi_b' = 12°$, a result which has already been found. If the strength along the shear plane AB is reduced to $c_r' = 0$ kN/m^2 and $\varphi_r' = 35°$, φ_b' has to increase to 14.7°, to reach equilibrium.

So far, equilibrium conditions have been used to find the mobilized strength parameters at failure. The condition of failure, when it is properly identified, which means, in particular, that slide geometry and pore water pressure distribution are known, is a procedure to find strength parameters or, better, a relationship among the strength parameters involved in the model selected to perform stability calculations. This procedure, illustrated above, is often described as a "back-analysis" of the failure.

In addition, one may be interested in knowing the safety factor for conditions other than failure. For instance, in the case of Vaiont, it makes sense to ask for the safety conditions of the slope before dam impoundment or at some particular elevation of the reservoir surface. These questions are addressed in the next section.

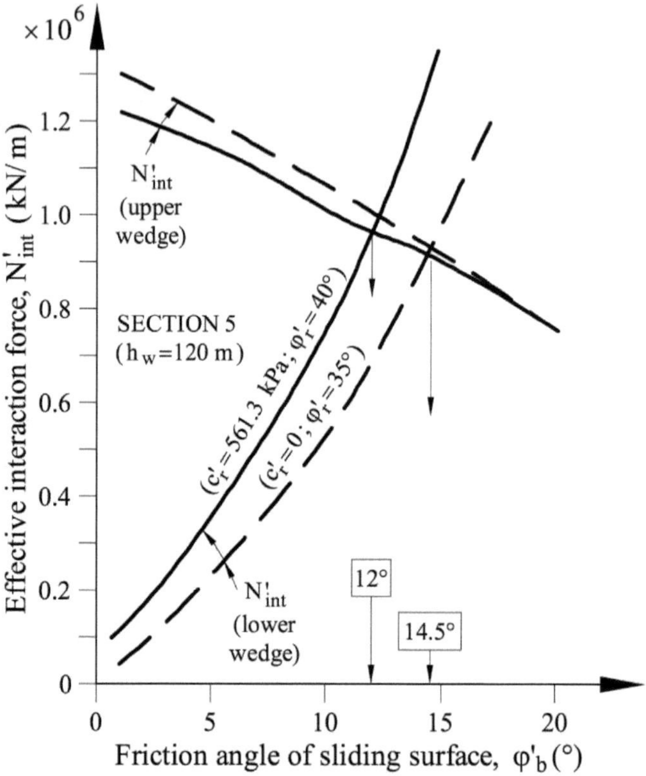

Figure 2.22 Effective interaction force, N'_{int} between upper and lower wedges. Section 5 of Vaiont slide.

2.6.5 Safety factors

In limit equilibrium methods (the analyses developed before belong to this class of methods) the safety factor is defined as the ratio between the available shear strength of the soil or rock and the shear stress necessary for strict equilibrium. Shear strength and shear stress are calculated on the failure surface. The model of two interacting wedges developed in 2.6.3 and 2.6.4 includes two failure surfaces: the "basal" surface that bounds the landslide and an internal shear surface (AB). which makes it kinematically possible. The nature of both surfaces is quite different: the former is located in a high plasticity clay in residual conditions, whereas the internal shear surface crosses sedimentary planes, distorts a competent rock and exhibits significant strength. However, it is quite possible that shear displacements will decrease to some extent the shear strength of this shear plane. For a particular situation of the slide (for instance, under natural conditions before dam construction), the two shearing surfaces will most probably not mobilize their shear strength in equal proportions. Likewise, if a change in external conditions takes place (reservoir impoundment, or rainfall), the available strength will not be mobilized at the same time among the two surfaces because the shear stiffness of the shearing surfaces and, indeed, of the whole rock mass, will also play a significant role.

Since the problem is complicated, let us accept, to initiate the discussion, that two different safety factors, F_b and F_r, are appropriate for the two surfaces. Then, the mobilized strength parameters will be defined as follows:

$$\tan\varphi'_{b\,\mathrm{mob}} = \tan\varphi'_b / F_b\,, \tag{2.29a}$$

$$\tan\varphi'_{r\,\mathrm{mob}} = \tan\varphi'_r / F_r\,, \tag{2.29b}$$

$$c'_{r\,\mathrm{mob}} = c'_r / F_r\,. \tag{2.29c}$$

A relevant question is to ask for the safety factor, F_r, of the Vaiont slide at the beginning of impoundment (i.e., $h_w = 0$), in the hypothesis that the mobilized stress at the basal sliding surface remained at residual conditions, $\varphi'_b = 12°$, (i.e., $F_b = 1$). It is also of interest to know how F_r would change, still under $F_b = 1$, if the slide moves forward following the mechanism described in Figure 2.15.

Alternatively, one may wish to maintain the classic approach and to find a unique and global safety factor, F, for the two situations mentioned, ($F = F_b = F_r$). The two possibilities will be examined here.

For Cross-section 5, it was found that the following set of strength parameters: $\varphi'_b = 12°$; $c'_r = 762.2$ kPa; $\varphi'_r = 38°$ leads to failure when $h_w = 120$ m. If these parameters are accepted as true strength parameters, then the equilibrium equations given in 2.6.3 are also valid, for conditions other than failure, if the reduced strength parameters (2.29a,b,c) are used instead of the true strength values (which are now assumed to be known). In other words, equilibrium conditions are now satisfied for the mobilized stresses prevailing at the shear surfaces. In fact, mobilized shear stresses are defined as those which satisfy equilibrium conditions. Therefore, in view of Equations (2.29), the overall equilibrium equation can be used to find the safety factor. However, the equilibrium equation will now be a function of F_b and F_r and therefore only **one** safety factor may be determined – either F if it is accepted that $F = F_b = F_r$, or F_r if F_b is fixed, for instance at $F_b = 1$, or any other alternative. This situation is similar to the already discussed determination of strength parameters at failure.

If the mobilized strength parameters (Eqs. (2.29a,b,c)) are substituted into the equilibrium Equation (2.26), the following expression is obtained.

$$\frac{c'_r}{F_r} = \frac{-W_1 t_1(F_r,F_b) - \left(W_2 - P_{w2} + P_{wf_y}\right) t_2(F_r,F_b) + P_{w\mathrm{int}} t_4(F_r,F_b) - P_{w1} t_5(F_r,F_b)}{AB' t_3(F_r,F_b)}, \tag{2.30}$$

where the dependence of the t_i expressions on the safety factors has been explicitly indicated in the Appendix 2.1. If Equation (2.30) is developed, it turns out to be a second-order algebraic equation for F_r (Eq. (A2.4) in the Appendix 2.1), which may be solved if F_b is assumed to be known. Details of the solution of Equation (2.30) are relegated to Appendix 2.1.

The safety factor F_r of Section 5 of the Vaiont slide was obtained for:

- Water pressure conditions prior to failure. As discussed before, pore water-pressure effects are integrated into the variable h_w, the reservoir level over the lower horizontal sliding plane.
- The changing geometry, as the slide moves forward and the water level maintains maximum elevation, h_w = 120 m. This is a purely static analysis performed on different geometries of the slide as it moves forward. The dynamics of the motion will be introduced in the next section and it will be discussed in more detail in Chapter 5.
- The effect of h_w on safety factor F_r, when $F_b = 1$, is plotted in Figure 2.23 (dashed line). The calculated value for $h_w = 0$ ($F_r = 1.2$) is not particularly high and it indicates that the mobilized strength in the rock mass before any impounding was quite substantial in order to maintain the slope in equilibrium.

The analysis of the changing geometry, shown in Figure 2.15, leads to the safety factor F_r plotted in Figure 2.24 (dashed line). The increase of F_r, again for $F_b = 1$, becomes more pronounced as slide displacement increases. The high values calculated for s = 150 m (F_r = 5), indicate that the mobilized resistance across shear plane AB is no longer necessary to maintain equilibrium. In fact, beyond s = 179 m, the residual friction angle at the main sliding surface is able to maintain the slope in equilibrium without any contribution from the sheared rock mass across the shear plane AB.

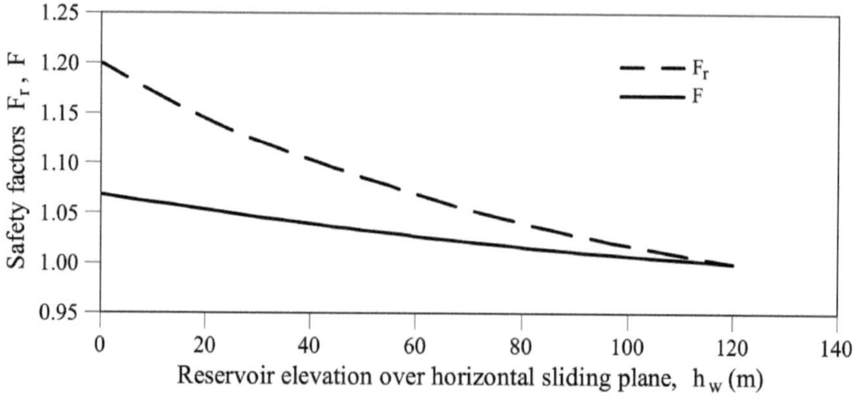

Figure 2.23 Section 5. Evolution of safety factor, F_r (if $F_b = 1$; see text) and global safety factor, F, when the water level increases in the reservoir.

Let us now consider the determination of a unique global safety factor F. The condition $F = F_b = F_r$ has to be introduced in Equation (2.30). The equilibrium

Equation (2.30) now becomes a fourth-order polynomial for the unknown F. A simple numerical procedure to solve the equation is described in Appendix 2.2.

Calculated global safety factors, with the help of Equation (A2.10), were plotted in Figures 2.23 and 2.24 (continuous line). Computed values of F are now significantly lower than the previously reported values of F_r.

One advantage of global safety factors is that geotechnical engineers have developed, over the years, a scale of numerical values that helps them to approximate the risk of failure. F values of 1.5 and above are generally regarded as indicators of a low risk of failure of slopes. A safety factor of 1.2 is probably close to the minimum that many would regard as an acceptable situation. Since different calculation procedures often result in changes in safety factor of ± 0.1 for a given slope stability problem, a safety factor of 1.1 conveys a clear message of risk.

However, one should distinguish between design situations and, on the other hand, the problem of analyzing an existing slide and its remedial measures. In the second case, the evidence of field instability, if properly interpreted, provides a robust reference value (F = 1 for failure conditions) which acts as a validation benchmark for any method of stability analysis. Then, calculated changes of safety factor over the reference situation (F = 1) are significantly more reliable than a pure predicting exercise based, for instance, on strength parameters determined in the laboratory or on estimated pore water pressures derived from flow calculations.

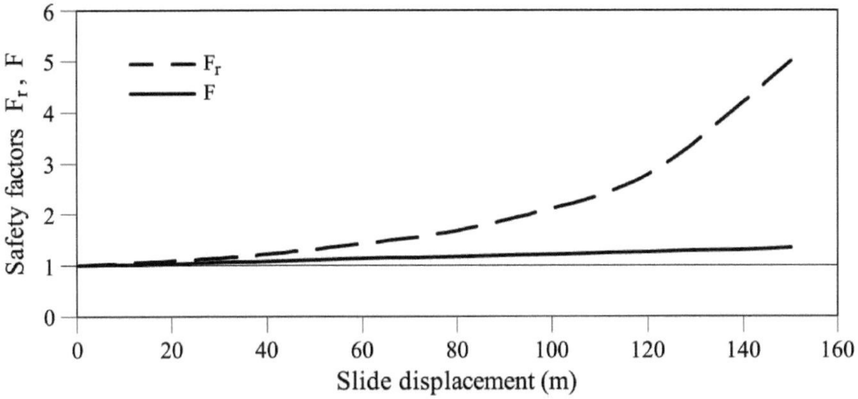

Figure 2.24 Section 5; h_w = 120 m. Evolution of safety factor, F_r (if F_b = 1; see text) and global safety factor, F, with slide displacement.

Vaiont obviously belongs to the second category. Nevertheless, the global safety factors calculated for changing water levels within a very large range (0 to 120 m of water column) (Fig. 2.23) look particularly low (F decreases from F = 1.07 for h_w = 0 m to F = 1 for h_w = 120 m). This is certainly a consequence of the very large size of the landslide but it also points out that the presence of the reservoir implied a relatively minor change in the safety of the slope, always within the perspective of risk associated with the classical definition of a global safety factor. Moreover, this result is also an indirect indication that in very large landslides, feasible remedial measures are expected to lead to relatively low increments of safety factor.

Figure 2.24 shows that the motion of the slide results in geometries with an increasing global safety factor. Given the preceding comments, changes are far from being negligible. In fact, displacements of 40, 100, and 150 m imply F values of 1.08, 1.22, and 1.36 respectively. (Interestingly, very similar changes were computed with the much simpler two-block model, Fig. 2.17b.) The increasing sophistication of the model did not change this basic result.

The relevant question in this case, already stated when discussing the two block model results, is to ask for the reasons for the accelerated motion of a landslide which seemed to move in a direction of increased stability. This aspect is essentially the subject of Chapter 5 but some additional discussion is offered in the next section.

2.6.6 Landslide run out

Equilibrium conditions, when inertia terms are included, results in the motion Equation (2.24). This equation, taking into account Equation (2.6) has the following form:

$$\frac{\mathrm{d}v}{\mathrm{d}t}=\frac{W_1t_1+\left(W_2-P_{w2}+P_{wf_y}\right)t_2+c_r'AB't_3-P_{w\mathrm{int}}t_4+P_{w1}t_5-s_5\dfrac{\mathrm{d}M_1}{\mathrm{d}t}-s_2\dfrac{\mathrm{d}M_2}{\mathrm{d}t}}{\left(s_5+s_2\right)}, \quad (2.31)$$

where time derivatives of M_1 and M_2 are known (Eq. (2.15)) and they depend on displacement and velocity. The weights (W_1 and W_2) and the length AB' also depend on the displacement. Therefore, Equation (2.31) can be written as

$$a=\frac{\mathrm{d}v}{\mathrm{d}t}=f(s,v). \quad (2.32)$$

At any given time of the motion, slide acceleration ($a=\mathrm{d}v/\mathrm{d}t$) is a function of slide displacement, s and velocity, v. Function f also includes information on geometry, specific weights, water pressures, and strength parameters. Finding a close-form solution for $v(t)$ is a hard task but the structure of (2.32) invites to develop a simple numerical algorithm of integration. If the following discrete approximation is adopted, the value of the acceleration and the velocity at time (t + 1) can be calculated as

$$a_{i+1}=\left(\frac{\mathrm{d}v}{\mathrm{d}t}\right)_i=\frac{v_{i+1}-v_i}{t_{i+1}-t_i}=f\left(s_i,v_i\right), \quad (2.33a)$$

$$v_{i+1}=v_i+f\left(s_i,v_i\right)\left(t_{i+1}-t_i\right), \quad (2.33b)$$

which are functions of known values evaluated in time t. In this way, an explicit time integration procedure is developed. Reducing $\Delta t=t_{i+1}-t_i$ leads to progressively more accurate results.

Displacements can be estimated from the following expression

$$v_i = \left(\frac{\mathrm{d}s}{\mathrm{d}t}\right)_i = \frac{s_{i+1} - s_i}{t_{i+1} - t_i} \tag{2.34a}$$

and therefore:

$$s_{i+1} = s_i + v_i\left(t_{i+1} - t_i\right). \tag{2.34b}$$

In view of the nature of the problem and the simplicity of the underlying mechanical model, it is probably not justified in this case to look for more sophisticated integration procedures. The integration algorithm was implemented in an Excel calculation sheet. Note that masses, weights and the length AB' should be updated at each time interval since they depend on the displacement.

It was argued in Section (2.6.3), when developing the model of two interacting wedges, that the effective rock cohesive intercept, c'_r, would be degraded during shear along plane AB. Since relative shear displacements along AB are controlled by displacement s, a simple degradation model will make c'_r dependent on s. For instance,

$$c'_r = c'_{r0} \exp(-\Gamma s), \tag{2.35}$$

where Γ is a constant (units: length^{-1}) that controls the rate of rock degradation and c'_{r0} is the initial cohesion intercept (c'_{r0} = 768.35 kPa for Cross-section 5, if φ'_r = 38°, and accepting that φ'_b = 12°). Expression (2.35) was also included in the motion equation in order to explore the effect of loss of shear strength on the dynamics of the motion. It is not reasonable, however, to expect a strong degradation of cohesion along AB′ and the reason is that the rock mass "crosses" the plane AB′ during the motion and therefore new – more or less undisturbed – rock is continuously sheared across AB.

Consider the following scenario: in a situation of strict equilibrium (reservoir elevation at h_w = 120 m in Cross-section 5) the water level is increased by a small amount (say h_w = 121 m), and it is maintained as constant thereafter. It is desirable to find the motion of the slide until a new situation of equilibrium is reached. Since the slide improves its static stability conditions as s increases – a result from the previous section – it should be expected that the slide will come to rest after some displacement.

The solution to this problem (which is the solution of Eq. (2.31) plotted as a relationship between the run out (s) and the velocity on the moving mass (v)) is shown in Figure 2.25 for no degradation of the rock strength (Γ = 0). The result shows that the slide stops after a displacement of 0.30 m and reaches a maximum velocity of 1.7 cm/s. If the water level is increased to h_w = 124 m and to h_w = 130 m, maximum displacements and velocities increase as shown in Figure 2.25, but the calculated values are far from the actual behaviour of the landslide, which reached velocities estimated in 30 m/s, more than two orders of magnitude higher than the maximum values found in this calculation.

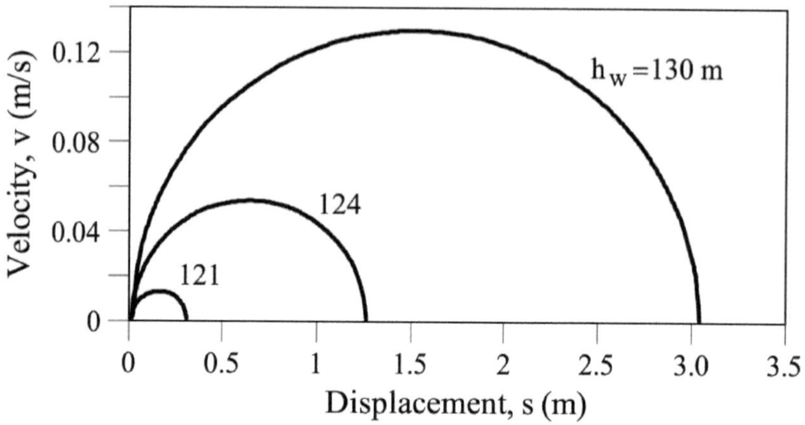

Figure 2.25 Cross-section 5. Calculated run outs and slide velocities for h_w = 121, 124, and 130 m. No rock strength degradation (Γ = 0).

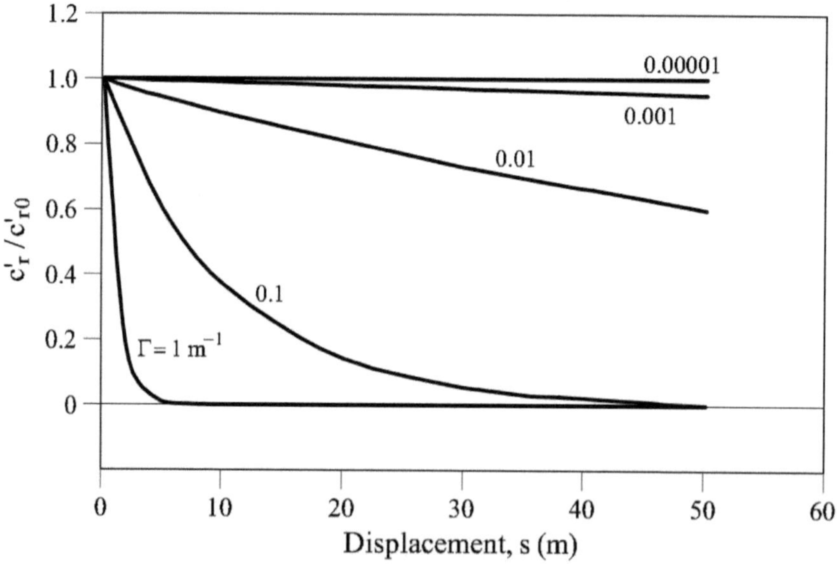

Figure 2.26 Assumed loss of effective cohesive strength parameter across shearing plane AB with slide displacement, for several values of parameter Γ.

The situation changes if some rock strength degradation is introduced into the analysis.

Figure 2.26 is a plot of Equation (2.35) for a few values of the degradation parameter Γ. It will be used as a reference for the results of run-out calculations.

Now the scenario is to start the slide motion by increasing the water level (to

h_w = 121 m) and to accept a certain degradation of the rock during the motion. The calculated response of the slide, again in terms of velocity vs. displacement, is shown in Figures 2.27 and 2.28. A moderate degradation of the effective strength parameter of the rock (Γ = 0.01 m^{-1}, Fig. 2.27) has a limited effect on the maximum sliding velocity and on the travelled distance. However, if the degradation of rock effective cohesion is more rapid (Γ = 0.1 m^{-1} and Γ = 1 m^{-1}; Fig. 2.28), the slide is able to travel long distances (60–70 m), although the maximum velocity does not increase beyond 3 m/s (16.2 km/h) even if a very rapid and complete destruction of the rock effective cohesion is imposed (for Γ = 1,see Fig. 2.28). Under the more realistic assumption of moderate rock degradation, $\Gamma \leq 1$ m^{-1}, the maximum slide velocity is quite small.

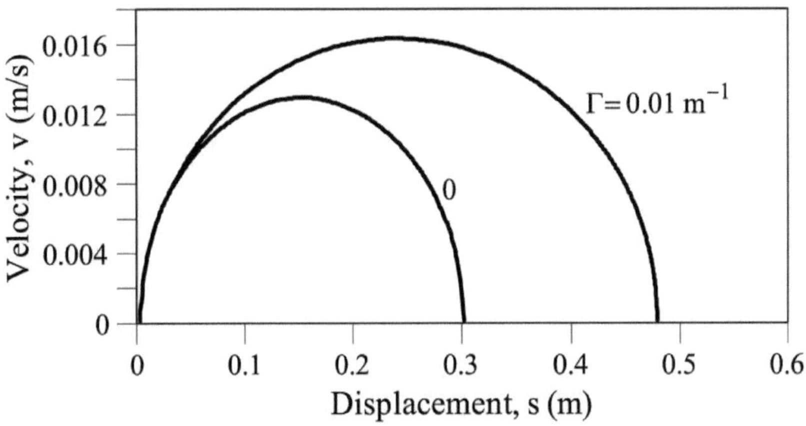

Figure 2.27 Cross-section 5. Calculated run outs and slide velocities for h_w = 121 m. Effect of rock strength degradation (Γ = 0 and Γ = 0.01 m^{-1}).

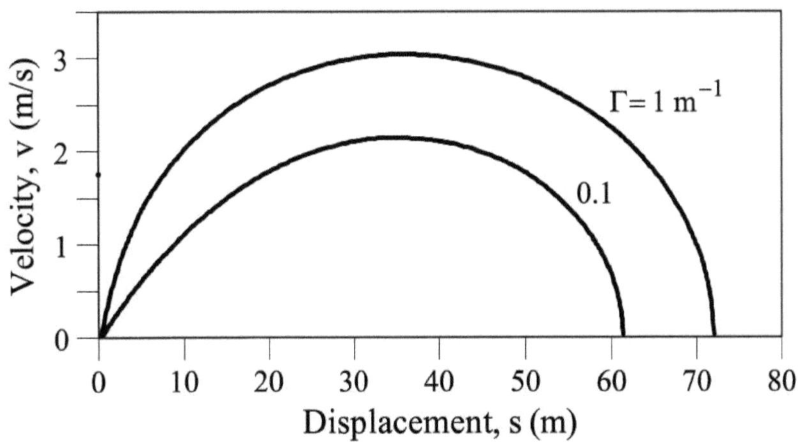

Figure 2.28 Cross-section 5. Calculated run outs and slide velocities for h_w = 121 m. Effect of rock strength degradation (Γ = 0.1 and Γ = 1 m^{-1}).

In all the cases analyzed, the mechanism leading to stopping the landslide motion is the change in geometry of the slide as it moves downwards.

The dynamic analysis developed here maintains, unanswered, the key question of the extremely high velocities reached by the slide. However, it indicates that a loss of internal rock strength, associated with the slide motion itself, is a potential mechanism to accelerate the slide.

2.7 Discussion

The investigations on the past history of the landslide by Semenza (2001), synthesized in Figure 2.7, and the work of Hendron and Patton (1987) highlight two fundamental aspects: Vaiont was a case of a slide reactivation and the sliding surface was located in fairly continuous layers of high plasticity clay. Taken together, the implication is that the basal sliding surface could not offer, against a new reactivation of the slide (essentially induced by an increase in pore water pressures in the lower massive passive wedge of the slide), an effective friction angle larger than, say, 10–12°. A good proportion of published back-analysis of Vaiont, which use conventional methods of limit equilibrium to find the actual friction angle prevailing at the sliding surface at the time of failure is not consistent with Vaiont past history. In fact, published back-analyses lead to friction angles in the range 18 – 28° (the simple two-block model of 2.6.2 is an example in this regard). Vaiont exhibits a safety factor significantly lower than one if a friction angle of 10 – 12° (and zero effective cohesion) is used in any of the currently available methods of slices. How to address this inconsistency?

Hendron and Patton (1987) argue that the side friction on the eastern edge of the slide provided the necessary resisting force to ensure equilibrium (however, some limited information on the direction of the displacements on this border, plotted in Fig. 2.2, tends to indicate that the moving mass was detaching from the stable rock massif). The alternative explanation developed here is that the kinematics of the motion, even in a two-dimensional cross-section, requires the relative shearing between the two large rock wedges defining the slide. Leonards (1987) also pointed out that the motion of the slide required such a rock shearing between the upper and the lower sliding blocks. The estimated shearing strength parameters across the common shearing plane are in reasonable agreement with the expected mass strength of cretaceous marls and limestones of Vaiont.

The acceleration of the motion during the catastrophic failure escapes the capabilities of the models presented here. A loss of strength is expected when rock masses are sheared, due to its inherent brittleness and the complex development of strains within the moving mass. "Progressive failure" is the term often used to describe these phenomena. The end result is a loss of the cohesive components of strength. Such a loss, when imposed on the strength available on the interacting shearing plane between the upper and lower wedges, results in an acceleration of the slide, which is unable to explain the high velocities reached by the landslide, even if a rapid and complete loss of rock cohesion is imposed (Section 2.6.6). Therefore, it becomes important to look for additional explanations for the apparent loss of strength experienced by the actual slide. If the mechanism of side friction proposed by Hendron and Patton (1985) is accepted as additional resisting phenomena, the need for a convincing mechanism for strength loss is even more pronounced. The discussion of this fundamental matter will continue in Chapter 5.

2.8 Lessons Learned

2.8.1 Slide reactivation

Ancient slides are rightly regarded as trouble-makers when they are affected by engineering works. Past sliding activity is responsible for the reduction of the strength available along "dormant" sliding surfaces to minimum values ("residual" strengths). In addition, if sliding surfaces are associated with high plasticity clays, the residual friction angles are particularly low. Typically, ancient slides in these circumstances maintain a low safety factor, which may be rapidly exhausted by engineering works. Vaiont is a good example.

2.8.2 Submerging the slide toe

Submerging the toe of slopes usually leads to a reduction of stability. The safety factor decreases as the water level increases. The reduction is first pronounced but, eventually, the negative effect associated with the reduction of effective normal stresses on the sliding surface is compensated by the beneficial hydrostatic forces acting against the exposed slope[(1)].

The safety factor reaches a minimum value for some intermediate water level and then increases again to reach values close to the initial safety factor of the "aerated" slope. The precise evolution of the safety factor when the reservoir water level increases depends also on the particular distribution of pore water pressures inside the slope, but a fundamental aspect of this problem is the geometry of both the slope and the sliding surface. Three examples are shown in Figure 2.29 to illustrate these comments. All of them were solved with a commercial slope stability program for soil slopes using the Morgenstern – Price method (Morgenstern and Price, 1965). In all cases the distribution of pore pressures inside the slide follows a horizontal water table. The first case (Fig. 2.29a) reproduces the geometry of Vaiont, Section 5. A uniform friction angle, $\varphi' = 12°$ (and zero effective cohesion) is assumed. The safety factor reaches a minimum for $h_w/H = 0.5$. A similar result is obtained if the lower sliding surface is inclined (Fig. 2.29b; now $\varphi' = 15°$). However, for a conventional slope (the upstream slope of an earthdam) and a critical circular failure surface, the

(1) An alternative explanation can be given in terms of submerged weights. When a lower part of the slope is flooded, its effective weight becomes the submerged weight (roughly equal to one half of the saturated total weight). Therefore, normal effective stresses on the sliding plane are reduced. But the (effective) weight also reduces. This weight has often a positive stabilizing effect when it is close to the toe. Therefore, reducing it also decreases the safety factor. But, as the water level increases, the upper parts of the slope, which contribute with unstabilizing weight, also reduce its effect and the calculated safety factor will increase again beyond some critical water level.

minimum is reached for $h_w/H = 0.3$.

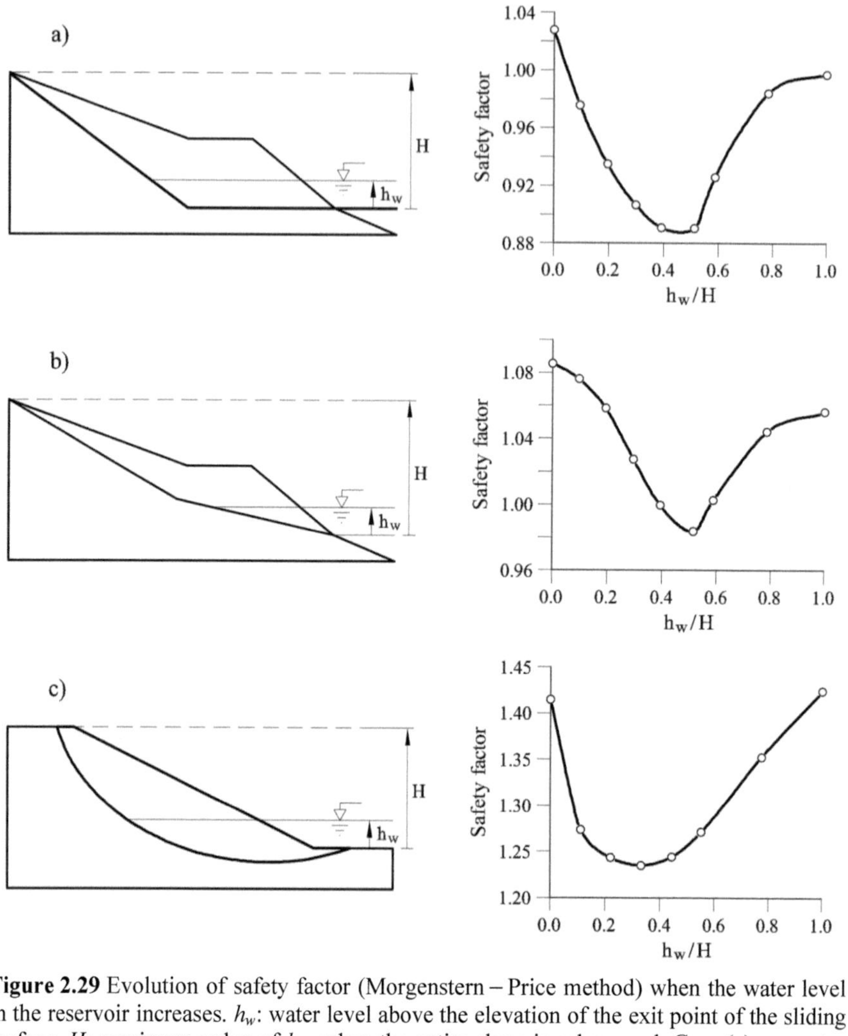

Figure 2.29 Evolution of safety factor (Morgenstern – Price method) when the water level in the reservoir increases. h_w: water level above the elevation of the exit point of the sliding surface. H: maximum value of h_w, when the entire slope is submerged. Case (a) geometry of Vaiont and ($c' = 0$; $\varphi' = 12°$); Case (b): geometry modified from Case (a) and ($c' = 0$; $\varphi' = 15°$); Case (c): conventional slope and circular failure surface ($c' = 0$; $\varphi' = 30°$).

The geometry of Vaiont is especially sensitive to the submergence of the toe, because the large toe passive wedge offers an ever decreasing resisting force when submerged. This is further illustrated in Figure 2.30, which shows a calculation of the global safety factor of Section 5, following the procedure described in 2.6.5. The height of the dam prevented h_w values higher than 145 m, approximately, in Section 5. Unfortunately, Vaiont slide never entered in a zone of increasing stability.

In conclusion, flooding slope toes is not good practice but it is an unavoidable situation in many circumstances, notably in valley slopes affected by dam reservoirs. If the initial safety factor is low (this was the case of Vaiont) flooding the toe implies trouble ahead. Note also that it takes a substantial water level elevation before the trend for lower safety, as the water level increases, is reversed.

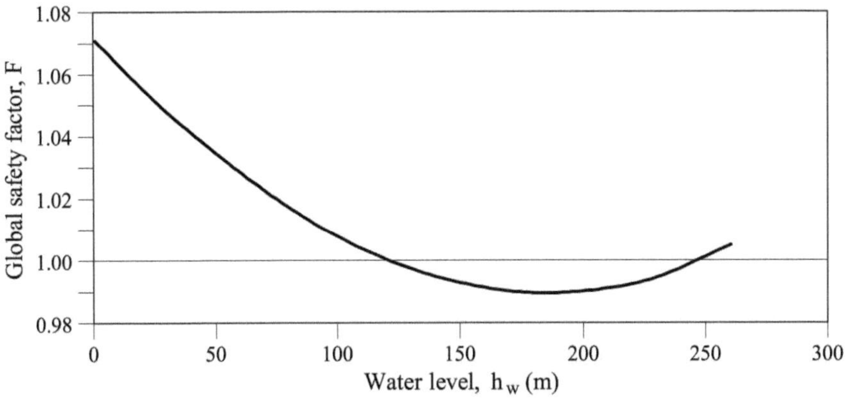

Figure 2.30 Variation of global safety factor with height of water above the horizontal basal plane. Two-wedge model. Section 5 of Vaiont.

2.8.3 Interpretation of field data

Interpretation of sliding risk was essentially made on the basis of reservoir elevation and surface displacements. There was also information on rainfall and on the levels of four piezometers. The "piezometers" were in fact open tubes which did not reach the level of the sliding surface and only provided average water pressures prevailing along their length. In addition, no direct information of the position of the failure surface and, in particular, on the type of material being sheared was available.

Identification of a landslide for the purposes of estimating its evolution and of defining any remedial measure requires information of a few key variables. Ideally, these key variables should also be used in the formulation of a mechanical model of the motion. In the case of Vaiont, early knowledge of the following data concerning the basal failure surface: geometry, pore water pressure, type of material, and drained strength parameters would have been fundamental to build a conceptual and mechanical model for the slide. This is a first step in understanding the problem, not only for Vaiont, but for any landslide. In the case of Vaiont, the observation that the slide velocity decreased when the reservoir level was reduced, irrespective of the absolute level of the water, provided a reservoir filling criterion which, finally led to the failure. In some sense, an "observational method" [(2)] was

[(2)] The observational method, described by Peck (1969), requires the following ingredients: a) direct observation of a key variable or property describing the essential nature of the problem; b) a proper conceptual, analytical or computational model able to provide an

applied: the conceptual model was essentially given by the preceding observation, illustrated in Figure 2.12. The key variables to be interpreted were the displacement rates of surface markers and the reservoir level. The action in mind, in case of excessive displacement rate, was to reduce the water level in the reservoir. It was accepted, despite this strategy, that a full slide was a likely event and that the expected height of the generated wave was even estimated by model studies. However, the conceptual model was not based on any mechanical analysis of the slide. In addition, the reservoir level did not necessarily provide the actual pore pressures on the failure surface and the remedial plans were too simple and weakly connected with the complex mechanisms taking place within the slide.

It should be borne in mind that these comments are made more than 50 years after the first investigations started in Vaiont. Their purpose is to learn from the case, not to criticize the involved individuals who had to work with the techniques and rules of practice available at that time.

Even today, managing a very large landslide is a daunting task. We are well equipped to extract field data (pore water pressures, absolute deformations, "in-situ" tests) in the first tens of meters of soil and rock. Going beyond 200 m requires sophisticated, not easily available, and time-consuming efforts. In addition, a very large landslide requires a vast site investigation. It is not a matter of only a few borings. Therefore, the difficulties to handle large landslides continue to be present and the words of Carlo Semenza, the dam designer, remain as a vivid testimony of the formidable challenge he was facing: "[...] *things are probably bigger than us and there are no adequate practical measures* [...] *After so many fortunate works and so many structures* [...] *I am in front of a thing which due to its dimensions seems to escape from our hands* [...]", (in a letter written in April 1961, quoted by Nonveiller, 1987; the full letter in Italian was published in Semenza, 2001).

2.8.4 Computational procedures

Most of the limiting equilibrium procedures commonly available to the geotechnical profession (methods of slices) do not include an internal shearing in the moving mass, which is described by strength parameters other than the parameters operating on the external bounding failure surface. Moreover, none of them may handle processes of stress redistribution induced by progressive failure mechanisms. In addition, they have no capability to approximate the initial stress state. Continuum models (finite elements) may reproduce better the stress state derived from a known history of slope development but modelling progressive failure is still a research subject with very little impact on current practice. It has to be accepted that, 47 years after the disaster, static methods to estimate the stability conditions of the Vaiont landslide still suffer from important limitations. To aggravate things, the dynamic behaviour of the slide is still being discussed

---

estimation of the risk, in a general sense, for some threshold values of the key variable(s) and c) a plan, defined in advance, to act in a specified manner when threshold values are exceeded.

and investigated. From a practical point of view, there are no reliable criteria to estimate the dynamic reaction of an impending landslide in the case of failure. More will be said in Chapter 5 on this aspect

One has to accept for the time being, by a simple comparison with Vaiont, that large landslides exhibiting symptoms of instability, sliding on (high plasticity?) clay levels and subjected to a definite reduction of stability conditions, may develop unexpected sliding velocities.

2.8.5 Could it have been avoided?

This has been a subject of much debate (see Leonards, 1987). It is probably fair to say that an improved knowledge of the field situation – based on a more comprehensive set of sensors, a detailed geology, and the strength properties of the sliding surface – and even a better conceptual and mechanical model of the slide would not have provided reliable criteria to stop the motion. Large engineering works also convey important pressures to be completed as planned. If this was the case of the Vaiont dam, there was probably not a reasonable procedure to avoid the slide. In fact, this risk was accepted by the designers, as mentioned before. There were also (abandoned) attempts to drain the failure surface by means of a drainage tunnel. Its potential effect remains unclear especially because the reservoir level marked an unavoidable minimum interstitial water pressure, which was already very high. Even if the operating water level of the reservoir was substantially reduced (more than 100 m), there remains the risk that an exceptional rainfall event (see Fig. 2.13) could have brought the water pressures to critical values. Perhaps a combination of a significant (no less than 100 m) reduction of the maximum reservoir level and an expensive tunnel-based drainage scheme of the failure surface could have achieved a sufficiently low risk of failure.

Appendix 2.1 Safety Factor F_r. Static Equilibrium

Equation (2.30) provides the condition of equilibrium of the entire slide in terms of mobilized strength parameters given in Equations (2.29). Coefficients t_i in Equation (2.30) are now written in more detail:

$$t_1 = s_1\left(F_b\right)s_5\left(F_b, F_r\right), \tag{A2.1a}$$

$$t_2 = \frac{\tan\varphi_b'}{F_b}s_2\left(F_b, F_r\right), \tag{A2.1b}$$

$$t_3 = s_3\left(F_b\right)s_5\left(F_b, F_r\right) - s_2\left(F_b\right)s_6\left(F_b, F_r\right), \tag{A2.1c}$$

$$t_4 = s_4\left(F_b\right)s_5\left(F_b, F_r\right) + s_7\left(F_b\right)s_2\left(F_b, F_r\right), \tag{A2.1d}$$

$$t_5 = \frac{\tan\varphi_b'}{F_b} s_5\left(F_b, F_r\right), \tag{A2.1e}$$

where

$$s_1\left(F_b\right) = \sin\alpha - \cos\alpha \frac{\tan\varphi_b'}{F_b}, \tag{A2.2a}$$

$$\begin{aligned} s_2\left(F_b, F_r\right) = \sin(\alpha/2)\frac{\tan\varphi_b'}{F_b} - \cos(\alpha/2)\frac{\tan\varphi_r'}{F_r}\frac{\tan\varphi_b'}{F_b} + \\ \cos(\alpha/2) + \sin(\alpha/2)\frac{\tan\varphi_r'}{F_r} = s_{a2}(F_b) + s_{b2}(F_b)/F_r, \end{aligned} \tag{A2.2b}$$

$$s_3\left(F_b\right) = \cos(\alpha/2)\frac{\tan\varphi_b'}{F_b} - \sin(\alpha/2), \tag{A2.2c}$$

$$s_4\left(F_b\right) = \sin(\alpha/2)\frac{\tan\varphi_b'}{F_b} + \cos(\alpha/2), \tag{A2.2d}$$

$$\begin{aligned} s_5\left(F_b, F_r\right) = \cos(\alpha/2) - \sin(\alpha/2)\frac{\tan\varphi_b'}{F_b} - \cos(\alpha/2)\frac{\tan\varphi_r'}{F_r}\frac{\tan\varphi_b'}{F_b} - \\ \sin(\alpha/2)\frac{\tan\varphi_r'}{F_r} = s_{a5}(F_b) + s_{b5}(F_b)/F_r, \end{aligned} \tag{A2.2e}$$

$$s_6\left(F_b\right) = \cos(\alpha/2)\frac{\tan\varphi_b'}{F_b} + \sin(\alpha/2), \tag{A2.2f}$$

$$s_7\left(F_b\right) = \sin(\alpha/2)\frac{\tan\varphi_b'}{F_b}, \tag{A2.2g}$$

where

$$s_{a2}\left(F_b\right) = \sin(\alpha/2)\frac{\tan\varphi_b'}{F_b} + \cos(\alpha/2), \tag{A2.3a}$$

$$s_{b2}\left(F_b\right) = \tan\varphi_r'\left(\sin(\alpha/2) - \cos(\alpha/2)\frac{\tan\varphi_b'}{F_b}\right), \tag{A2.3b}$$

$$s_{a5}\left(F_b\right)=\cos(\alpha/2)-\sin(\alpha/2)\frac{\tan\varphi_b'}{F_b}, \tag{A2.3c}$$

$$s_{b5}\left(F_b\right)=\tan\varphi_r'\left(-\sin(\alpha/2)-\cos(\alpha/2)\frac{\tan\varphi_b'}{F_b}\right). \tag{A2.3d}$$

If Equations (A2.1) to (A2.3) are substituted into Equation (2.30), the following algebraic equation for F_r is obtained:

$$aF_r^2+bF_r+c=0, \tag{A2.4}$$

where

$$a=As_{a5}-Cs_{a2}, \tag{A2.5a}$$

$$b=As_{b5}+Bs_{a5}-Cs_{b2}-Ds_{a2}, \tag{A2.5b}$$

$$c=Bs_{b5}-Ds_{b2}, \tag{A2.5c}$$

and

$$A=W_1s_1(F_b)-P_{w\,\mathrm{int}}s_4(F_b)+P_{w1}\frac{\tan\varphi_b'}{F_b}, \tag{A2.6a}$$

$$B=c_r'm's_3(F_b), \tag{A2.6b}$$

$$C=P_{w\,\mathrm{int}}s_7(F_b)+\frac{\tan\varphi_b'}{F_b}(P_{wf_y}-P_{w2}+W_2), \tag{A2.6c}$$

$$D=c_r'm's_6(F_b). \tag{A2.6d}$$

The valid root of Equation (A2.4) is

$$F_r=\left(-b+\sqrt{b^2-4ac}\right)/2a. \tag{A2.7}$$

Appendix 2.2 Global Safety Factor F

Equation (A2.7), when $F = F_r = F_{b,}$ is, in fact, the static equilibrium equation. Therefore, the value of F should satisfy

$$G(F)=F-\left(-b+\sqrt{b^2-4ac}\right)/2a=0, \tag{A2.8}$$

where the terms a, b, c in (A2.8) should now be calculated for F.

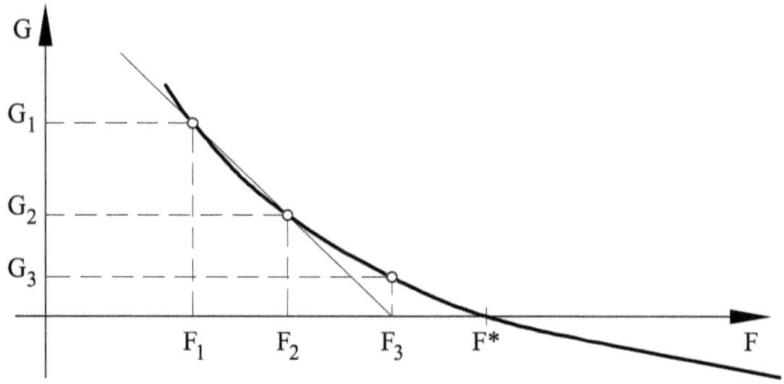

Figure A2.1 Numerical determination of global safety factor, *F*.

Consider in Figure A2.1 the function $G(F)$. The solution sought is marked as F^*. Consider now two F values (F_1 and F_2) and the corresponding G values given by (A2.8). The straight line through (F_1, G_1) and (F_2, G_2) intersects the F axis at F_3, given by

$$F_3 = F_2 - \frac{G(F_2)(F_1 - F_2)}{G(F_1) - G(F_2)}, \tag{A2.9}$$

which could be generalized, as F_i approaches F^*

$$F_i = F_{i-1} - \frac{G(F_{i-1})(F_{i-2} - F_{i-1})}{G(F_{i-2}) - G(F_{i-1})}. \tag{A2.10}$$

Equation (A2.10) provides a recurrent expression to find the target F^*. The success of the procedure depends on the shape of function G around F^*. The upwards concave shape illustrated in Figure A2.1 is the actual shape of (A2.8) in the vicinity of the solution (which, actually, is quite close to the critical value F = 1). For all the global safety factors calculated the two initial values of F_i to initiate the calculation through Equation (A2.10) were $F_1 = 1$ and $F_2 = 1.01$.

References

Barton, N. (2007) *Rock Quality, Seismic Velocity, Attenuation and Anisotropy*. Taylor and Francis. London.

Belloni, L.G. and Stefani, R. (1987) The Vajont slide: Instrumentation – Past experience and the modern approach. *Engineering Geology* 24, 445–474.

Broili, L. (1967) New knowledge on the geomorphology of the Vaiont Slide slip surface. *Rock Mechanics and Engineering Geology* 5, 38–88.

Hendron, A.J. and Patton, F.D. (1985) The Vaiont slide, a geotechnical analysis based on new geologic observations of the failure surface. *Technical Report GL–85–5*. Department of the Army US Army Corps of Engineers, Washington, DC.

Hendron, A.J. and Patton, F.D (1987) The Vaiont slide. A geotechnical analysis

based on new geologic observations of the failure surface. *Engineering Geology* 24, 475–491.
Hoek, E. (2007) Practical Rock Engineering http://www.rocscience.com/hoek/PracticalRockEngineering.asp
Jaeger, C. (1965) The Vaiont rockslide. Part I. *Water Power*. March, 110–111.
Jaeger, C. (1965) The Vaiont rockslide. Part II. *Water Power.* April, 142–144.
Kenney, T.C. (1967) Stability of the Vajont valley slope, discussion of a paper by L. Müller (1964) on the rock slide in the Vajont valley. *Rock Mechanics and Engineering Geology* 5, 10–16.
Leonards, G.A. (1987) Overview and personal commentary. *Engineering Geology* 24, 577–612.
Lupini, J.F., Skinner, A.E. and Vaughan, P.R. (1981) The drained residual strength of cohesive soils. *Géotechnique* 31 (2), 181–213.
Mencl, V. (1966) Mechanics of landslides with non-circular slip surfaces with special reference to the Vaiont Slide. *Géotechnique* 19 (4), 329–337.
Morgenstern, N.R. and Price, V.E. (1965) The analysis of the stability of general slip surfaces. *Géotechnique* 15, 79–93.
Müller, L. (1964) The rock slide in the Vajont Valley. *Rock Mechanics and Engineering Geology* 2, 148–212.
Nonveiller, E. (1967) Shear strength of bedded and jointed rock of landslides with noncircular surfaces with special reference to Vajont slides. *Proceedings of the Geotechnical Conference*. Oslo, Norway. Norwegian Geotechnical Institute. 1, 289–294.
Nonveiller, E. (1987) The Vajont reservoir slope failure. *Engineering Geology* 24, 493–512.
Peck, R.B. (1969) Advantages and limitations of the observational method in applied soil mechanics. Ninth Rankine Lecture. *Géotechnique* 19 (2), 171–187.
Semenza, E. (2001) *La Storia del Vaiont Raccontata del Geologo che ha Scoperto la Frana. Tecomproject.* Editore Multimediale. Ferrara.
Semenza, E. and Ghirotti, M. (2000) History of the 1963 Vaiont slide: the importance of geological factors. *Bulletin of Engineering Geology and the Environment* 59, 87–97.
Skempton, A.W. (1966) Bedding-plane slip, residual strength and the Vaiont landslide. *Correspondence to Géotechnique* 16 (2), 82–84.
Tika, Th.E. and Hutchinson, J.N. (1999) Ring shear tests on soil from the Vaiont slide slip surface. *Géotechnique* 49 (1), 59–74.
Valdés Díaz-Caneja, J.M. (1964) Meditaciones sobre la Catástrofe del Vaiont. *Boletín nº 20*. Servicio Geológico. Ministerio de Obras Públicas. Madrid.

PART II

EMBANKMENTS AND DAMS

Chapter 3

Collapse of Compacted Soil: Girona Road Embankments, Spain

TABLE OF CONTENTS

E.E. Alonso et al., *Geomechanics of Failures. Advanced Topics*,
DOI 10.1007/978-90-481-3538-7_3, © Springer Science+Business Media B.V. 2010

Chapter 3

Collapse of Compacted Soil: Girona Road Embankments, Spain

3.1 Case Description

The eastern coast of the Iberian Peninsula experiences intense rainfall events every autumn when the warm and humid Mediterranean air masses are hit by Northern and Atlantic colder winds. One of these events, which caused extensive damage to a new road in the province of Girona, NE Spain, is shown in Figure 3.1. The accumulated rainfall in a 38-day period reached 374 l/m^2. The highest intensities were measured the 10 and 11 October, 1994 (123 and 56 l/m^2, respectively).

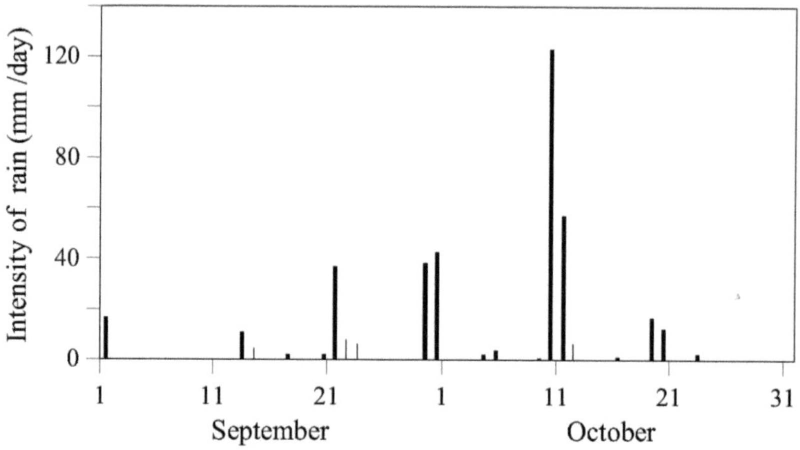

Figure 3.1 Rainfall record in autumn, 1994.

The road embankments were heavily eroded and the pavement lost support at some points in the vicinity of the embankment shoulders. In addition, shallow translational slides modified the original geometry. However, the most serious damage was attributed to the pronounced and systematic volume loss of the embankments. The most serious situation occurred in the access to bridges, because transition slabs were damaged. Bridge abutments had the "wing" design shown in Figures 3.2 and 3.3. In order to smooth the transition from the deformable embankment to the bridge structure, concrete slabs were placed on top of the access embankments. The volume loss experienced by the compacted soil as a result of heavy rains resulted in generalized settlements that reached 30 cm in some locations (Figs. 3.2, 3.3 and 3.4). Fortunately, no vehicle accidents were reported and the road was closed to traffic during the subsequent repair works.

Figure 3.4 shows the nature of the embankment collapse. The fill in contact with the bridge abutment not only settled (the original position of the fill is shown in the concrete wall) but it also retracted horizontally, leaving a wide, empty space which was later filled with an asphalt mixture (the black band adjacent to the abutment wall in the photograph).

Figure 3.2 Collapsed fill around a bridge abutment.

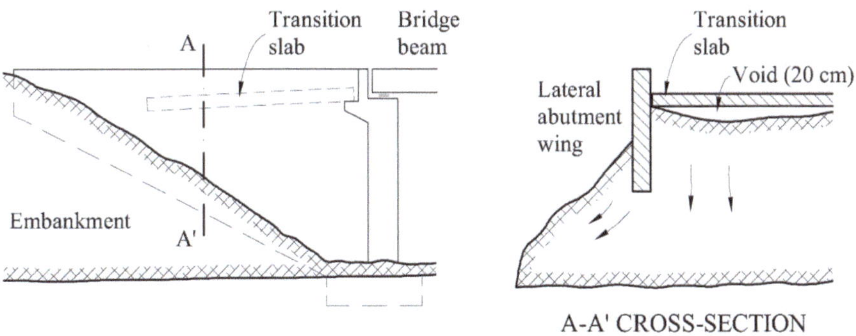

Figure 3.3 Scheme of voids formation under the access slabs to bridge abutments.

3.1.1 Questions asked

After the rains, the Road Administration had two main concerns:

- Identifying the causes of the damage suffered by the recently built road and

- determining if future rainfalls of similar or higher intensity would induce additional damage.

Figure 3.4 Collapse of embankments after rainfall.

3.1.2 Soil properties

Project data, the results of construction control tests, and some additional laboratory experiments performed on samples recovered on borings perforated after the rainfall events, allowed identifying the relevant soil properties. Embankments were built during the summer of 1994 and they were made of compacted decomposed granite, a sandy clay of low to medium plasticity (% < 74 μm = 13–83% (45%); w_L: 31–46% (36%); PI: 7–24% (13%); average value in parentheses). Granite feldspars and micas, when exposed to atmospheric action, decompose into clay minerals (mainly kaolin) while quartz grains remain stable. The end result is a mixture of sand grains and clay minerals.

Compaction specifications were defined on the basis of the Standard Proctor test. Dry density and water content values determined during embankment construction are shown in Figure 3.5 in a compaction plane (γ_d vs. w). The figure also shows the position of the Proctor Optimum, the average "point" of several (γ_d, w) determinations during construction and the contours of equal degree of saturation for S_r = 1, 0.8 and 0.6. The plot shows a significant heterogeneity but it clearly indicates that most of the embankment volume was compacted dry of optimum. Very often the degree of saturation did not reach the value S_r = 0.6, which is a low value, below acceptable specifications for compaction, even if the dry density matches the established target (in this case, the Standard Proctor Optimum).

After the heavy rainfalls it should be expected that the degree of saturation would increase. This was indeed the case, as shown in Figure 3.6. Specimens recovered from borings showed a systematic increase in S_r, especially in the proximity of embankment shoulders and slopes.

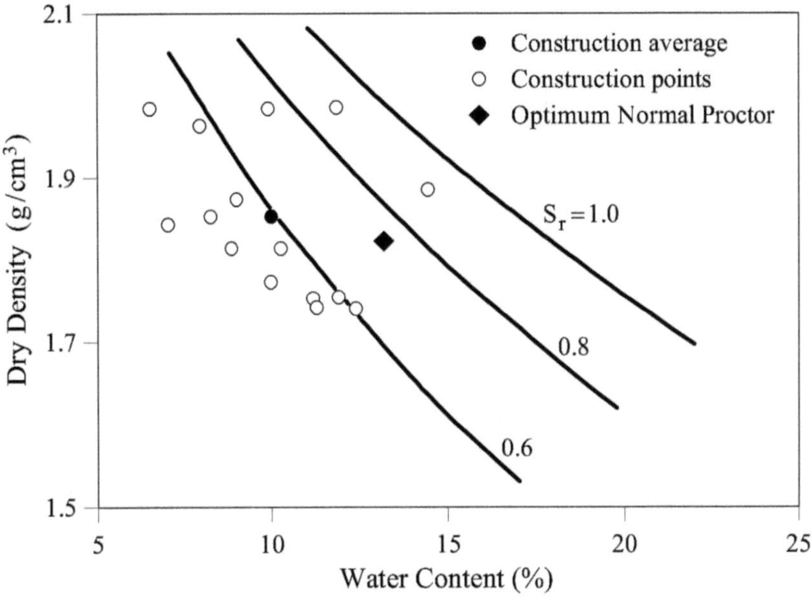

Figure 3.5 Compaction data.

Some of the recovered specimens were also subjected to a "stress path" similar to expected conditions "in situ". A point such as "A" within the embankments, in central positions (Fig. 3.7) was first compacted to a given dry density and water content. It was loaded by the overlying layers to a vertical stress,

$$\sigma_v = \gamma_{nat} z. \tag{3.1}$$

Then, infiltration from rain will increase the degree of saturation of Point A. It would become saturated under extreme conditions.

Oedometer tests were performed in order to reproduce this simple and approximate stress and wetting path. The result of one of these tests is shown in Figure 3.8. When the soil was wetted under a vertical total stress of 0.7 MPa, a compression was observed: the void ratio decreased from 0.65 to 0.59, which implies a volumetric deformation close to 4%. In fact, this is one of the highest volumetric compressions measured. In most of the recovered specimens, the measured "collapse" under the estimated overburden pressure ranged between 0.5 and 2%. The specimen represented in Figure 3.8 was loaded under saturated conditions and finally unloaded.

These results indicate that some collapse potential was still existing in the embankment after the heavy rains. Note also that the test in Figure 3.8 does not provide the initial collapse behaviour, which will be discussed later on the basis of the observed total deformation experienced by the embankments since no collapse tests for the initial as-compacted conditions were available.

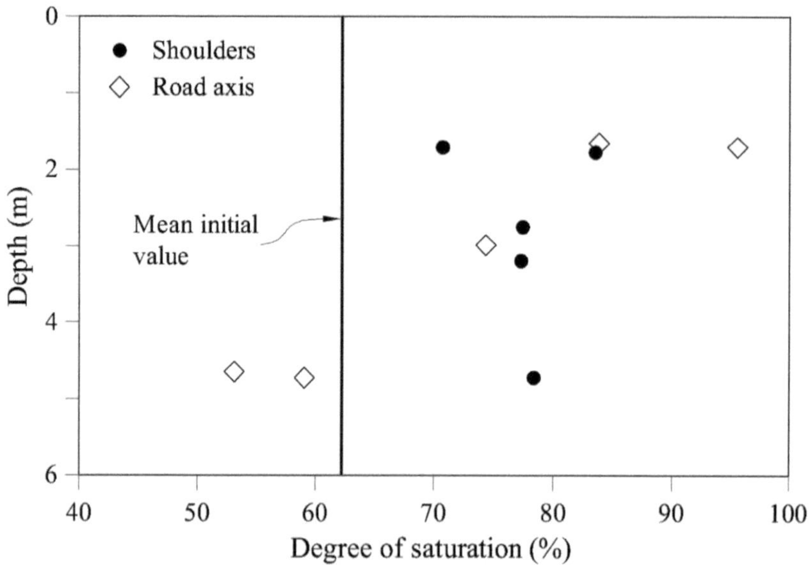

Figure 3.6 Degree of saturation in embankments after the rains.

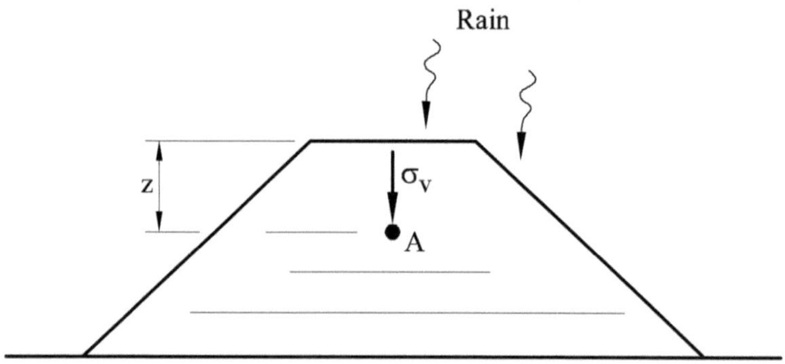

Figure 3.7 Scheme of the embankment.

Water infiltration increases also the natural unit weight of the soil and therefore the vertical total stress σ_v. This effect is not reproduced in the oedometer test performed (wetting at constant stress) but it will be introduced in the analysis described later.

The water retention characteristics of the soil were also investigated. The water retention capacity depends on the type of soil, mineralogy, particle size distribution, and soil structure (arrangement of particles). The water retention capacity is expressed by means of a curve which provides the amount of stored water in the soil when it is subjected to varying values of suction. Variables such as the degree of saturation (ratio of the volume of water and volume of voids), water content (ratio of the weights – or masses – of the water and the solids) or water ratio (volume of water over volume of solids) are measures of the amount of stored water.

In order to obtain the water retention curve of the soil the initial suction of one of the recovered specimens was measured (a value s = 0.4 MPa was obtained). This initial suction was reduced in steps until full saturation and changes in water content were registered. The result, a water retention curve for wetting conditions, is given in Figure 3.9. This test was performed in an oedometer cell under a constant net vertical stress of 0.07 MPa. The role of net stress (difference between total and air pressure in the specimen) in unsaturated soil mechanics will be discussed below. In many practical application p_a = 0 (the reference atmospheric pressure) and the net stress is identical to the total stress if the soil remains unsaturated.

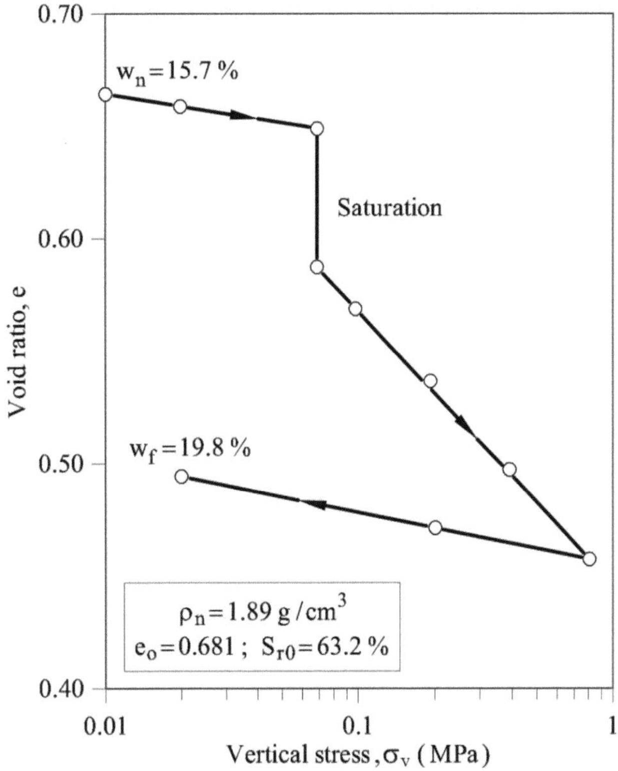

Figure 3.8 Oedometer collapse test. Natural (w_n) and final (w_f) water contents are indicated.

The oedometer response given in Figure 3.8 indicates one fundamental aspect of partially saturated soil behaviour, namely the possibility of experiencing a volumetric compression when subjected to wetting under a given stress. This behaviour is known as "collapse" and it will be further examined in the next section.

3.2 Collapse in Engineering Practice

3.2.1 Collapse of natural and compacted soils

Natural unsaturated soils of low density are capable of significant collapse when wetted. In principle, some applied confining stress is required to trigger the collapse of the soil structure when water is added. The self-weight is often sufficient to cause collapse. A widespread class of soils known to collapse is loess. Their open structure derives from its aeolian origin. Loess particles are fine (silt size) and include quartz and limestone but clay particles are also present. They are found in arid regions of Europe, Asia, and America.

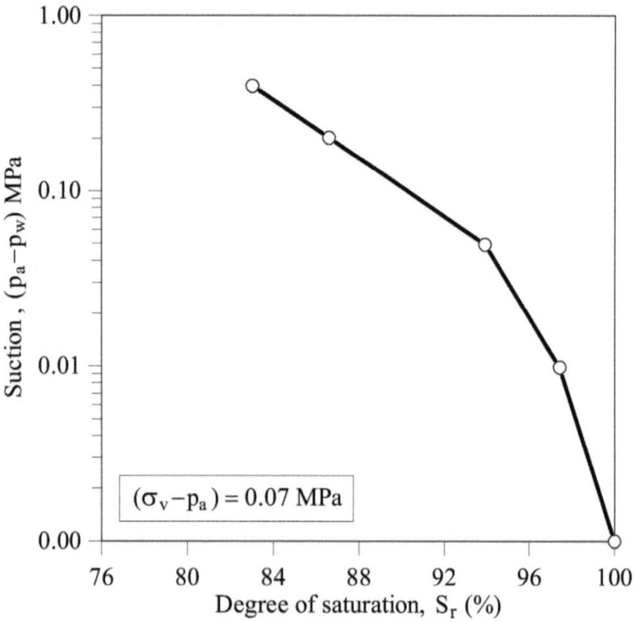

Figure 3.9 Water retention curve (under wetting) of a recovered specimen of the compacted fill.

The photograph in Figure 3.10 shows the granular structure of loess from northern France (Delage *et al.*, 2005).

Their low natural saturation provides an explanation for the collapse mechanism. Consider two spherical particles in contact in Figure 3.11. Water is held at the particle contacts and, due to capillary effects, it remains in tension. The contact menisci introduce a net compression force, F_c, normal to the tangent plane between the two spheres at the contact point.

This is a stabilizing force which helps to resist external stress induced forces (F) which, in general, introduce a given shear at contacts. Water flooding destroys the stabilizing force F_c and leads to a rearrangement of the soil microstructure which is externally perceived as a volumetric compression.

Figure 3.10 SEM photograph of a loess from northern France (Delage *et al.*, 2005).

Not just loess but also other natural alluvial or residual soils of low density are known to collapse when wetted under load. Canals in low density alluvial soils or loess are particularly delicate structures because the presence of water is guaranteed. Two catastrophic canal failures in collapsible silts are shown in Figure 3.12. The Figure 3.12a shows a severe canal dislodgement induced by generalized collapse. The second picture shows a longitudinal crack of one of the side slabs of the canal, attributed to the collapse settlement of the supporting soil, as illustrated in the sketch included in the figure.

Compacted soils may also collapse when compacted on the dry side to relatively low densities. In fact, virtually any soil may collapse if its density, applied confining stress, and initial water content reach an appropriate range of values. When the density is low (man-made fills) collapse upon wetting may lead to large settlements, absolute and differential. Damage associated with collapse phenomena tends to be high because of the magnitude of the settlements often recorded. Case records and other practical aspects of collapse in man-made fills and embankments are reported in Lawton *et al.* (1992) and Skinner *et al.* (1999).

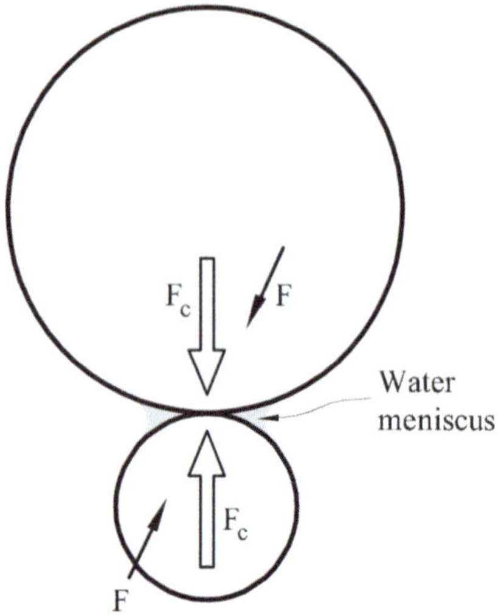

Figure 3.11 Water meniscus at the contact between two spherical grains.

(a) (b)

Figure 3.12 Two canal failures on collapsible soils: (a) Terreu canal; (b) Algerri-Balaguer canal. Ebro Valley, Spain.

3.2.2 Rockfill collapse

Rockfill and coarse granular soils are also known to collapse. In this case, the capillary explanation given before does not make sense because capillary forces at contacts result in a very low value of equivalent stress (sum of forces per unit area).

(a)

(b)

Figure 3.13 Specimen of hard sandstone crushed gravel tested in a 30 cm oedometer cell: (a) grain breakage after testing involving loading and wetting; (b) detail of broken particle.

A fundamental mechanism of rockfill collapse is the breakage of particles heavily loaded at grain to grain contacts. Particles break because a dormant crack suddenly propagates and splits the original particle. It has been shown (Atkinson, 1985; Oldecop and Alonso, 2001) that crack propagation velocity depends on the prevailing relative humidity (RH) in big pores around rock particles. Wetting a rockfill increases RH and this change induces a higher rate of crack propagation, which may eventually lead to the breakage of some particles. The subsequent rearrangement of the structure results in a volumetric compression or collapse.

Figure 3.13 is a picture of broken particles of sandstone after loading and wetting in a 30 cm diameter oedometer cell.

The effects of rockfill collapse have been often observed in rockfill dams. Reservoir impounding and rainfall result in significant deformations which are of concern in dam engineering design. The integrity of the upstream impervious face in some designs (concrete or asphalt face rockfill dams) is of particular concern. Figure 3.14 shows the good correlation between the rate of settlement and rainfall intensity for a 40 m high shale embankment on the high-speed railway line between Madrid and Sevilla.

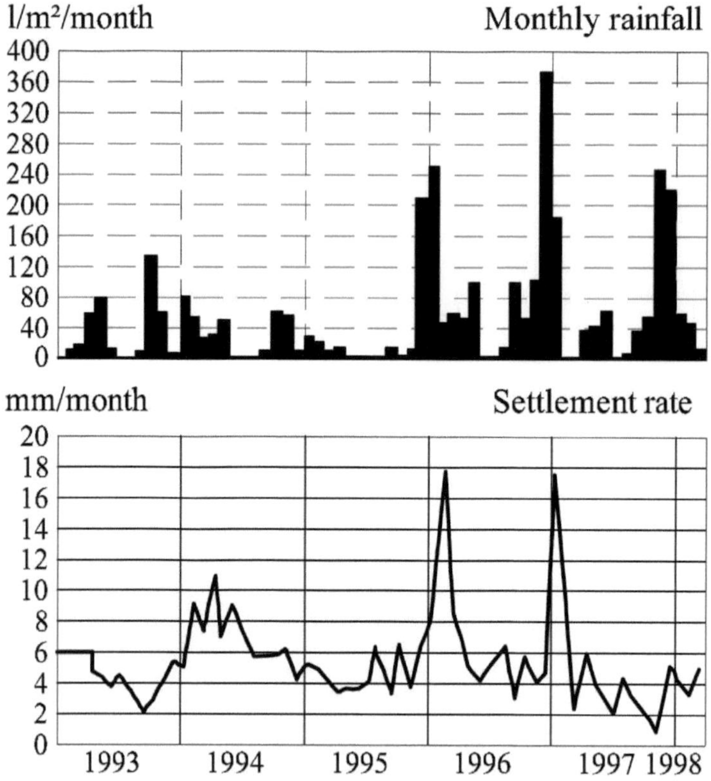

Figure 3.14 Rainfall record and surface settlement rate of a 40 m high rockfill embankment (Soriano and Sánchez, 1999).

3.3 Description of Collapse and its Modelling

3.3.1 Effective stress

Let us maintain isotropic stress conditions. Under saturated conditions, the effect of pore-water pressure on the soil mechanical behaviour is included through the Terzaghi effective stress. There is no need for independent consideration of total stress and pore water pressure.

The capillary interpretation of the state of water in an unsaturated granular soil (see Fig. 3.11) explains that the water is under tension (if the air pressure is taken as the reference zero pressure: in general, the atmospheric pressure). The absolute value of the negative (capillary) pressure is called "matric" suction. The term suction is also used in a more general sense to describe the state of energy of the water. It has units of stress (energy per unit volume).

Unfortunately, no single expression for a single effective stress has been found, in the case of unsaturated soils, to be consistent with experimental observations. Consider, as a first trial, that Terzaghi's expression ($\sigma'_{ij} = \sigma_{ij} - p_w\delta_{ij}$ where δ_{ij} is the Kronecker delta) remains valid for unsaturated conditions. Then, the effective stress would be found by adding total stress and suction. Now consider a collapse phenomenon. Wetting implies a reduction in suction and, therefore, a reduction in effective stress. One would expect a soil expansion if effective stress holds. The observed collapse (volumetric compression) in the above mentioned cases invalidates the proposed definition of effective stress. The same argument applies if only a fraction of the suction is added to the total stress to find an effective stress for unsaturated soils. This is the Bishop (1959) proposal.

One alternative is to accept two independent stress states: total stress (for instance, mean stress, p, under isotropic conditions) and suction s (which is an isotropic value). In order to be general, air pressure, p_a, is introduced as a reference pressure. In this way, the two "effective" stresses are ($\sigma_{ij} - p_a\delta_{ij}$), which is also known as "net stress" and $s = p_a - p_w$ which is the suction. In an isotropic case, if air pressure is the reference zero value, the stress plane (p, s) may conveniently be used to investigate the behaviour of unsaturated states. In the remainder of the chapter, the air pressure will be taken as constant and equal to the reference atmospheric value. This is a reasonable assumption in most engineering applications because of the high gas permeability of unsaturated soils. In this case, the net stress becomes the total stress and suction is equal, in absolute value, to the negative pore water pressure.

3.3.2 Isotropic yielding of unsaturated soils

A convenient starting point for understanding collapse is to consider the response of an unsaturated soil subjected to isotropic loading at different applied suctions ($s = 0$, $s = s_1$, $s = s_2$) (stress paths are given in Fig. 3.15a). Take the soil response under saturated conditions ($s = 0$) as a reference. The soil behaves elastically before a yielding point Y_0 is reached (yield stress p_0^*). The elastic compressibility (the slope of the elastic compression line in the plane e–ln$p$) is defined by κ. Beyond the yield point, the soil will follow a virgin compressibility line with a stiffness defined by a compressibility coefficient $\lambda(0)$ (for $s = 0$).

The effect of suction is to allow the soil maintaining higher void ratios than those corresponding to the saturated virgin compression line. Experiments may be interpreted in the sense that suction increases the yield stress to values p_{01} for $s = s_1$, p_{02} for $s = s_2$, etc. Beyond the yield stress, the unsaturated response will be characterized by a compressibility coefficient $\lambda(s)$ that, in general, will depend on suction. If points p_0^*, p_{01}, p_{02} are collected in the (p,s) space (Fig. 3.15c) a yield locus will be defined which has been named Loading-Collapse (LC) curve for the reasons which will be given below. Points on the left of LC will remain in an elastic state. In Figure 3.15b, all compression lines for stresses below the yield points are plotted parallel with a common elastic compression index κ. Note that they start at a different void ratio ($e_0 > e_1 > e_2$) because the application of an increasing suction (drying) results in a volumetric compression (shrinkage).

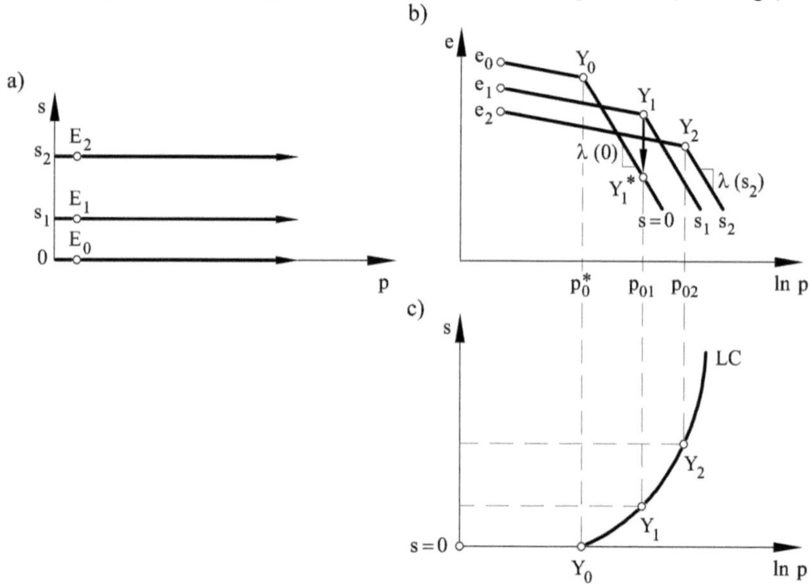

Figure 3.15 (a) Stress paths for isotropic loading at constant suction; (b) compression curves; (c) yielding points and LC yield envelope.

Wetting the soil from a given state on the yield curve such as Y_1 (for suction s_1) to $s = 0$ will imply a reduction in volume because the equilibrium state under stress p_{01} and zero suction ($s = 0$) is the state Y_1^* in Figure 3.15b. The change in void ratio from point Y_1 to point Y_1^* is a collapse deformation.

This idea is used again in Figure 3.16. Consider a Point 1 ($p_1, \bar{s}$) on the yield curve LC; and its corresponding state in the e vs. ln$p$ plot. If wetted at constant $p =$

p_1, the void ratio will evolve from Point 1 to 3 in Figure 3.16b. The yield curve LC will also be dragged to its new position LC_f. The path 1–3 in Figure 3.16 is a collapse path. The soil will experience the same irreversible compression if it is loaded at constant suction from Point 1 to 2. It will deform following the compression line for suction $\bar{s}$. In both cases, a hardening process takes place. These plots indicate that loading and collapse behaviour are intimately related. This is the reason for the name LC given to the isotropic yield locus on the (p,s) plane.

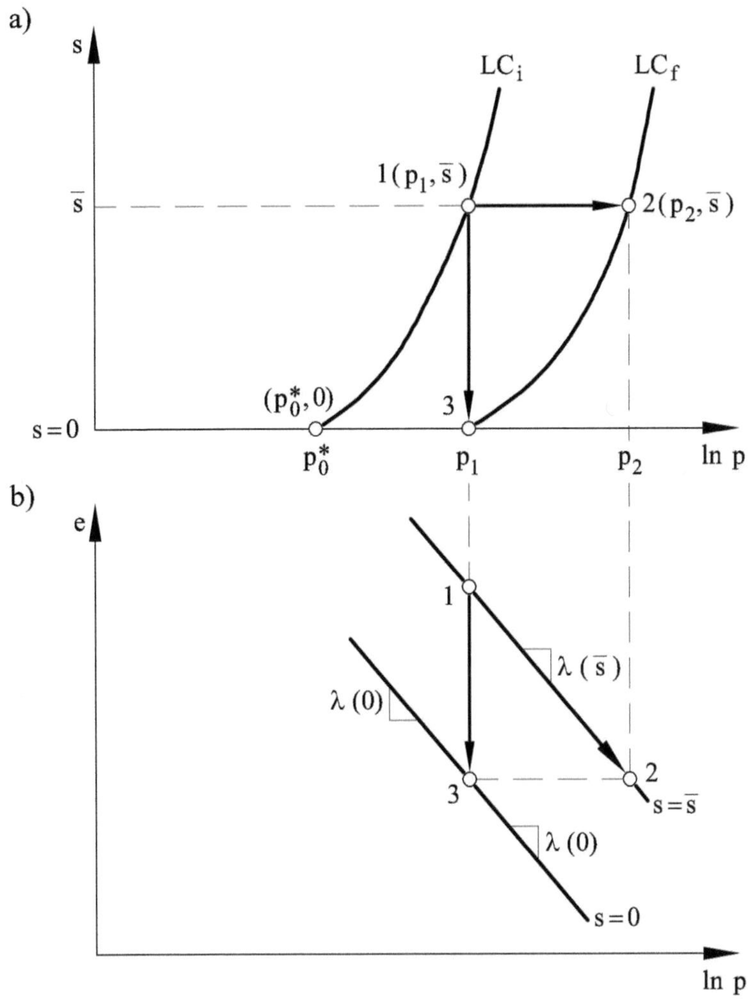

Figure 3.16 (a) Loading (1–2) and collapse (1–3) paths; (b) expected soil response.

The sketch in Figure 3.17 shows the two alternative stress suction paths leading to the same volumetric deformation. The state of the soil at Point 1 is represented by an "open" structure. The arrangement of clay aggregates and sand or silt particles is adapted to large pore sizes. The soil in this state is in equilibrium under a confining stress p and a suction s. The existing suction and its associated set of internal stabilizing forces help to maintain the open structure. If suction is reduced to zero (lower path in the figure), capillary forces vanish and grains are no longer in equilibrium under the applied stress. They fall to new, denser positions and this is indicated in the assumed evolution of the microstructure. But increasing the load at a maintained suction (upper path in the figure) essentially induces the same mechanism of deformation. The suction is maintained but now it is unable to guarantee the initial soil structure under increasing stress. Therefore, the LC yield curve represents the limit arrangement of particles and clay aggregates able to stay in equilibrium under a given combination of p and s. This is the meaning of the yield curve. Any increase in p, decrease in s, or any combination thereof, implies a new equilibrium, a more compressed state, and an associated displacement of LC yield locus towards the right.

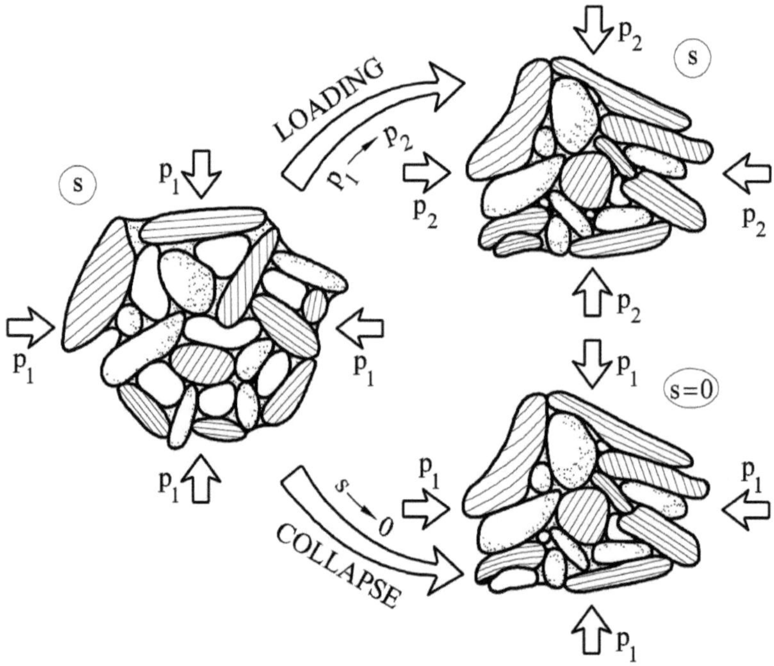

Figure 3.17 Sketch showing an interpretation of collapse and loading in unsaturated soils. (Alonso and Gens (1994) ©1994 Taylor and Francis Group. Used with permission).

3.3.3 Developing a simple model for collapse calculations

In order to proceed, a simple model will be built following a standard procedure:

- Propose a suitable shape for curve LC.
- Propose a hardening rule for LC

For the sake of simplicity, only plastic loading from states on the yield curve will be considered. Therefore, the elastic response will not be discussed.

Experiments indicate that the yield stress does not increase indefinitely with suction. It seems to reach an asymptotic value. This idea is introduced in the following equation

$$p_0(s) = p_0^*\left[1 + a\left(1 - \exp(-\alpha s)\right)\right], \tag{3.2}$$

which provides the preconsolidation or yield stress p_0 as a function of the saturated yield stress p_0^* and the suction s. Coefficients a and α are material parameters: α controls the rate of increase of p_0 with s, and a provides the limiting value of $p_0(s)$ for infinite suction,

$$p_0(\infty) = p_0^*(1 + a). \tag{3.3}$$

The family of yield curves (3.2) was plotted in Figure 3.18 for a given set of material parameters starting at a given point under yielding conditions, i.e. on the LC curve. Additional loading at constant suction (this would represent the accumulation of layers in an embankment, on top of a reference position) would displace the LC curves towards the right and they would change in shape, as shown in the figure.

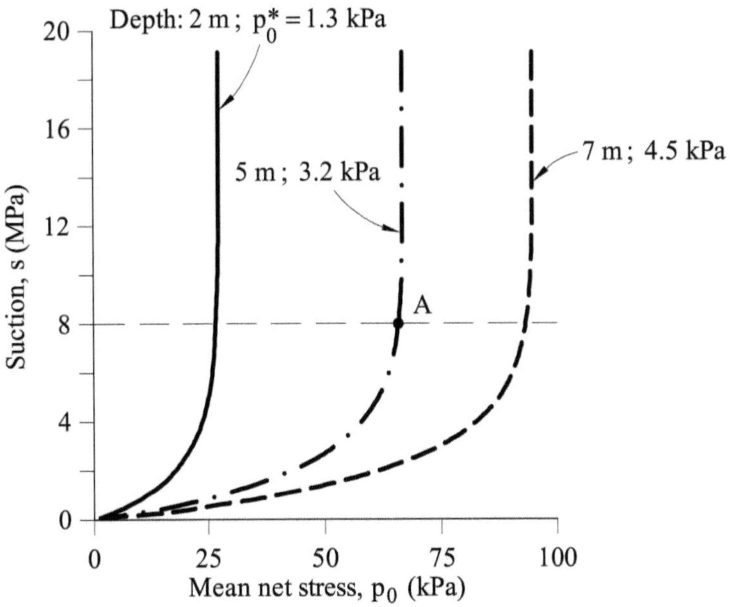

Figure 3.18 LC yield curves.

The three initial stress states considered in plotting Figure 3.18 correspond to a common suction $s = 8$ MPa and the estimated mean total stresses at depths of 2, 5 and 7 m. For an at-rest earth pressure coefficient $K_0 = 0.5$, and for a natural unit weight γ_{nat} = 20 kN/m^3, the mean net stress at a depth of 5 m will be $5\times(20+2\times0.5\times20)/3 = 66.6$ kPa (Point A). Then, for the following material constants: a = 20, α = 0.5 MPa^{-1}, a value $p_0^* = 3.2$ kPa is found from Equation (3.2) for the saturated mean preconsolidation stress. The material constants indicated are derived from the analysis of the collapse of the Girona embankments as shown below.

The yield stress $p_0(s)$, given in Equation (3.2), will be assumed to harden when plastic volumetric strains (ε_{vol}^p) accumulate. Hardening will be controlled through p_0^*, which becomes the hardening parameter in (3.2): $p_0^*(\varepsilon_{vol}^p)$.

The following logarithmic expression is proposed for the evolution of p_0^* with ε_{vol}^p:

$$\mathrm{d}\varepsilon_{vol}^p = \frac{\lambda(0)}{1+e}\frac{\mathrm{d}p_0^*}{p_0^*}. \tag{3.4}$$

Equation (3.4) simply states a linear logarithmic relationship between the plastic volumetric deformations and the applied mean stress under saturated conditions. This is a common behaviour of soils. Equations (3.2) and (3.4) allow calculating the plastic volumetric strains due to any loading and wetting from a state on a given yield curve. In fact, recognizing that plastic strains result from changes in mean stress and suction, one may write

$$\mathrm{d}\varepsilon_{vol}^p = \frac{\partial\varepsilon_{vol}^p}{\partial p}\mathrm{d}p + \frac{\partial\varepsilon_{vol}^p}{\partial s}\mathrm{d}s, \tag{3.5a}$$

where, in view of Equation (3.4),

$$\frac{\partial\varepsilon_{vol}^p}{\partial p} = \frac{\lambda(0)}{1+e}\frac{1}{p_0^*}\frac{\partial p_0^*}{\partial p}, \tag{3.5b}$$

$$\frac{\partial\varepsilon_{vol}^p}{\partial s} = \frac{\lambda(0)}{1+e}\frac{1}{p_0^*}\frac{\partial p_0^*}{\partial s}. \tag{3.5c}$$

(Stress variables p and p_0 are equivalent since stress points remain on the yield surface: they describe the mean net stress.)

Isolating p_0^* from Equation (3.2) and taking the derivative of this variable with respect to p and s, we obtain

$$\frac{\partial p_0^*}{\partial p} = \frac{p_0^*}{p_0}, \tag{3.6a}$$

$$\frac{\partial p_0^*}{\partial s} = \frac{-p_0^* \, a \alpha \exp(-\alpha s)}{\left[1 + a\left(1 - \exp(-\alpha s)\right)\right]}. \tag{3.6b}$$

Then

$$\frac{\partial \varepsilon_{\text{vol}}^p}{\partial p} = \frac{\lambda(0)}{(1+e)} \frac{1}{p_0}, \tag{3.7a}$$

$$\frac{\partial \varepsilon_{\text{vol}}^p}{\partial s} = \frac{-\lambda(0)}{(1+e)} \frac{a \alpha \exp(-\alpha s)}{\left[1 + a\left(1 - \exp(-\alpha s)\right)\right]}, \tag{3.7b}$$

which can now be introduced in (3.5a) to find the increment in plastic volumetric strain for any change in mean stress and suction. Integrating (3.5a) along a given stress and suction path will provide the soil deformation. This is shown in the next section.

3.3.4 Calculating loading and wetting strains

Consider an imposed loading path at constant suction ($\overline{s}$) as the path 1–2 indicated in Figure 3.16. Equation (3.2) provides the variation of p_0^* due to the imposed increment of yield isotropic stress p_0 ($\Delta p = p_2 - p_1$) at constant suction:

$$\Delta p_0^* = p_{0_2}^* - p_{0_1}^* = \frac{\Delta p}{1 + a\left(1 - \exp(-\alpha \overline{s})\right)}. \tag{3.8}$$

Integrating Equation (3.4), the volumetric plastic strain due to the increment in mean stress $p_1 \rightarrow p_2$ becomes

$$\Delta \varepsilon_{\text{vol}}^p = \frac{\lambda(0)}{1+e} \ln\left(\frac{p_{0_2}^*}{p_{0_1}^*}\right) = \frac{\lambda(0)}{1+e} \ln\left(\frac{p_2}{p_1}\right). \tag{3.9}$$

Similarly, volumetric plastic strains accumulated during a collapse path (wetting at constant loading, path 1 to 3 in Fig. 3.16) can be calculated. In this case, Equations (3.5a) and (3.7b) for dp = 0 result in

$$\mathrm{d}\varepsilon_{\text{vol}}^p = -\frac{\lambda(0) a \alpha}{1+e} \frac{\exp(-\alpha s)}{\left[1 + a\left(1 - \exp(-\alpha s)\right)\right]} \mathrm{d}s, \tag{3.10}$$

which can now be integrated between two suction values $s_1 \rightarrow s_2$:

$$
\begin{aligned}
\Delta\varepsilon_{\mathrm{vol}}^{p} &= \int_{s_1}^{s_2} \mathrm{d}\varepsilon_{\mathrm{vol}}^{p} = -\frac{\lambda(0)a\alpha}{1+e}\int_{s_1}^{s_2}\frac{\exp(-\alpha s)}{\left[1+a\left(1-\exp(-\alpha s)\right)\right]}\mathrm{d}s = \\
&= -\frac{\lambda(0)}{1+e}\ln\left[\frac{1+a-a\exp(-\alpha s_1)}{1+a-a\exp(-\alpha s_2)}\right].
\end{aligned}
\tag{3.11}
$$

As an example, for a compressibility coefficient for saturated conditions, $\lambda(0)$= 0.024, a = 20, α = 0.5 MPa^{-1}, an initial void ratio e_0 = 0.60, and a change in suction from s_1 = 10 MPa to s_2 = 1 MPa, a collapse strain of $\varepsilon_{\mathrm{vol}}^{p}$ = 0.013 = 1.3% is calculated.

The next necessary step is to link the volumetric behaviour of the soil to the flow in the embankment.

3.3.5 Flow and collapse modelling

In the previous section, a simple formulation for modelling plastic volumetric strains, and, in particular, the collapse due to reduction in suction, has been developed. In this section, the formulation required to calculate the variation of suction in an unsaturated/saturated porous media is given.

In an unsaturated deformable porous soil, changes in suction will mainly depend on

- water flow
- water retention of the soil
- changes in volume of the soil skeleton, water and solid particles.

In order to take into account all these phenomena in the calculation of pore pressures, mass balance equations of solid particles and water are required.

Solid mass balance

In order to formulate the solid mass balance equation, we consider a representative volume V (Fig. 3.19) fixed in space (Eulerian description). S is the surface closing this volume, $\mathbf{n}$ is the unit outward normal to the boundary surfaces, and $\mathrm{d}\mathbf{S} = \mathbf{n}\mathrm{d}S$ is a differential element of area. The volume contains a porous media whose solid particles have a density ρ_s and a porosity n. Voids are partially occupied with water of density ρ_w.

The total mass of solid particles per unit volume is $\rho_s(1-n)$. In volume V, the total mass of solid particles will be

$$
\int_V \rho_s(1-n)\mathrm{d}V. \tag{3.12}
$$

Transfer of solid mass in or out of volume V results in the following rate of change of the total solid mass:

$$\frac{\partial}{\partial t}\left[\int_V \rho_s(1-n)\,\mathrm{d}V\right]=\int_V \frac{\partial}{\partial t}\left[\rho_s(1-n)\right]\mathrm{d}V. \tag{3.13}$$

The velocity of solid particles is **v**. Therefore, the outward solid mass flux through a differential element of area is $\rho_s(1-n)\mathbf{v}\cdot\mathbf{n}$ and the total outward solid mass flux through surface *S* is

$$\int_S \rho_s(1-n)\mathbf{v}\cdot\mathbf{n}\,\mathrm{d}S. \tag{3.14}$$

Conservation of mass requires the rate of mass within the volume to be equal to the net flux. Therefore

$$\int_V \frac{\partial}{\partial t}\left[\rho_s(1-n)\right]\mathrm{d}V=-\int_S \rho_s(1-n)\mathbf{v}\cdot\mathbf{n}\,\mathrm{d}S. \tag{3.15}$$

The negative sign of the flux term indicates that the increase of mass, as well as the outward flux, were accepted as positive.

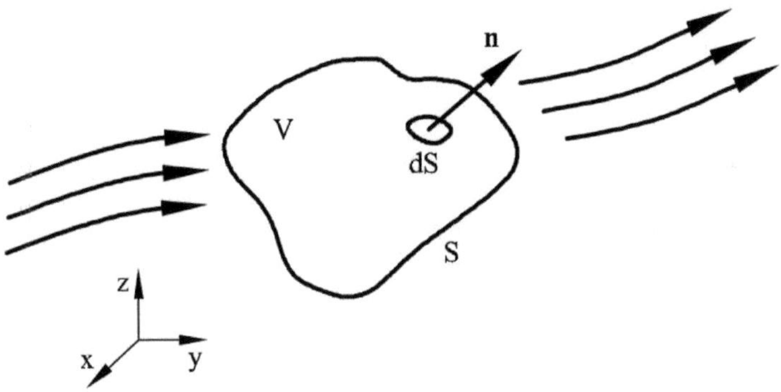

Figure 3.19 Fixed volume of soil, *V*, bounded by a surface *S* in a Cartesian space. A flow of mass crosses the volume.

In order to convert this integral equation of mass conservation into a differential equation, the Gauss divergence theorem is applied to the surface integral:

$$\int_S \rho_s(1-n)\mathbf{v}\cdot\mathbf{n}\,\mathrm{d}S=\int_V \mathrm{div}\left[\rho_s(1-n)\mathbf{v}\right]\mathrm{d}V. \tag{3.16}$$

When the divergence operator is applied to $\rho_s(1-n)\mathbf{v}$ the result is

$$\mathrm{div}\left[\rho_s(1-n)\mathbf{v}\right]=\frac{\partial\left(\rho_s(1-n)v_x\right)}{\partial x}+\frac{\partial\left(\rho_s(1-n)v_y\right)}{\partial y}+\frac{\partial\left(\rho_s(1-n)v_z\right)}{\partial z}.$$

Equations (3.15) and (3.16) result in

$$\int_V \frac{\partial}{\partial t}\left[\rho_s(1-n)\right]\mathrm{d}V + \int_V \operatorname{div}\left[\rho_s(1-n)\mathbf{v}\right]\mathrm{d}V = 0 \tag{3.17a}$$

or

$$\int_V \left(\frac{\partial}{\partial t}\left[\rho_s(1-n)\right] + \operatorname{div}\left[\rho_s(1-n)\mathbf{v}\right]\right)\mathrm{d}V = 0. \tag{3.17b}$$

Since Equation (5.9b) is valid for any representative volume, the expression under the integral sign must vanish. This condition provides the conservation of mass expression in local form:

$$\frac{\partial}{\partial t}\left[\rho_s(1-n)\right] + \operatorname{div}\left[\rho_s(1-n)\mathbf{v}\right] = 0, \tag{3.18a}$$

where the first term is the time variation of solid mass stored per unit volume and the second term provides the net flux of solid particles. The preceding equation can be expanded to give

$$\begin{aligned}&(1-n)\frac{\partial\rho_s}{\partial t} - \rho_s\frac{\partial n}{\partial t} + (1-n)\,\mathbf{grad}(\rho_s)\cdot\mathbf{v} - \\ &\rho_s\,\mathbf{grad}(n)\cdot\mathbf{v} + \rho_s(1-n)\operatorname{div}(\mathbf{v}) = 0,\end{aligned} \tag{3.18b}$$

where the vector gradient operator, **grad(·)**, has been introduced. In detail, Equation (3.18b) is written:

$$\begin{aligned}&(1-n)\frac{\partial\rho_s}{\partial t} - \rho_s\frac{\partial n}{\partial t} + (1-n)\left(\frac{\partial\rho_s}{\partial x},\frac{\partial\rho_s}{\partial y},\frac{\partial\rho_s}{\partial z}\right)\begin{pmatrix}v_x\\ v_y\\ v_z\end{pmatrix} - \\ &\rho_s\left(\frac{\partial n}{\partial x},\frac{\partial n}{\partial y},\frac{\partial n}{\partial z}\right)\begin{pmatrix}v_x\\ v_y\\ v_z\end{pmatrix} + \rho_s(1-n)\left(\frac{\partial v_x}{\partial x} + \frac{\partial v_y}{\partial y} + \frac{\partial v_z}{\partial z}\right) = 0.\end{aligned} \tag{3.18c}$$

The definition of material derivative of a property of a material point moving with velocity **v**,

$$\frac{\mathrm{D}(\bullet)}{\mathrm{D}t} = \frac{\partial(\bullet)}{\partial t} + \mathbf{v}\cdot\mathbf{grad}(\bullet) \tag{3.19}$$

allows the writing of (3.18b) in a compact form:

$$(1-n)\frac{\mathrm{D}\rho_s}{\mathrm{D}t} - \rho_s\frac{\mathrm{D}n}{\mathrm{D}t} + \rho_s(1-n)\operatorname{div}(\mathbf{v}) = 0. \tag{3.20}$$

The divergence of the solid velocity can be transformed into the volumetric strain rate as follows:

$$\operatorname{div}(\mathbf{v}) = \operatorname{div}\left(\frac{\partial \mathbf{u}}{\partial t}\right) = -\frac{\partial \varepsilon_{\text{vol}}}{\partial t}, \tag{3.21}$$

where compressive volumetric deformation are considered to be positive. This result is obtained by a simple interchange of time and spatial derivative.

Equation (3.20) provides the rate of change of the porosity (Dn/Dt) as a function of the stiffness of the soil skeleton and of the changes in density of solid particles,

$$\frac{Dn}{Dt} = \frac{(1-n)}{\rho_s}\frac{D\rho_s}{Dt} + (1-n)\operatorname{div}(\mathbf{v}), \tag{3.22}$$

a result which will be immediately used.

In many applications the compressibility of the skeleton is large compared with the compressibilility of solid particles. Then Equation (3.22) can be reduced to

$$\frac{Dn}{Dt} = -(1-n)\frac{\partial \varepsilon_{\text{vol}}}{\partial t}. \tag{3.23}$$

Water balance

Water flow velocity ($\mathbf{v}_w$) will be expressed as the sum of two terms: the velocity of solids ($\mathbf{v}$) already introduced, and the relative velocity of the fluid with respect to solids ($\mathbf{q}^*$). Consider now a cross-section of the porous medium of unit area. Only a portion of this area (approximately given by porosity multiplied by the degree of saturation, nS_r) will be available for fluid flow. Therefore, the flow rate of water through a cross-section of unit area will be given by $nS_r\,(\mathbf{v} + \mathbf{q}^*)$. The term $nS_r\mathbf{q}^* = \mathbf{q}$ is generally known as the Darcy flow rate of a fluid filtrating through an unsaturated porous medium. It is the relative flow rate of the fluid with respect to the soil skeleton.

The previous derivation of the balance equation for the solid mass can now be followed step by step. But it may be directly written from the solid mass conservation equation by formally replacing $(1-n)$ by nS_r, ρ_s by ρ_w and $(1-n)\mathbf{v}$ by $nS_r\mathbf{v} + \mathbf{q}$. Therefore, from Equation (3.18a) it follows

$$\frac{\partial}{\partial t}(\rho_w n S_r) + \operatorname{div}\left[\rho_w\left(nS_r\mathbf{v} + \mathbf{q}\right)\right] = 0. \tag{3.24}$$

The first term in Equation (3.24) is the rate of change of water mass stored in a unit volume of soil. The second term is the net water mass flow entering or leaving the unit soil element. It has two components: the flux due to the displacement of the porous media and the relative flux ($\mathbf{q}$) between the water and solid particles.

Equation (3.24) can now be expanded as follows:

$$
\begin{aligned}
&nS_r \frac{\partial \rho_w}{\partial t} + nS_r \mathbf{v} \cdot \mathbf{grad}(\rho_w) + \rho_w S_r \frac{\partial n}{\partial t} + \rho_w S_r \mathbf{v} \cdot \mathbf{grad}(n) + \\
&n\rho_w \frac{\partial S_r}{\partial t} + n\rho_w \mathbf{v} \cdot \mathbf{grad}(S_r) + \rho_w n S_r \operatorname{div}(\mathbf{v}) + \operatorname{div}(\rho_w \mathbf{q}) = 0.
\end{aligned} \tag{3.25}
$$

If the definition of material derivative (Eq. (3.19)) is introduced, the water mass balance equation becomes

$$
nS_r \frac{\mathrm{D}\rho_w}{\mathrm{D}t} + n\rho_w \frac{\mathrm{D}S_r}{\mathrm{D}t} + S_r \rho_w \frac{\mathrm{D}n}{\mathrm{D}t} + \rho_w n S_r \operatorname{div}(\mathbf{v}) + \operatorname{div}(\rho_w \mathbf{q}) = 0. \tag{3.26}
$$

In most practical situations except for hard soils or soft rocks, the compressibility of the water can be neglected and equation (3.26) can be simplified to

$$
n \frac{\mathrm{D}S_r}{\mathrm{D}t} + S_r \frac{\mathrm{D}n}{\mathrm{D}t} - nS_r \frac{\partial \varepsilon_{\mathrm{vol}}}{\partial t} + \operatorname{div}(\mathbf{q}) = 0, \tag{3.27}
$$

where Equation (3.21) was used.

Integrated mass balance equation. The field differential equation

Solid and water mass balance equations can now be combined substituting the porosity variation in time given by Equation (3.23) into Equation (3.27):

$$
n \frac{\mathrm{D}S_r}{\mathrm{D}t} - S_r \frac{\partial \varepsilon_{\mathrm{vol}}}{\partial t} + \operatorname{div}(\mathbf{q}) = 0. \tag{3.28}
$$

If the spatial variation of the degree of saturation is neglected, the balance equation becomes

$$
n \frac{\partial S_r}{\partial t} - S_r \frac{\partial \varepsilon_{\mathrm{vol}}}{\partial t} + \operatorname{div}(\mathbf{q}) = 0. \tag{3.29}
$$

Equation (3.29) identifies the two terms of stored water in an unsaturated soil. Water can be stored by varying the volume of water in the voids (degree of saturation) or by deforming the soil skeleton.

In order to proceed, the terms of Equation (3.29) should be related to the unknown variables: (net) stress and suction.

The retention curve provides the degree of saturation in terms of suction. Therefore, it is necessary to fit a mathematical expression for the water retention curve (WRC) obtained in the laboratory. Different expressions can be found in the literature. A popular one is the expression published by Van Genuchten (1980):

$$
S_r = S_{r\min} + \left(S_{r\max} - S_{r\min}\right) \left[1 + \left(\frac{s}{P_0}\right)^{\frac{1}{1-\lambda}}\right]^{-\lambda}, \tag{3.30}
$$

where λ is related to the slope of the WRC in its central part and P_0 is related to the "air entry value", i.e. the necessary suction to desaturate an initially saturated soil. The preceding equation is defined between two saturation limits (a maximum (S_{rmax}), close to one, and a minimum (S_{rmin}), close to zero). Figure 3.20 provides the shape of water retention curves and shows the effect of varying parameters P_0 and λ.

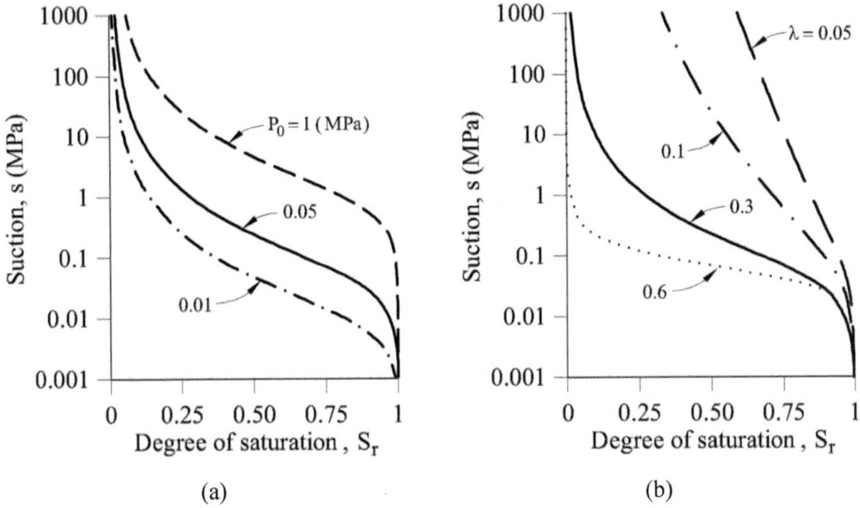

Figure 3.20 Van Genuchten retention curves for (a) $\lambda = 0.3$ and different values of P_0 and (b) $P_0 = 0.05$ MPa and different values of λ.

The term $\partial S_r/\partial t$ in Equation (3.29) can be expressed as the product of $(\partial S_r/\partial s)(\partial s/\partial t)$ where $\partial S_r/\partial t$ is obtained by differentiating Equation (3.30):

$$\frac{\partial S_r}{\partial s} = \left(S_{r\max} - S_{r\min}\right)\frac{1}{P_0}\frac{(-\lambda)}{(1-\lambda)}\left[1+\left(\frac{s}{P_0}\right)^{\frac{1}{1-\lambda}}\right]^{-\lambda-1}\left(\frac{s}{P_0}\right)^{\frac{\lambda}{1-\lambda}}. \tag{3.31}$$

Volumetric strains depend on changes in net mean stress and suction. They have been described in Equations (3.5), (3.6) and (3.7). Because of the simple case analyzed here, only isotropic changes in stress have been considered. For a general case, the stress tensor should be introduced since not only isotropic but also deviatoric stress variation may cause volumetric strain.

The final term in Equation (3.29) refers to the flow through pores due to the gradient of head (Darcy's law). A generalized Darcy law for a compressible fluid describes the relative flow velocity **q** in terms of gradients of pore-water pressure and the gradient of elevation as follows:

$$\mathbf{q} = -\frac{k}{\gamma_w}\left[\mathbf{grad}(-s) + \gamma_w\,\mathbf{grad}(z)\right], \tag{3.32}$$

where γ_w is the unit weight of water, z is the vertical coordinate and k is the hydraulic conductivity (the term permeability will also be used). In an unsaturated soil, the permeability to water decreases when the degree of saturation decreases because of the reduction of pore volume filled with water and the difficulty of the water adhered to particles to displace. A common approach is to express the unsaturated permeability in terms of the saturated value, k_{sat} through a relative permeability term (k_{rel}) which decreases fast with S_r according to the following widely used expression:

$$k = k_{\text{sat}} k_{\text{rel}} = k_{\text{sat}} \left(S_r\right)^m . \tag{3.33}$$

The power m typically takes values in the range 2–4. Figure 3.21 shows the effect of m on the relative permeability coefficient.

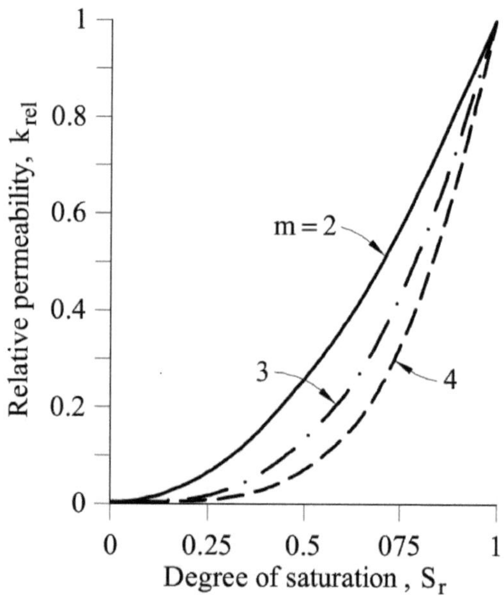

Figure 3.21 Relative permeability curve for different values of parameter m (Eq. (3.33)).

For a one-dimensional analysis in the z-direction, **grad(•)** and div(•) are reduced to $\partial(\bullet)/\partial z$. Then, the flow velocity **q** of Equation (3.32) becomes

$$\mathbf{q} = \frac{k}{\gamma_w}\frac{\partial s}{\partial z} - k\frac{\partial z}{\partial z} = \frac{k}{\gamma_w}\frac{\partial s}{\partial z} - k. \tag{3.34}$$

Equation 3.34 is consistent with the common formulation of Darcy's law for saturated soils provided the concept of suction is formally extended to saturated states as the negative value of the water pressure. Then if hydrostatic conditions are considered $s = -p_w = -\gamma_w z$ and Equation (3.34) would provide a zero flow.

The last term of Equation (3.29) can be developed as follows:

$$\operatorname{div}(\mathbf{q}) = \frac{\partial q_z}{\partial z} = \left(\frac{1}{\gamma_w}\frac{\partial s}{\partial z} - 1\right)\frac{\partial k}{\partial z} + \frac{k}{\gamma_w}\frac{\partial^2 s}{\partial z^2}, \tag{3.35}$$

where $\dfrac{\partial k}{\partial z} = \dfrac{\partial k}{\partial S_r}\dfrac{\partial S_r}{\partial s}\dfrac{\partial s}{\partial z}$ and $\partial k/\partial S_r$ are obtained differentiating Equation (3.33) and $\partial S_r/\partial s$ is given in Equation (3.31).

Finally, using the previous equations, Equation (3.29) can be rewritten as follows:

$$\begin{aligned}&\left(n\frac{\partial S_r}{\partial s} - S_r\frac{\partial \varepsilon_{vol}}{\partial s}\right)\frac{\partial s}{\partial t} - S_r\frac{\partial \varepsilon_{vol}}{\partial p}\frac{\partial p}{\partial t} + \frac{k}{\gamma_w}\frac{\partial^2 s}{\partial z^2} + \\ &\left(\frac{1}{\gamma_w}\frac{\partial s}{\partial z} - 1\right)\frac{\partial k}{\partial S_r}\frac{\partial S_r}{\partial s}\frac{\partial s}{\partial z} = 0.\end{aligned} \tag{3.36}$$

This is a parabolic partial differentiated equation. The terms $\partial S_r/\partial s$, $\partial \varepsilon_{\mathrm{vol}}/\partial p$ and $\partial \varepsilon_{\mathrm{vol}}/\partial s$ are given by Equations (3.31), (3.7a), and (3.7b), respectively.

The integration of Equation (3.36) with the appropriate boundary and initial conditions will provide the suction of any point of the integration domain at any time. In the next section, this equation will be solved for the case of the collapsed Girona road embankments.

3.4 Modelling the Collapse of Girona Road Embankments

The geometry is sketched in Figure 3.22. A central column of soil will be analyzed under one-dimensional oedometric conditions. Several embankments of different sizes were affected by the rainfall but only one case, an embankment 8 m high, will be considered here. Only vertical displacements will be calculated. Note, however, that collapse strains are volumetric and therefore the expected deformed shape of a collapsed embankment will be given by a decrease in size in all directions, as shown in Figure 3.22. This was also the observation in the field (see the photograph in Fig. 3.4).

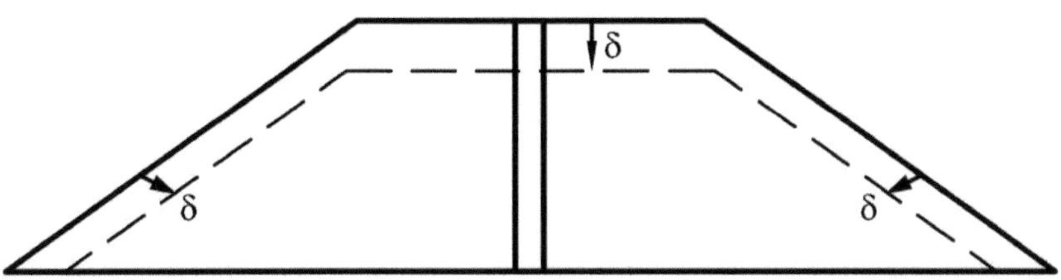

Figure 3.22 Sketch of the collapsed embankment. The coupled flow-deformation problem will be solved in a central column. Also indicated is the expected deformed shape of the embankment after collapse.

It will be assumed that the soil within the embankment is normally consolidated and that the initial stress distribution depends on the weight of the soil, which, since it is in unsaturated condition, can be expressed as follows:

$$\gamma_{\text{nat}} = (1-n)\gamma_s + nS_r\gamma_w . \tag{3.37}$$

Under oedometric conditions, horizontal stresses at any point are equal to $K_0\sigma_v$ where K_0 is the at-rest earth pressure coefficient. Then the mean stress ($p = (\sigma_v + 2\sigma_h)/3$) for a given depth z is equal to

$$p = \int_0^z \frac{1}{3}(1+2K_0)\left[(1-n)\gamma_s + nS_r\gamma_w\right] z\,\mathrm{d}z \tag{3.38}$$

The initial suction of the embankment can be derived from the water retention curve (Fig. 3.9) and the data on compaction (Fig. 3.5). It appears that the average degree of saturation after compaction of embankments was 0.6. The water retention data were again plotted in Figure 3.23 and the Van Genuchten expression (3.30) was adapted to the experimental points (Fig. 3.23). The derived parameter values are $S_{r\max} = 1$, $S_{r\min} = 0$, $\lambda = 0.09$ and $P_0 = 0.05$ MPa. For $S_r = 0.6$, an initial average suction $s = 8$ MPa is obtained.

The virgin compressibility index (a unique value, independent of suction, in the model developed) is calculated from the oedometer test for saturated conditions shown in Figure 3.8. The value of C_c is calculated as follows:

$$C_c = \frac{e-e_0}{\log(\sigma_v/\sigma_{v0})} = \frac{0.133}{\log(0.8/0.07)} = 0.054 \tag{3.39}$$

and, therefore, $\lambda = C_c/\ln 10 = 0.024$.

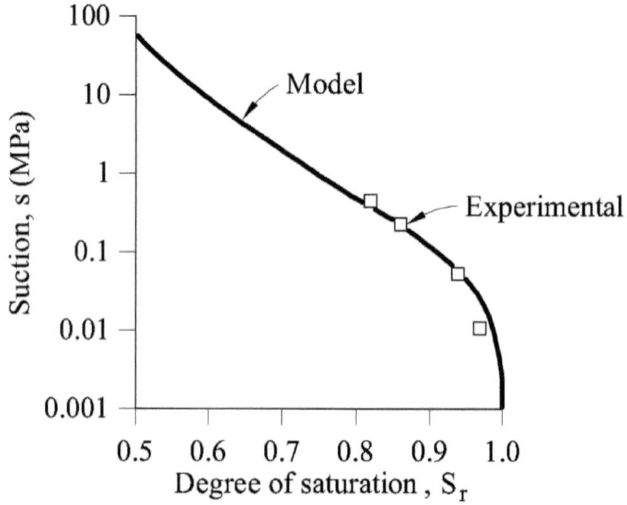

Figure 3.23 Water retention curve for calculation compared with experimental data.

The top of the embankment will be in contact with the atmosphere and it will be infiltrated by rainfall. At the base, it will be in contact with a saturated natural soil, having the water table on the surface. Therefore, an upward flow of water will cross the lower embankment boundary due to suction gradients (s = 8 MPa inside the embankment and s = 0 MPa at the lower boundary). The rain may be simulated by imposing a water inflow at the top of the embankment as a boundary condition. To do that, it is necessary to estimate the fraction of rain that seeps into the ground and the fraction that leaves the embankment as a surface flow. In order to simplify the problem, a boundary condition at the top of the embankment imposing a suction equal to zero is defined. It implies the presence of a thin film of liquid water on the embankment surface capable of delivering the necessary infiltration flow. This condition is also imposed in the bottom of the embankment, as mentioned before.

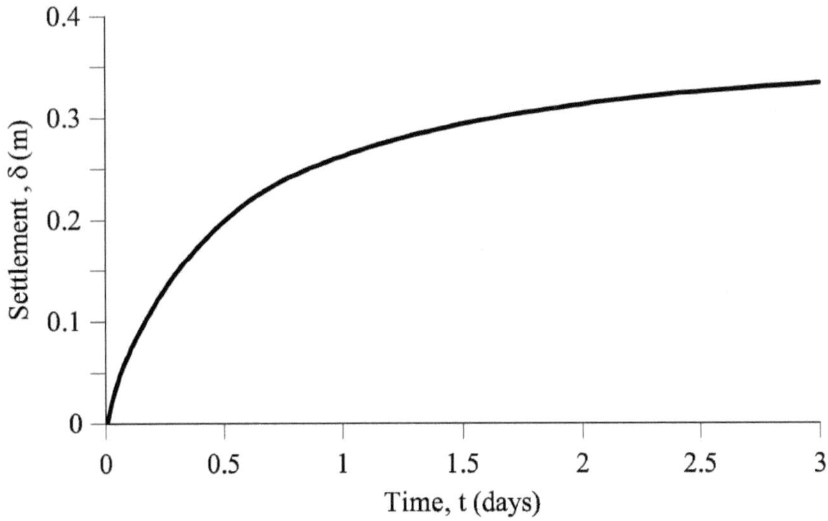

Figure 3.24 Calculated evolution of crest settlement of an 8 m high embankment under a top and bottom infiltration.

The inward flow of water will induce increments in the degree of saturation and collapse plastic strains in the embankment. The increment in the degree of saturation also involves changes in the total mean stress because of the increase of soil unit weight

$$\frac{\partial p}{\partial t}=\frac{\partial}{\partial t}\left[\int_0^z \frac{1}{3}(1+2K_0)[(1-n)\gamma_s+nS_r\gamma_w]z\,\mathrm{d}z\right]= \\ =\frac{1}{3}(1+2K_0)\gamma_w\int_0^z n\frac{\partial S_r}{\partial t}z\,\mathrm{d}z. \tag{3.40}$$

Note that total stresses change in time due to the modification of soil natural specific weight as the degree of saturation changes. The porosity and time variation of the degree of saturation depend on coordinate z. Equation (3.40) is used in the field Equation (3.36).

The problem of flow in a (deformable and unsaturated) soil can be solved through the formulation just given. The differential Equation (3.36) has to be integrated. Since the terms preceding $\partial s/\partial t$ depend on suction, an analytical solution of this equation cannot be found in general. It will be solved by a finite difference approximation developed in the Appendix 3.1.

3.5 Results

3.5.1 Collapse settlements

Experimental data on the shape of the LC yield curve (Fig. 3.18) were not available in this case. There was information on initial soil compaction and some collapse tests performed on samples recovered after the heavy rains (Fig. 3.8). The test in Figure 3.8 provided a value for the compression index $\lambda(0) = 0.024$. The water retention curve, $S_r(s)$, was also measured (Figs. 3.9 and 3.23). This information provided a value for the initial soil suction which was $s = 10$ MPa (for $S_r = 0.6$; see Fig. 3.23). A further hypothesis was that the soil was under normally consolidated conditions and this information provides a direct relationship between the field mean confining stresses and the saturated mean stresses (Fig. 3.18), provided parameters a and α of the yield locus LC are known. But the field observation that the higher embankments (heights of 7–8 m) had settled 30–40 cm after the heavy rainfall period could be used to derive, by means of a trial and error procedure, the parameters of the LC Equation (3.2).

Figure 3.24 shows the calculation of the crest settlement of an 8 m high embankment, following the procedure detailed in the Appendix 3.1. The top and bottom of the unsaturated soil column were flooded ($s = 0$) at $t = 0$. For a permeability $k = 10^{-7}$ m/s, $\alpha = 0.5$ MPa^{-1}, $a = 20$, and the remaining data as indicated previously, crest settlements of 30 – 40 cm were calculated for 1.5 – 3 days of continuous wetting. The settlement rate is highest at the start of the process and decreases continuously. The calculated suction isochrones for the first day of infiltration are shown in Figure 3.25. The time evolution of suction of two points located at depths of 2 and 5 m, representatives of the core of the embankment, are given in Figure 3.26. After one day of wetting, and despite of the relatively low permeability, the embankment is almost fully wetted. Infiltration (Darcy) flow rates are not only controlled by permeability but also by the suction gradients which, in our case, are very large. The development of plastic collapse strains follows the dissipation of suction, as shown in Figure 3.27.

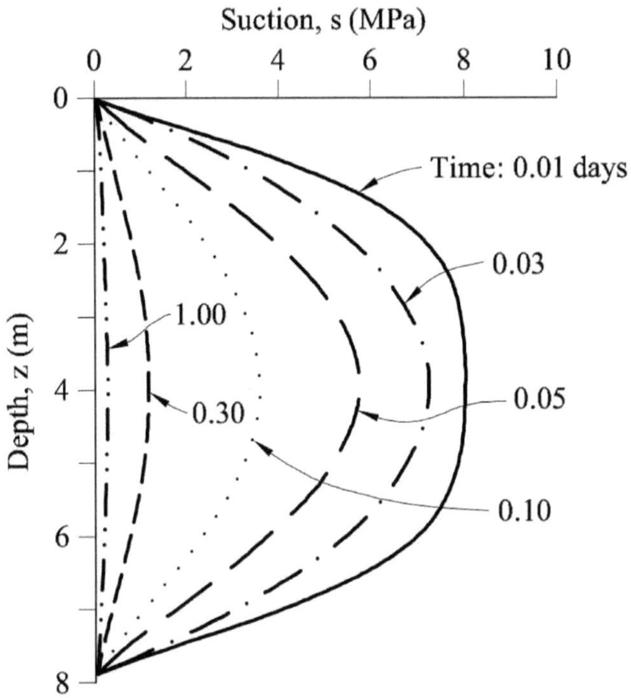

Figure 3.25 Calculated suction isochrones.

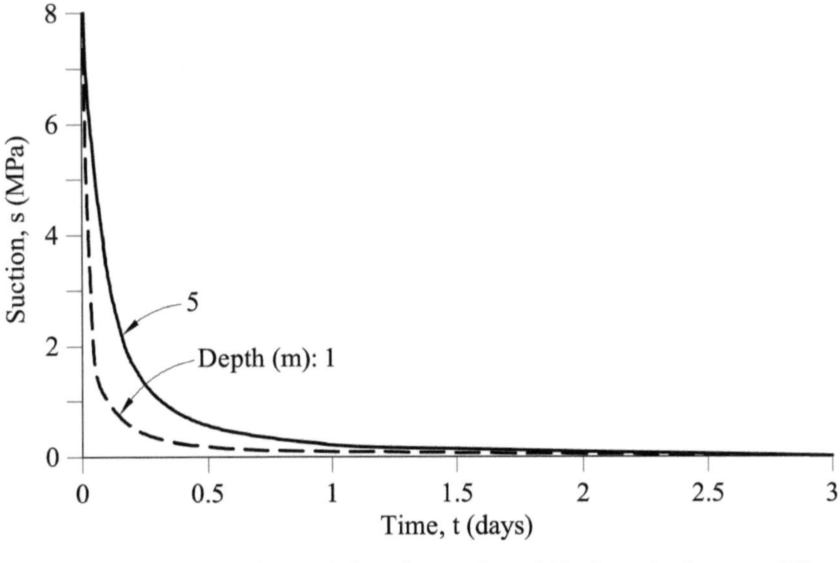

Figure 3.26 Calculated suction evolution of two points within the embankment at different depths.

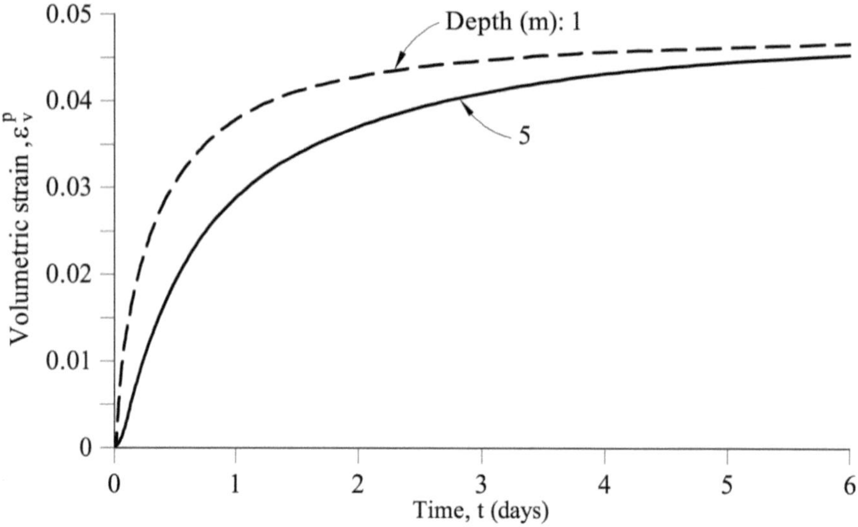

Figure 3.27 Calculated evolution of collapse strains for two points within the embankment.

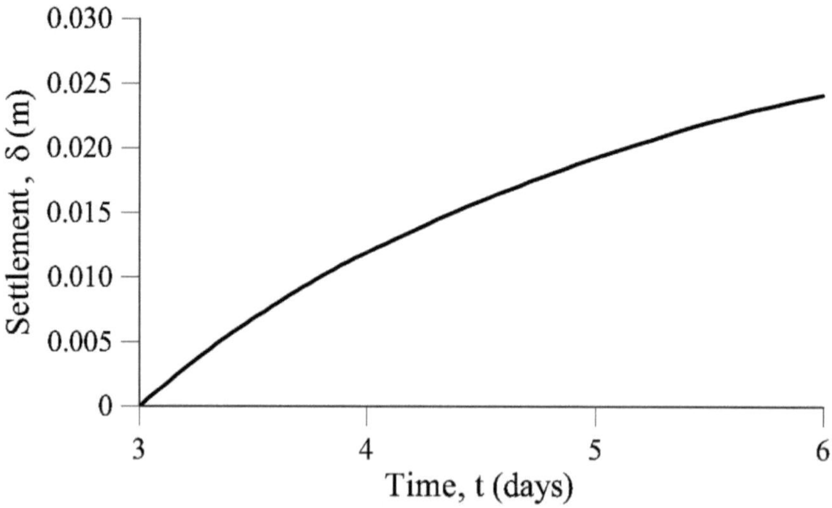

Figure 3.28 Calculated evolution of crest settlement of an 8 m high embankment under a top and bottom infiltration, after an initial stage of wetting lasting three days.

3.5.2 Possibility of additional collapse settlements

The question of the danger of suffering additional damage, should heavy rains occur again, is answered in Figure 3.28, which shows the calculated crest settlements for a maintained infiltration after the first three days. Since collapse strains are irreversible, only suction values smaller than the suction values attained previously are capable of inducing additional collapse strains. In other words, subsequent drying and wetting cycles, which do not take the suction of the soil to a lower value than the suction remaining immediately after the initial rains, are not capable of inducing collapse strains. The simulation performed in Figure 3.28, starts at t = 3 days after the beginning of the assumed infiltration period and therefore takes the soil to yet lower suction values, as shown in Figure 3.29. The calculated collapse settlements are small: a few centimetres after a continuous wetting lasting for a few more days.

3.5.3 Discussion

The collapse plastic model developed was deliberately simple and formulated under isotropic stress conditions. Yet, it incorporates fundamentals aspects of unsaturated soil behaviour. It may be easily generalized by the reader by changing and improving some of the assumptions made. The embankment analysis was performed under one-dimensional conditions, but a fairly general derivation of the basic coupled flow-deformation phenomena in unsaturated soils was given. Again, a more precise analysis would require a two-dimensional or a three-dimensional approach. But the model computations were generally consistent with the laboratory data available and with field observations. In the second case, the observed total overall settlement was used, in a back-analysis procedure, to approximate two model parameters. Those parameters could be obtained by oedometer tests on the as-compacted samples, although information was not available. But the overall analysis was able to provide an answer to one of the pressing questions asked by the Road Administration, namely if there were further risks of dangerous embankment collapse.

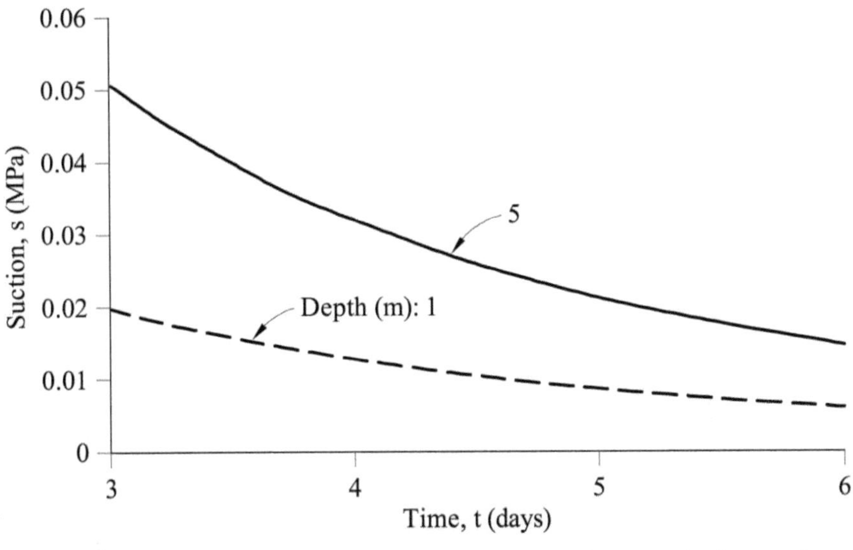

Figure 3.29 Calculated suction evolution of two points within the embankment after an initial stage of wetting lasting three days.

3.6 Mitigation Measures

For a given applied confining stress, collapse decreases when porosity and initial suction decrease. When compacting wet of optimum the initial suction is close to zero and wetting has no effect (the soil is almost saturated). In practice wet of optimum compaction is difficult or not advisable to achieve for a variety of reasons: trafficability is impaired; dry density decreases when water content is increased; in dry regions and summer months, water evaporates fast; and water shortage. Increasing dry density is achieved by increasing the energy of compaction. For many suitable soils for compaction, the target dry density and water content should be located in the vicinity of the Normal and Modified Proctor test optimums. In plastic soils, Modified Proctor energy may lead to soil expansion when wetted.

Girona embankments collapsed for two main reasons: the initial water content was very low and, therefore, suctions were high. This leads to "open" and unstable soil microstructures. In addition, the density achieved was heterogeneous and insufficient, with a significant proportion of specimens having densities lower than Proctor Optimum (Fig. 3.5). The compaction data on this figure corresponds to specimens recovered after collapse; initial densities should be lower than the values plotted.

Concerning the future of the embankments, it was calculated that the remaining collapse potential may result in some limited deformations only on the occasion of extreme rainfall events more intense than the rainfalls of October 1994. Such extreme rainfall events have a low probability but, even in this case, the calculated settlements may cause some pavement cracking but no serious damage to pavements and bridge structures, as in the first wetting. It was concluded that this limited damage could be assumed without any further action after repairs were made to the affected roads. The transition slabs to bridge abutments were repaired. Cavities below them were filled with lean concrete. Pavement shoulders were rebuilt and the embankment slopes were taken to its initial geometry. Fast growing vegetation was installed on slopes to limit erosion to facilitate evapotranspiration (which helps to maintain high suctions within the embankment) and to reduce water infiltration.

3.7 Lessons Learned

3.7.1 Compaction on the dry side

Soils compacted dry of optimum may experience significant compressive volumetric strains when wetted under load. An additional condition required to experience collapse is to compact the soil to a relatively low density. Collapse behaviour is easy to identify in practice by means of oedometer tests. They may be designed to follow the actual stress path experienced in the field. In the case of embankments, a point within the soil experiences, once compacted, an increase in confining stress and later, once the embankment is finished, an increase in water content at essentially constant stress. This simple history of loading and wetting is easily reproduced in an oedometer test.

3.7.2 Natural collapsible soils

Collapse behaviour is also common in natural low density clayey and silty soils that are often weakly cemented. They are found in arid climates (loess) and in tropical environments (residual soils).

3.7.3 Suction and stress variables

The state of the water in a partially saturated soil is conveniently characterized by its suction. Under a capillary interpretation, suction is a positive stress equal, in absolute value, to the pore water pressure. Wetting leads to a reduction in suction. Because of the difficulty in finding a single effective stress, unsaturated soil behaviour is defined in terms of two independent stress variables which combine total stress and suction. A simple choice was made in the chapter: the two selected independent stress variables are total stress and suction.

3.7.4 The nature of collapse

Collapse strains are irreversible, i.e., plastic. In the limit, if collapse is induced by a full wetting, the soil will not experience further collapse deformations if it is later dried and wetted again. Collapse deformations are essentially volumetric, conceptually similar to a temperature induced contraction. A collapsed embankment not only settles. It reduces isotropically in size and therefore it may separate from neighbouring structures such as bridge abutments.

3.7.5 Capillary rise

Rainfall or reservoir impounding, in the case of embankment dams, is a common situation leading to the increase of water content of unsaturated compacted or natural soils. However, capillary rise from shallow water tables may also lead to wetting. In periods of heavy rain water tables are likely to rise. This is an additional reason for the accelerated wetting of embankments because under these circumstances, they are subjected to surface as well as to base infiltration.

3.7.6 Modelling collapse

A simple elastoplastic isotropic model has been developed to quantify collapse deformations. It is based on a fundamental experimental observation, namely that suction increases the apparent preconsolidation stress of soils. The model is built from a proposed variation for this increase in yielding stresses and from some assumed compression behaviour of the soil under increasing suction.

3.7.7 Coupled flow-deformation

Collapse deformations require an increase in water content. In a field situation, water content changes as a result of infiltrating flow, which is, in turn, induced by changes in boundary conditions. Suction provides the link between flow and mechanical behaviour because suction gradients control flow and suction changes control volumetric deformations. This chapter describes a step-by-step procedure to derive the field equation for the coupled flow and collapse deformations under one-dimensional conditions. This case is yet a further example of coupled problems involving flow and deformation. In addition to the classical consolidation analysis under saturated conditions, Chapter 5 will also provide a case of coupled flow, deformation, and heat under one-dimensional conditions, in the context of rapid landslide analysis. The procedure developed was applied to reproduce the collapse deformations observed in embankments subjected to rainfall wetting.

3.7.8 Predicting the future behaviour of embankments

The coupled model developed is useful to investigate the future behaviour of embankments under weather action. The analysis performed on Girona embankments suggests that most of the collapse potential of the embankments was released during the first heavy rains after construction. The analysis shows that further collapse deformations are possible under extreme rainfall events, but they are small and unlikely to cause additional damage.

3.8 Advanced Topics

Collapse upon wetting is a distinct and fundamental feature of the mechanical behaviour of unsaturated soils. It is closely related to the discussion of effective stress. The notion of collapse strain could not be reconciled with the concept of a single effective stress and models for unsaturated soil behaviour were eventually formulated in terms of two independent stress components which combine total stress, gas pressure and liquid pressure. The model developed here uses a particular choice (net stress and suction) but other alternatives have also been used (Gens, 1995).

The mechanics of unsaturated soils has received increasing attention during the past two decades. A comprehensive description of the state of knowledge before 1987 is found in Alonso *et al.* (1987). The first proposal to model unsaturated soil behaviour through the concepts of hardening plasticity is also described in this reference. Constitutive model development has evolved rapidly since 1990. The isotropic collapse model described in this chapter has some similarities with the so-called Barcelona Basic Model (BBM) (Alonso *et al.*, 1990).

Appendix 3.1 Solving the Coupled Flow-Deformation Equation of the Collapsing Embankment

Changing the notation, Equation (3.36) is written here as

$$f\frac{\partial s}{\partial t}-g\frac{\partial p}{\partial t}+\frac{k}{\gamma_w}\frac{\partial^2 s}{\partial z^2}+\left(\frac{1}{\gamma_w}\frac{\partial s}{\partial z}-1\right)h\frac{\partial s}{\partial z}=0, \tag{A3.1}$$

where

$$f=n\frac{\partial S_r}{\partial s}-S_r\frac{\partial \varepsilon_{\text{vol}}}{\partial s},\quad g=S_r\frac{\partial \varepsilon_{\text{vol}}}{\partial p}\quad\text{and}\quad h=\frac{\partial k}{\partial S_r}\frac{\partial S_r}{\partial s}.$$

Note that γ_w is constant and

$$\frac{\partial \varepsilon_{\text{vol}}}{\partial p}, \quad \frac{\partial \varepsilon_{\text{vol}}}{\partial s}, \quad \frac{\partial S_r}{\partial s} \quad \text{and} \quad \frac{\partial k}{\partial S_r}$$

are known functions given by Equations (3.5), (3.31) and, from Equation (3.33),

$$\frac{\partial k}{\partial S_r} = mk_{\text{sat}}\left(S_r\right)^{m-1}. \tag{A3.2}$$

S_r is calculated with Equation (3.30) and n varies with the volumetric strains according to the integration of Equation (3.23) as follows:

$$\int_{n_0}^{n} d\varepsilon_{\text{vol}} = \int_{n_0}^{n} -\frac{\mathrm{d}n}{1-n} \Rightarrow \varepsilon_{\text{vol}}\left(t\right)-\varepsilon_{\text{vol}}\left(t_0\right) = \ln\left(\frac{1-n}{1-n_0}\right) \Rightarrow$$
$$n = 1-(1-n_0)\exp\left(\varepsilon_{\text{vol}}\left(t\right)-\varepsilon_{\text{vol}}\left(t_0\right)\right). \tag{A3.3}$$

A forward finite difference procedure will be used to solve the system of equations. The one-dimensional spatial domain is subdivided into n small elements of thickness Δz. The z_i coordinate of any point is defined by an index i such that $z_i = i\Delta z$. The following indices define singular points: $i = 1$ corresponds to $z = 0$. The far boundary is located at a distance $z = L = 8$ m (it corresponds to the depth of the embankment) where $i = n_z$.

Time derivatives of a general function y at any time t, for $z = z_i$ can be approximated by (Forward Euler Method):

$$\left.\frac{\partial y}{\partial t}\right|_{z_i} \approx \frac{y_{z_i,t+\Delta t} - y_{z_i,t}}{\Delta t}. \tag{A3.4}$$

where the subscript (z_i, t) indicates that the function y is evaluated in the point z_i and time t.

The first and second derivatives with respect to z will be approximated by a central difference:

$$\left.\frac{\partial y}{\partial z}\right|_{t} \approx \frac{y_{z_{i+1},t} - y_{z_{i-1},t}}{\Delta z}, \tag{A3.5a}$$

$$\left.\frac{\partial^2 y}{\partial z^2}\right|_{t} \approx \frac{y_{z_{i+1},t} - 2y_{z_i,t} + y_{z_{i-1},t}}{\left(\Delta z\right)^2}. \tag{A3.5b}$$

The term $\partial p/\partial t$ given by Equation (3.40) should be also discretized. Because of the complexity of this equation it will be approximated by a backward scheme:

$$\frac{\partial p}{\partial t} = \frac{p_{z_i,t} - p_{z_i,t-\Delta t}}{\Delta t}. \tag{A3.6}$$

Since changes in total stress induced by increasing soil saturation are moderate and occur slowly in time, this approximation is sufficiently accurate.

Once the numerical approximations of derivatives are substituted into the governing Equation (A3.1), the following discrete equation is obtained:

$$s_{z_i,t+\Delta t} = s_{z_i,t} + \Delta s_{z_i,t+\Delta t}, \tag{A3.7}$$

where

$$\Delta s_{z_i,t+\Delta t} = \frac{\Delta t}{f_{z_i,t}} \left[-\frac{k_{z_i,t}}{\gamma_w} \frac{s_{z_{i+1},t} - 2s_{z_i,t} + s_{z_{i-1},t}}{(\Delta z)^2} + g_{z_i,t} \frac{p_{z_i,t} - p_{z_i,t-\Delta t}}{\Delta t} - \frac{1}{\gamma_w} h_{z_i,t} \left(\frac{s_{z_{i+1},t} - s_{z_i,t}}{\Delta z} \right)^2 + h_{z_i,t} \frac{s_{z_{i+1},t} - s_{z_i,t}}{\Delta z} \right]. \tag{A3.8}$$

Equations (A3.7) and (A3.8) allow calculating the suction at any point within the soil column knowing its value at the previous time in the same point and the points just above and below. Initial and boundary conditions should be defined in order to initiate the calculation procedure. According to the previous description, the initial condition can be written as

$$s(z_i, t_0) = 8 \text{ MPa} \quad \text{for } i = 1, \ldots, n_L \tag{A3.9}$$

and the boundary conditions for any time as

$$s(z_{n_1}, t) = s(z_{n_L}, t) = 0 \text{ MPa}. \tag{A3.10}$$

A Fortran program was developed to solve the numerical procedure described above for the calculation of the collapse in one dimension and for the boundary and initial condition described before. The complete code is given below.

```
!*****************************************************************!
!                    COLLAPSE CALCULATION - PARAMETERS
!*****************************************************************!

      IMPLICIT NONE
!--------------------------WORKING VARIABLES----------------------!
      INTEGER :: nz,nt,nw
      REAL(8) :: tmax,Dz,Dt,Dtw,znt,ff
      REAL(8) :: param(100),suct0
      REAL(8) :: gamma_w,gamma_s,Srmax,Srmin
      REAL(8) :: kappa,lambda0
      REAL(8) :: K0,p0_RC,lambda_RC,A_krel,b_krel,perm_sat
      REAL(8) :: alpha,aaa
      REAL(8) :: height,nn0,ee0,Q,suctBC,Depth_w
!-----------------------------------------------------------------!

!PARAMETERS:
!Specific weight
      gamma_w = 0.01      !MN/m2
      gamma_s = 0.027     !MN/m2
!At rest earth pressure coefficient
```

```
      K0 = 0.5
!Water retention (van Genuchten)
      Srmax = 1.
      Srmin = 0.
      lambda_RC = 0.09
      p0_RC = 0.05        !MPa
!Permeability
      A_Krel = 1.
      b_krel = 3.
      perm_sat = 1.e-7*86400     !m/day
!Compressibility
      lambda0 = 0.024
!LC curve
      alpha = 0.5       !MPa-1
      aaa = 20.
!Geometry
      height = 8.       !m

!INITIAL CONDITIONS
      !Suction
      suct0 = 8.        !MPa
      !Porosity and void ratio
      nn0 = 0.35
      ee0 = nn0/(1-nn0)

!BOUNDARY CONDITIONS
      suctBC = 0.0      !MPa. Suction will be imposed in z=0 and z=8

!DISCRETIZATION
      nz = 80.
      tmax = 6.
      Dz = height/nz
      Dt = 1.e-6        !days
      znt=tmax/Dt
      nt=int(znt)

!PRINT RESULTS
      Dtw=0.05
      ff=Dtw/Dt
      Depth_w = 5       !m Depth at which results in time are printed
      nw = int(1.*nz/height)

!CALCULATION, PROCEDURE STARTS
      param(1) = gamma_w
      param(2) = gamma_s
      param(3) = K0
      param(4) = Srmax
      param(5) = Srmin
      param(6) = lambda_RC
      param(7) = p0_RC
      param(8) = A_Krel
      param(9) = b_krel
      param(10) =  perm_sat
      param(11) = kappa
      param(12) = lambda0
      param(13) = rr
      param(14) = beta
      param(15) = pc
      param(16) = alpha
      param(17) = aaa

       call Integration (nz,Dz,nt,Dt,nw,
     .                   param,suct0,nn0,ee0,Q,suctBC,ff)
       end

!*******************************************************************!
```

```
!            INTEGRATION OF ONE-DIMENSIONAL COLLAPSE
!*******************************************************************!
      subroutine  Integration (nz,Dz,nt,Dt,nw,
     .                         param,suct0,nn0,ee0,suctBC,ff)
      IMPLICIT NONE
!------------------------IN VARIABLES--------------------------------!
      INTEGER,INTENT(IN)  :: nz,nt,Model_coeff,BC,nw
      REAL(8),INTENT(IN)  :: Dz,Dt,param(100),suct0,nn0,ee0,suctBC,ff
!--------------------------------------------------------------------!
!---------------------WORKING VARIABLES------------------------------!
      INTEGER     ::  i,j,k,m,mm,iw,ifile
      REAL(8)     ::  z,time,pp
      REAL(8)     ::  gamma0,gamma_w,gamma_s,Srmax,Srmin,DSr
      REAL(8)     ::  kappa,lambda0
      REAL(8)     ::  K0,C_K0,p0_RC,lambda_RC,C_RC
      REAL(8)     ::  A_krel,b_krel,perm_sat,d_perm_dSr(nz)
      REAL(8)     ::  pc,alpha,aaa
      REAL(8)     ::  height
      REAL(8)     ::  Dsuct(nz),suct(nz),gamma(nz)
      REAL(8)     ::  p0(nz),p0old(nz)
      REAL(8)     ::  Sr0,Sr(nz),perm(nz)
      REAL(8)     ::  d_Sr_ds(nz),p0ast(nz)
      REAL(8)     ::  d_p0ast_ds(nz),d_p0ast_dp(nz),d_p_t(nz)
      REAL(8)     ::  epsvol(nz),d_epsvol_ds(nz)
      REAL(8)     ::  d_epsvol_dp(nz),Inc_epsvol(nz)
      REAL(8)     ::  funct_f(nz),funct_g(nz),funct_h(nz)
      REAL(8)     ::  ee(nz),nn(nz),displacement(nz)
      REAL(8)     ::  aux(nz),aux2(nz)
!--------------------------------------------------------------------!
      !FILE TO PRINT RESULTS
      open (unit=12, file='AA_Suction.dat', status='unknown')
      open (unit=13, file='AA_p0ast.dat', status='unknown')
      open (unit=14, file='AA_p0.dat', status='unknown')
      open (unit=15, file='AA_displacement.dat', status='unknown')
      open (unit=16, file='AA_Sr.dat', status='unknown')
      open (unit=17, file='AA_eps_vol.dat', status='unknown')

      !INITIALIZE
      iw=0
      do i=1, nz
       epsvol(i)=0.0
       Dsuct(i) = 0.0
      enddo

      !PARAMETERS
      gamma_w = param(1)
      gamma_s = param(2)
      K0 = param(3)
      Srmax = param(4)
      Srmin = param(5)
      lambda_RC = param(6)
      p0_RC = param(7)
      A_Krel = param(8)
      b_krel = param(9)
      perm_sat = param(10)
      kappa = param(11)
      lambda0 = param(12)
      rr = param(13)
      beta = param(14)
      pc = param(15)
      alpha = param(16)
      aaa = param(17)
      !Constant coefficients
      DSr = Srmax-Srmin
      C_RC = lambda_RC/(1.-lambda_RC)
      C_K0 = (1.+2.*K0)/3
```

```
      !Initial values
      gamma0 = gamma_w*Sr0*nn0+gamma_s*(1-nn0)
      do i=1,nz
        z = Dz*i
        suct(i) = suct0
        ee(i) = ee0
        nn(i) = nn0
        p0old(i) = gamma0*C_K0*z
        p0(i) = p0old(i)
      enddo
      if (BC.eq.1) then
        suct(1) = suctBC
        suct(nz)= suctBC
      endif
!--------------------CALCULATION PROCEDURE STARTS-------------------!
      DO k=1,nt   !Iteration in time
     !Calculation of variables at each point
      do i=1,nz
        z = Dz*i
        aux(i) = (suct(i)/p0_RC)**(1/(1-lambda_RC))
        Sr(i) = DSr*(1.+aux(i))**(-lambda_RC)+
     .          Srmin
        perm(i) = perm_sat* A_krel*Sr(i)**b_krel
        d_perm_dSr(i) = b_krel*perm(i)/Sr(i)
        aux(i) = (1+aaa*(1-exp(-alpha*suct(i))))
        aux2(i) = lambda0/(1+ee(i))
        p0ast(i) = p0(i)/aux(i)
        d_epsvol_dp(i) = aux2(i)/p0(i)
        d_epsvol_ds(i) = -aux2(i)*aaa*alpha*exp(-
     .                   alpha*suct(i))/aux(i)
        aux(i) = (suct(i)/p0_RC)**(1/(1-lambda_RC))
        d_Sr_ds(i) = (DSr/p0_RC)*(-C_RC)*(1+aux(i))**(-lambda_RC-1)*
     .               (suct(i)/p0_RC)**C_RC
        d_p_t(i) = (p0(i)-p0old(i))/Dt
        funct_f(i) = nn(i)*d_Sr_ds(i)-Sr(i)*d_epsvol_ds(i)
        funct_g(i) = Sr(i)*d_epsvol_dp(i)
        funct_h(i) = d_perm_dSr(i)*d_Sr_ds(i)
       enddo
       !Calculation of increment of suction
        Dsuct(1) = 0.0
        Dsuct(nz) = 0.0
       do i=2,nz-1
          Dsuct(i)=(Dt/funct_f(i))*(
     .    -(perm(i)/gamma_w)*(suct(i+1)-2*suct(i)+suct(i-1))/(Dz*Dz)-
     .    funct_g(i)*d_p_t(i)-(1/gamma_w)*funct_h(i)*
     .    ((suct(i+1)-suct(i))/Dz)**2+funct_h(i)*(suct(i+1)-
     .    suct(i))/Dz)
       enddo
       !Update variables
       do i=1,nz
          suct(i)=suct(i)+Dsuct(i)
          Inc_epsvol(i) = d_epsvol_ds(i)
     .                    *Dsuct(i)+d_epsvol_dp(i)*d_p_t(i)
          epsvol(i)=epsvol(i)+Inc_epsvol(i)
          gamma(i) = gamma_w*Sr(i)*nn(i)+gamma_s*(1-nn(i))
          p0(i) = 0.
          do j=1,i
            p0(i) = p0(i)+gamma(j)*C_K0*Dz
          enddo
          ee(i) = -1.+(1.+ee(i))*exp(-Inc_epsvol(1))
          nn(i) = ee(i)/(1+ee(i))
          p0old(i) = p0(i)
       enddo
     !Calculation of settlements
       do i=1,nz
        displacement(i) = 0.
```

```
        do j=i,nz
         displacement(i) = displacement(i)+epsvol(j)*Dz
        enddo
       enddo
!-------------------------- PRINT RESULTS ------------------------!
       time=k*Dt
       if ((k.eq.2).or.(k.eq.1).or.(mm.eq.k)) then
         mm=int(ff)
         mm=mm*iw
         iw=iw+1
         ifile=iw+100
         write (ifile,*) 'Time(days)      ',time
         do m=1,nz
          z=(m-1)*Dz
          write (ifile,'(2(e15.5,1x))') z, suct(m)
         enddo
         write (12,'(2(e15.5,1x))') time, suct(nw)
         write (13,'(2(e15.5,1x))') time, p0ast(nw)
         write (14,'(3(e15.5,1x))') time, p0(nw)
         write (15,'(3(e15.5,1x))') time, displacement(1)
         write (16,'(3(e15.5,1x))') time, Sr(nw)
         write (17,'(3(e15.5,1x))') time, epsvol(nw)
        endif
        ENDDO
       close(12)
       close(13)
       close(14)
       close(15)
       close(16)
       close(17)
     return
     end
```

References

Alonso, E.E., Gens, A. and Hight, D.W. (1987) Special Problem Soils. General Report. *Proceedings of the 9th European Conference on Soil Mechanics and Foundation Engineering*. Dublin, Ireland. Hanrahan, Orr & Widdis (eds). 1087 – 1146.

Alonso, E.E. Gens, A. and Josa, A. (1990) A constitutive model for partially saturated soils. *Geotechnique* 40 (3), 405 – 430

Alonso, E.E. and Gens, A. (1994). On the mechanical behaviour of arid soils. Keynote Lecture. *Engineering Characteristics of Arid Soils,* 173-205

Atkinson, B.K. (1984) Subcritical crack growth in geological materials. *Journal of Geophysical Research* 89 (B6), 4077–4114.

Bishop, A.W. (1959). The principle of effective stress. *Tecknisk Ukeblad* 106 (39), 859 – 863.

Delage, P., Cui, Y.J. and Antoine, P. (2005) Geotechnical problems related with loess deposits in Northern France. *Proceedings of the International Conference on Problematic Soils*. Famagusta, Cyprus. Bilsel and Nalbantoglu (eds). 2, 517 – 540

Gens, A. (1995) Constitutive modelling: Application to compacted soils. Unsaturated soils. *Proceedings of the 1st International Conference on Unsaturated Soils.* Paris, France. Alonso and Delage (eds). 3, 1179 – 1201.

Lawton, E.C., Fragaszy, R.J. and Hetherington, M.D. (1992) Review of wetting-induced collapse in compacted soil. *Journal of Geotechnical Engineering* 118

(9), 1376–1394.

Oldecop, L.A. and Alonso, E.E. (2001) A model for rockfill compressibility. *Géotechnique* 51 (2), 127–139.

Skinner, H.D., Charles, J.A. and Watts, K.S. (1999) Ground deformations and stress redistribution due to a reduction in volume of zones of soil at depth. *Géotechnique* 49 (1), 111–126

Soriano, A. and Sánchez, F.J. (1999) Settlements of railroad high embankments. *Proceedings of the 12th European Conference on Soil Mechanics and Geotechnical Engineering*. Amsterdam, The Netherlands. Balkema. 3, 1885–1890.

Van Genuchten, M.Th. (1980) A closed-form equation for predicting the hydraulic conductivity of unsaturated soils. *Soil Science Society of America Journal* 44, 892–898.

Chapter 4

Earth Dam Sliding Failure: Aznalcóllar Dam, Spain

TABLE OF CONTENTS

E.E. Alonso et al., *Geomechanics of Failures. Advanced Topics*,
DOI 10.1007/978-90-481-3538-7_4,

Chapter 4

Earth Dam Sliding Failure: Aznalcóllar Dam, Spain

4.1 The Failure

Mines in the area of Aznalcóllar, a town in the province of Sevilla, southwest Spain, have exploited from ancient times a number of metallic minerals (zinc, lead, silver) associated with pyritic formations. The process of mineral extraction produces large volumes of pyritic tailings which, in the Aznalcóllar mine, were stored under water in a large pond area. The pond, whose layout is shown in the air photograph of Figure 4.1, is maintained by a perimeter dyke, which was conceived as a homogeneous rockfill dam made impervious by an upstream mantle of clay. The pond evolved in volume during the lifetime of mining operations in order to accommodate the increasing amounts of waste. The increasing demand of waste storage capacity was resolved (in the original design) by increasing the height and size of the dam in the manner schematically indicated in Figure 4.2.

A small embankment was first built and a small cut-off wall was installed to avoid leakages through the upper thin alluvium indicated in the cross-section (Fig. 4.2). Then, the dam height was increased in a "forward" or "downstream" procedure maintaining the position of the upstream face.

The toe of the downstream slope moved forward and the horizontal size of the dam base increased. In the original design, which dates back to 1977, the successive downstream slopes were supposed to be parallel to each other and a constant width of the dam crest was maintained (Fig. 4.2).

The initial pond was built in 1978. Twenty-one years later, the dyke had reached a height of 28 m. However, the design was somewhat changed, as shown in Figure 4.10. The implications of the changes introduced will be discussed later.

On April 25, 1998, some time during the night, a breach opened in the long perimeter dam located in the east side of the tailing's pond and a catastrophic flood of liquefied mine tailings invaded the valleys of nearby rivers (first the Agrio river and then the Guadiamar river – a tributary of the Guadalquivir river).

The acid nature of the tailings and the presence of heavy metals created deep concern. Doñana National Park, an emblematic and protected natural environment, famous for its unique Iberian fauna, was at risk.

Figure 4.3 shows a cross-section of the tailings pond in a west-east direction. The dyke sits on granular and pervious alluvium (a terrace of the Agrio river), 4–5 meters thick, which overlies a deep stratum of marine over-consolidated clays of tertiary age. The clay, known locally as Guadalquivir blue marl, dips gently (2 to 4°) towards the SSE.

The lower boundary of the blue clay is located at a depth of 60 m below the dam foundation. The thick clay deposit sits on a pervious aquifer whose piezometric level is located at the surface. Therefore, the clay layer is bounded by two pervious strata having essentially the same piezometric head. In addition to the sub-horizontal sedimentation planes, the over-consolidated blue clays are dissected by vertical joints having smooth surfaces. Three vertical joint families were identified, the most prominent one being in the direction NE–SW.

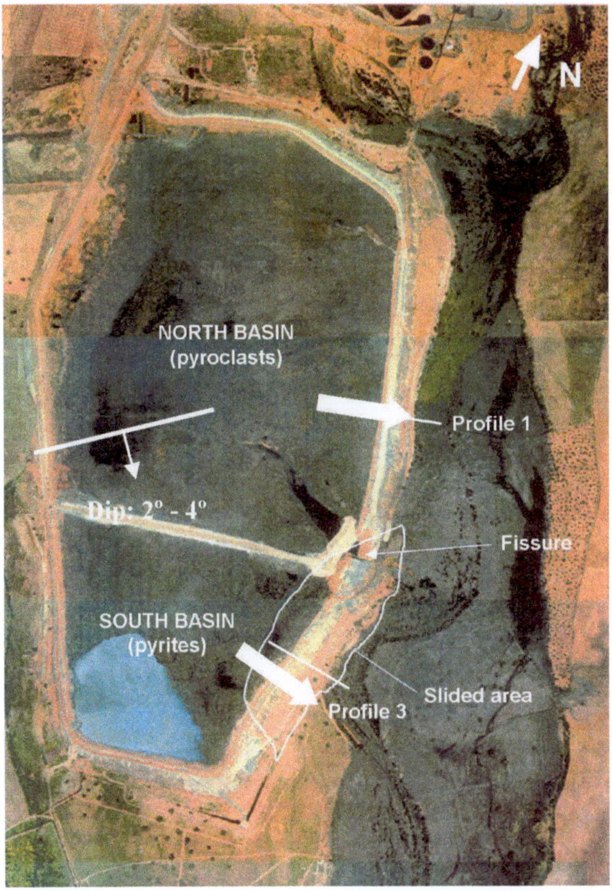

Figure 4.1 Aerial view of the pond a few weeks after the failure showing the direction and dip of stratification planes.

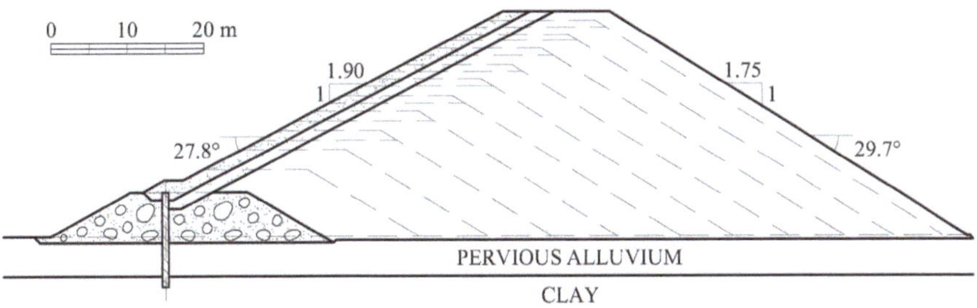

Figure 4.2 Aznalcóllar dam according to original project (1977) (Alonso and Gens, 2006).

Figure 4.3 Breach open in the dam. The foreground shows the valley of the Agrio river, covered by a thick deposit of mine waste (acid liquefied pyrites) (Alonso and Gens, 2006).

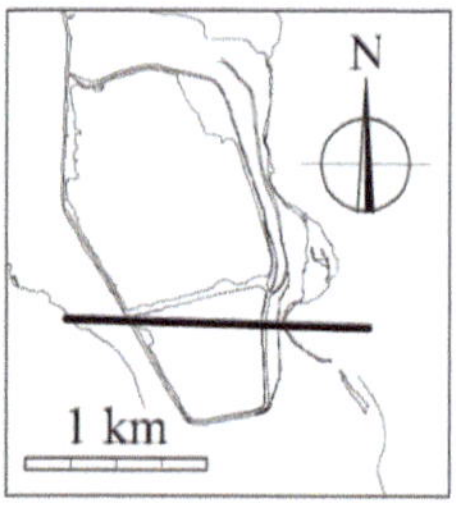

(a)

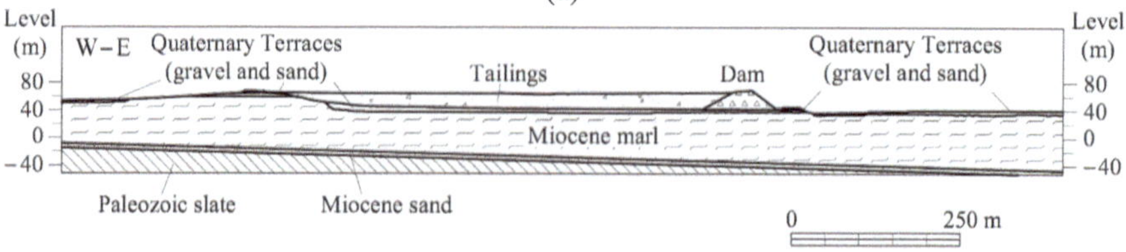

(b)

Figure 4.4 (a) Plan view of the pond; (b) representative cross-section in a west-east direction (Alonso and Gens, 2006).

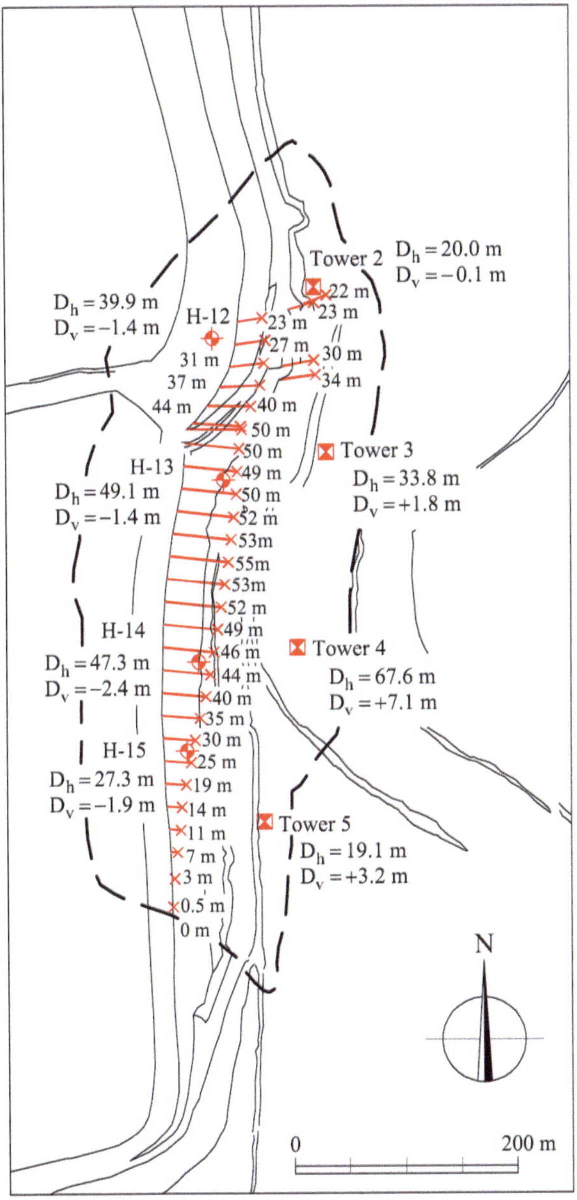

Figure 4.5 Displacement of the dam, indicated by the plotted segments, which join the same points of the dam crest before and after failure. The discontinuous line indicates the contour of the slide (Alonso and Gens, 2006).

A comparison of the topography before and after the failure (Fig. 4.5) was useful to identify the origin of the failure. It was discovered that a 600 m long stretch of the southern portion of the dyke facing the Agrio river had displaced forward, an average distance of 50 m. The plot in Figure 4.5 shows a series of segments which connect two equivalent points in the dam crest, before and after failure. The written number indicates the travelled distance.

For some particular marks, two numbers are indicated (for instance, for mark H-13, D_h = 49.1 m and D_v = -1.4 m). The two numbers show the horizontal distance (D_h) travelled by the dam and the vertical distance (D_v). The negative sign implies a fall and it indicates that the dam slid forward in a plane dipping gently towards the east. The calculated dip of the segments (atan(D_v/D_h)) is close to 2–3°. This information was interpreted as a strong indication that the failure plane was actually a sedimentation plane. Downstream, a few electricity towers were also displaced by the slide. Horizontal and vertical displacements of these towers were also measured. The positive D_v numbers indicate now a ground heave. These towers were located in the passive, resisting side of the slide.

The cross-section of the southern dyke after the failure, determined by the information provided by borings and exploration pits, is indicated in Figure 4.6a.

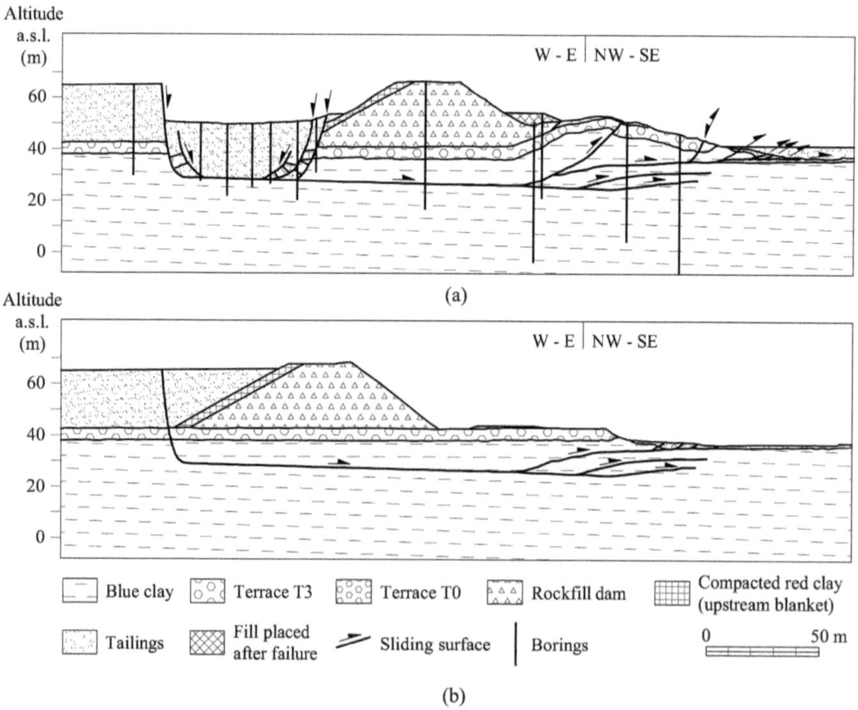

Figure 4.6 Cross section of Profile 3 (location in Fig. 4.1): (a) after failure (the position of the borings drilled is shown); (b) reconstructed position of the initial failure and sliding planes, at the start of the sliding motion (after Moya, 2004).

The section corresponds to the cross-section identified as Profile 3 in Figure 4.1. The figure shows that the dam moved forward, sliding on a sedimentary plane located approximately at a depth of 14 m below the axis of the dam. The upper figure was interpreted by Moya (2004), who was able to plot the original geometry of the incipient failure and the shape of the sliding surface as shown in Figure 4.6b.

The dam motion originated a large upstream void, which was limited by an essentially vertical plane located in the position of the toe of the upstream dam slope. This observation implies that the stored tailings were capable of maintaining a stable vertical "cliff", a result which has interesting consequences regarding the motion of the dam, discussed further in Chapter 6. The failure surface daylights downstream in more than one sliding surface. The motion created an accumulation of layers that resulted in the elevation of the ground, immediately downstream of the dam (recorded also by the displacement of the electric towers in Fig. 4.5). Maximum elevations of 8–10 m were measured. It was also noticed that the motion of the dam had a slight rotation towards the south.

The pond (Fig. 4.4) was divided in two parts (or "basins"), separated by a jetty. The southern one stored fine pyritic tailings. In the northern one, somewhat coarser pyroclastic granular waste was stored. The upper level of the waste in the pond was approximately the same in both basins. The direction of the dam limiting the northern basin is oriented in a direction close to north-south. The dyke direction in the southern basin changes approximately 20° with respect to the northern dyke direction and faces an ESE direction.

Interestingly, despite being essentially identical in terms of geometry and geotechnical conditions, the northern dyke remained still. As a result, a breach was opened at the junction between the two dykes. An explanation for this behaviour will be given later in this chapter.

The vertical jointing of the clay also controlled the geometry of the slide in its upstream end: the vertical limiting surface within the clay in Figure 4.6b is probably a consequence of the well-developed system of vertical discontinuities. On the other hand, the orientation of the failure surface crossing the dyke (at the position of the breach) follows a NE–SW orientation which agreed with the direction of the dominant family of vertical discontinuities.

4.2 Geotechnical Properties of Tailings and Foundation Clay

The tailings of the southern basin may be described as a non-plastic fine and uniform silt. The main mineral is pyrite (iron disulphide, S_2Fe). Other metallic minerals and chemical compounds in minor proportions complete the composition of the tailings. The silt is very homogeneous (coefficient of uniformity = 4.7) around an average grain diameter of 10 μm. Pyrite is heavy (solid unit weight γ_s = 43 kN/m^3) and it leads to a high natural density of tailings (average saturated unit weight γ_t = 31 kN/m^3) significantly higher than most natural soils.

Triaxial tests on undisturbed specimens resulted in a drained friction angle of 37°. It was also found that these pyritic tailings had a significant cementation (the unconfined compression strength of saturated specimens ranged between 100 and 200 kPa in most of the tests performed). Its permeability is low (10^{-7} to 10^{-6} cm/s), a value consistent with the grain size distribution.

The lower blue clay is a uniform deposit of high plasticity clay (w_L= 63–67%;

IP = 32–35%; clay content = 47–58%) which classifies as CH or MH. The water content (30–35%) is close to the plastic limit (it indicates high consistency). The void ratio ranges between 0.8 and 1. Clay minerals (70% of the total) include calcic smectite (35%), illite, and kaolinite. The remaining non-clay minerals are calcite and quartz.

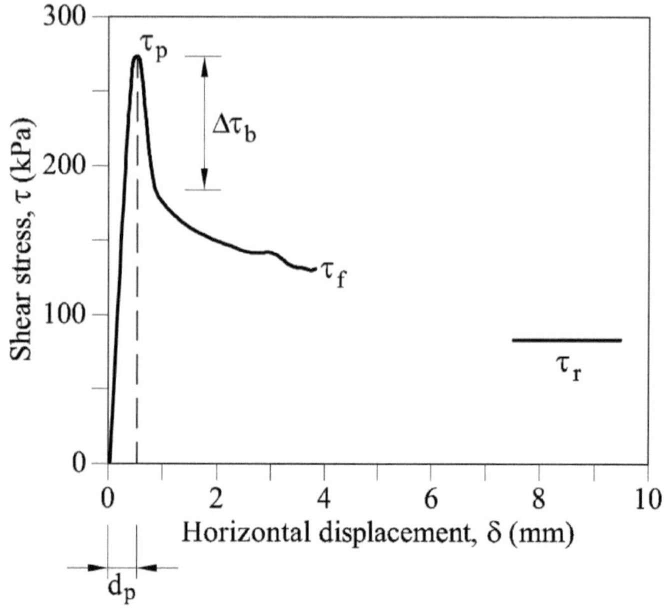

Figure 4.7 Drained direct shear test on the blue clay. Effective normal stress: 400 kPa (Alonso and Gens, 2006).

This clay exhibits a characteristic brittle behaviour. This is shown in Figure 4.7. Peak strength (τ_p) is reached for a small relative displacement ($d_p < 1$ mm) in drained direct shear tests. The loss of strength immediately after peak ($\Delta\tau_b$) is also rapid. The strength continues to drop as relative displacement accumulates. Shear tests on natural discontinuities and ring shear tests indicated a residual friction of 11°, a value consistent with the mineralogy of the clay.

A synthesis of strength tests is given in Figure 4.8. The plot highlights the brittle nature of the clay and the rapid loss of strength as the relative displacement between shear planes, δ, increases. The peak envelope is characterized by $c'_p = 65$ kPa and $\phi'_p = 24.1°$. It takes a small relative displacement (6 mm) to eliminate any apparent drained cohesion and to bring the friction angle down to 18–20°. Residual conditions require higher relative displacements, in the order of a few centimeters.

Oedometer tests provided the following range of values for the coefficient of consolidation (c_v = 0.5 to 1.5×10^{-3} cm^2/s) and permeability (K = 2 to 7×10^{-9} cm/s), a very low value.

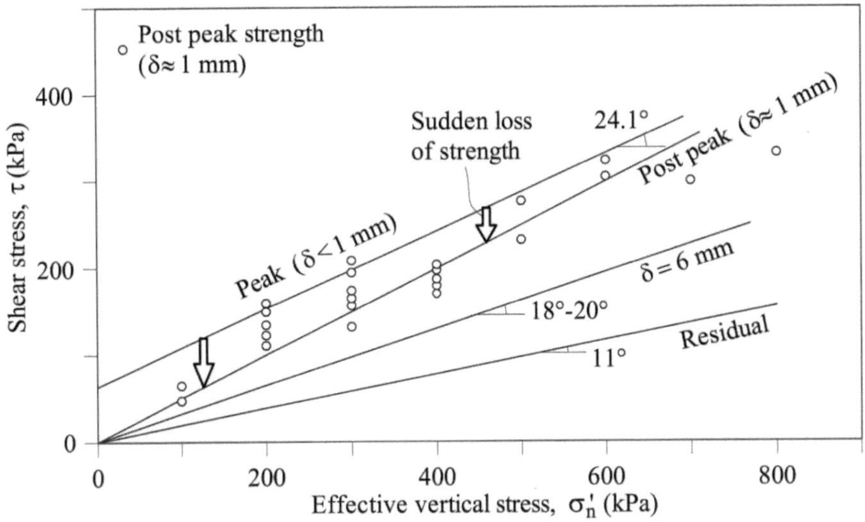

Figure 4.8 Strength envelopes of Guadalquivir blue clay based on direct shear tests. Strength is controlled by the relative shear displacement, δ (after Alonso and Gens, 2006).

4.3 Water Pressures and Stresses in the Foundation

4.3.1 Water pressure measurements

After the failure, borings were performed at different locations to investigate subsoil conditions. Borings were arranged in "profiles" in order to provide representative cross sections of the dam. Two of them (Profiles 1 and 3) are shown in Figure 4.1. Profile 1 was located in the northern dyke, which remained stable. Therefore, it provides, as a first approximation, the subsoil conditions existing in the failed dam before the accident. Vibrating wire piezometers were located in three borings S1–1, S1–2, S1–3 at two elevations (piezometers P1 and P2 in Fig. 4.9a). Measured pore pressures after stabilization are also shown in Figure 4.9a. Two representations are given: a vertical segment equivalent to the column of water and a plot in a horizontal scale in order to provide a better picture of pore water pressure variations with depth. Also indicated in the figure is the hydrostatic distribution of pressures associated with the upper boundary conditions. The measured pressures indicate a marked vertical gradient within the upper 20 m of blue clay. Excess pressures dissipate towards the pervious granular layer underlying the dam where a phreatic surface is permanently established. However, upstream of the cut-off wall, which was built at the start of dam construction, the boundary water pressure is controlled by the level of the reservoir (the tailings were permanently submerged in the deposit). The estimated profiles of hydrostatic pressures at the position of the three boreholes represented in Figure 4.9a (S1–1, S1–2, S1–3) are also given.

It was realized that existing pore pressures at the elevation of the failure surface were unexpectedly high. This is a very relevant result which provided a clue for the instability of the dam. The high existing pore pressures could have reduced the available strength on potential failure surfaces to low values. However, Profile 1 is not in the failed zone. The reported pore pressures were measured a few months after the failure and the modifications of the pond conditions since the time of the failure had been significant (outflow and associated erosion of tailings and a reduction of the water level in the pond).

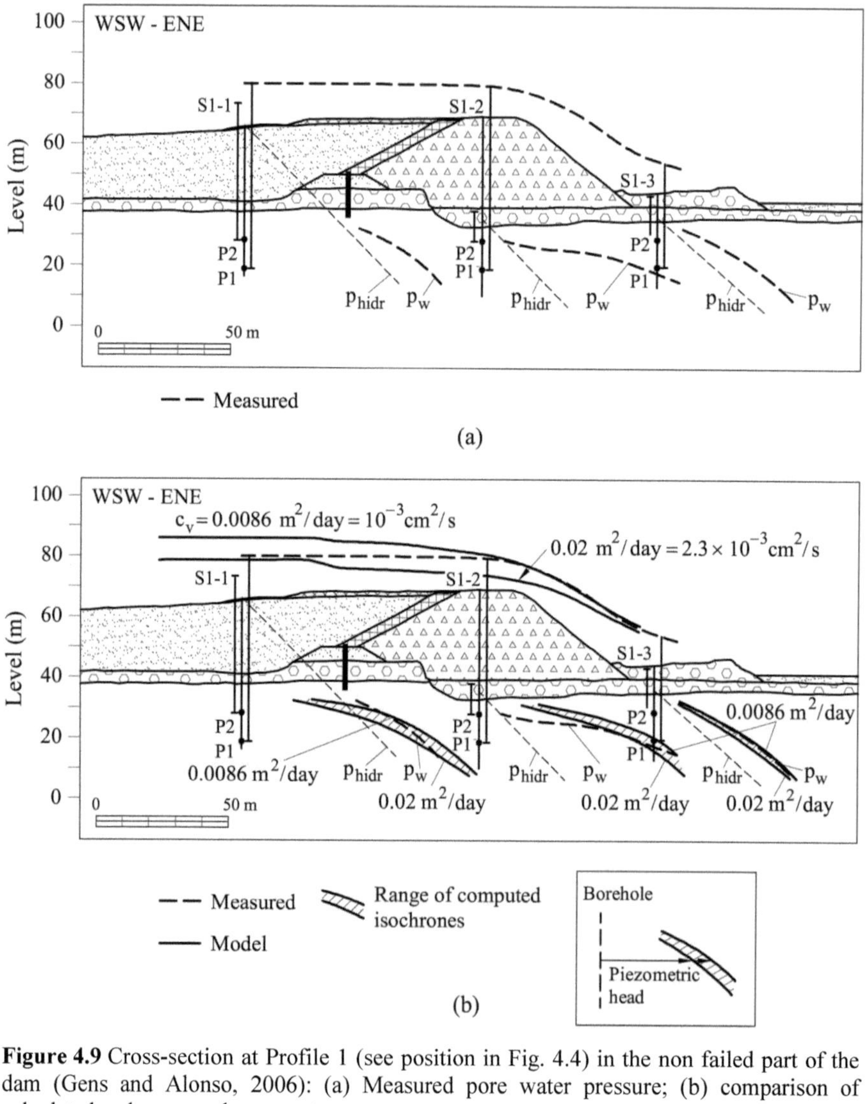

Figure 4.9 Cross-section at Profile 1 (see position in Fig. 4.4) in the non failed part of the dam (Gens and Alonso, 2006): (a) Measured pore water pressure; (b) comparison of calculated and measured pore water pressure.

But data on Profile 1 are very useful to validate any calculation procedure for water pressures, a key issue in Aznalcóllar failure. Once validated, the procedure could be applied to estimate the conditions of the failed dam immediately before the rupture.

A simple calculation method may be devised by combining some classical solutions of the theory of elasticity and the one-dimensional solution of the consolidation problem (Terzaghi, 1943).

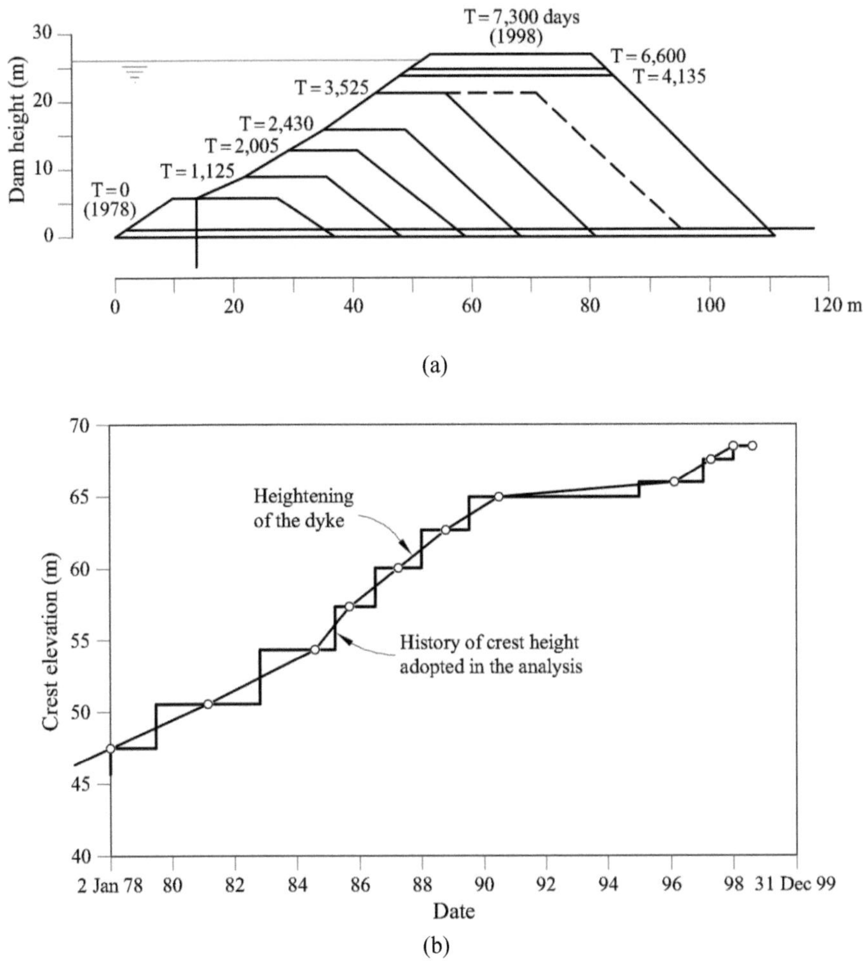

Figure 4.10 (a) Evolution of the dam cross-section; (b) increase in dam height (Gens and Alonso, 2006).

4.3.2 Evolution of the dam height

The actual evolution of cross section geometry and dam height along the years is shown in Figure 4.10. The plot (Fig. 4.10a) shows that the original dam cross section (Fig. 4.1) was somewhat changed. The dam lost its symmetry and a downstream steeper slope (39° against 30° in the original design) was introduced. In addition, the crest width was substantially increased in 1998, 10 years after the initiation of construction in 1988. The height of the dam increased at an approximately constant pace (1.5 m/yr) in the period 1978–1990 (Fig. 4.10b). The rate of increase of dam height slowed thereafter for a few years to increase again in the two years previous to the rupture.

4.3.3 A simple calculation model and its implications

A simple procedure to calculate stress and pore water pressures in the foundation soil under the evolving geometry of the dam was developed. Total stress will be first calculated. Then it will be accepted that the increase in pore pressure induced by the (sudden) application of a total stress is given by the increment of mean stress (this is the case in an elastic porous media). Excess pore water pressures will then be dissipated towards the upper pervious boundary.

a) Calculation of total stresses

Consider the elastic solution for an embankment loading of infinite lateral extent in one direction (plane strain conditions) shown in Figure 4.11.

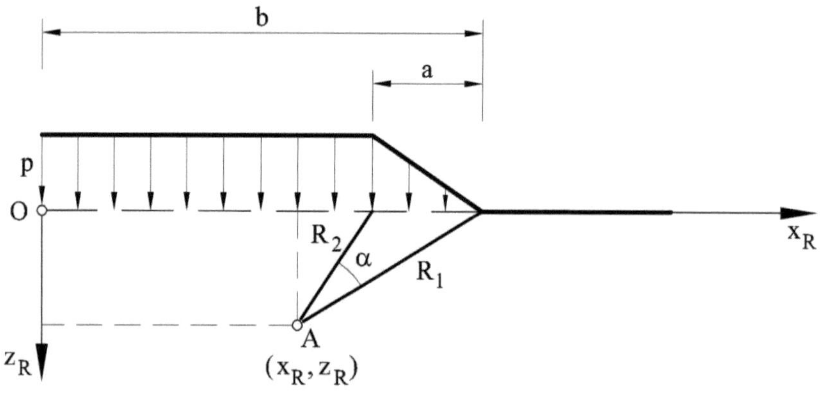

Figure 4.11 Embankment loading of semi-infinite extent.

The state of stress ($\sigma_z, \sigma_x, \sigma_{xz}$) at arbitrary Point A ($x_R$, z_R) is given by (Poulos and Davis, 1974)

$$\sigma_z = \frac{p}{\pi}\left[\left(\frac{\pi}{2}+\beta_2\right)+\alpha\left(\frac{x_m}{a}\right)\right]=\frac{p}{\pi}f_1\left(x_R, z_R, a, b\right), \tag{4.1a}$$

$$\sigma_x = \frac{p}{\pi}\left[\left(\frac{\pi}{2}+\beta_2\right)+\alpha\left(\frac{x_m}{a}\right)+2\left(\frac{z_m}{a}\right)\ln\left(\frac{R_2}{R_1}\right)\right]=\frac{p}{\pi}f_2\left(x_R, z_R, a, b\right), \tag{4.1b}$$

$$\tau_{xz} = \frac{-p}{\pi}\alpha\left(\frac{z_m}{a}\right) = \frac{p}{\pi} f_3\left(x_R, z_R, a, b\right), \tag{4.1c}$$

where

$$x_m = b - x_R,$$

$$z_m = z_R,$$

$$R_2 = \sqrt{\left(x_m - a\right)^2 + z_m^2},$$

$$R_1 = \sqrt{x_m^2 + z_m^2},$$

$$\beta_1 = \text{atan}(x_m / z_m),$$

$$\beta_2 = \text{atan}\left(\left(x_m - a\right)/ z_m\right),$$

$$\alpha = \beta_1 - \beta_2$$

and p is the maximum stress applied by the embankment.

Consider now, in Figure 4.12, the situation of the first stage of construction of the Aznalcóllar dam. Tailings and rockfill have significantly different total unit weights (γ_t and γ_r). It was mentioned that γ_t = 31 kN/m^3. The rockfill may have a natural unit weight of the order of γ_r = 19 kN/m^3.

The load applied by dam and tailings (Fig. 4.12a) may be viewed, with sufficient accuracy, as indicated in Figure 4.12b. This interpretation facilitates the justification of the superposition indicated in Figure 4.12: the external load is obtained by superimposing two semi-infinite embankment loadings, those shown in Figures 4.12c and 4.12d. The first one (Fig. 4.12c) has a uniform specific weight of γ_r and the second one (Fig. 4.12d), located in a different position with respect to the reference system (x_R, z_R), introduces the excess of stress applied by the heavy tailings, not accounted for in Figure 4.12c.

Symbolically, it could be written, with reference to Figure 4.12

$$(a) = (b) = (c) + (d). \tag{4.2}$$

Equations (4.1a) to (4.1c) are now applied to the geometry of Figures 4.12c and 4.12d and the results are added.

Note that the value of p in the first case is $p_1 = \gamma_r h$ and in the second case, $p_2 = \left(\gamma_t - \gamma_r\right)h$. Therefore, the stress in point A will be given by

$$\sigma_z = \frac{\gamma_r h}{\pi}\left[f_1\left(x_R, z_R, a_1, b_1\right)\right] + \frac{\left(\gamma_t - \gamma_r\right)h}{\pi}\left[f_1\left(x_R, z_R, a_2, b_2\right)\right], \tag{4.3a}$$

$$\sigma_x = \frac{\gamma_r h}{\pi}\left[f_2\left(x_R, z_R, a_1, b_1\right)\right] + \frac{\left(\gamma_t - \gamma_r\right)h}{\pi}\left[f_2\left(x_R, z_R, a_2, b_2\right)\right], \tag{4.3b}$$

$$\tau_{xz} = \frac{\gamma_r h}{\pi}\left[f_3\left(x_R, z_R, a_1, b_1\right)\right] + \frac{\left(\gamma_t - \gamma_r\right)h}{\pi}\left[f_3\left(x_R, z_R, a_2, b_2\right)\right]. \tag{4.3c}$$

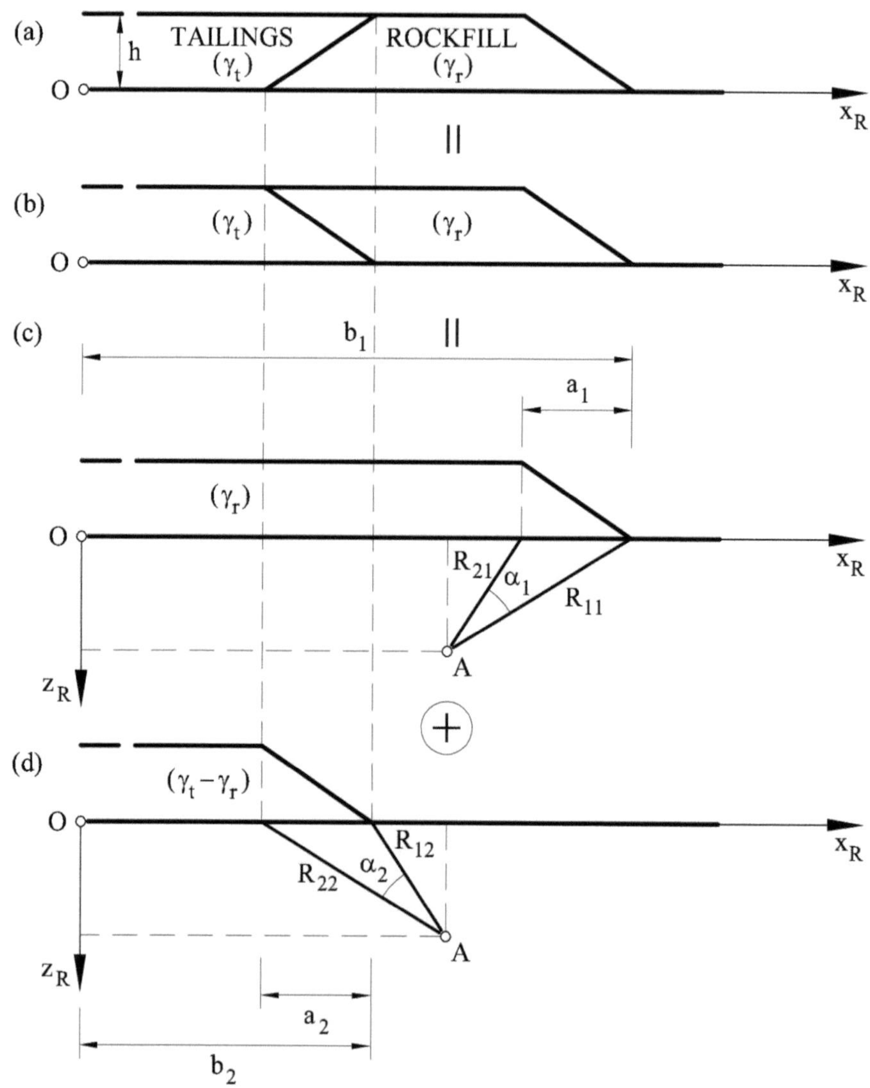

Figure 4.12 Superposition of solutions to solve the stress distribution for the dam and pond.

At this point, it is convenient and sufficiently accurate to assume that the "continuous" evolution of dam geometry was actually made in a number of finite jumps. This hypothesis facilitates the calculations of pore water pressure dissipation. The actual sequence of height jumps assumed is also indicated in Figure 4.10. Jumps are made more frequent when the rate of loading is higher. Consider now two successive dam geometries (I and II) in Figure 4.13.

The total stress, when II is built, will be a function of geometry II through Equations (4.3a) to (4.3c). However, the analysis of pore pressures requires the calculation of the loading increments because each one of them marks the beginning of a consolidation process and their origin is changing, as shown in Figure 4.10.

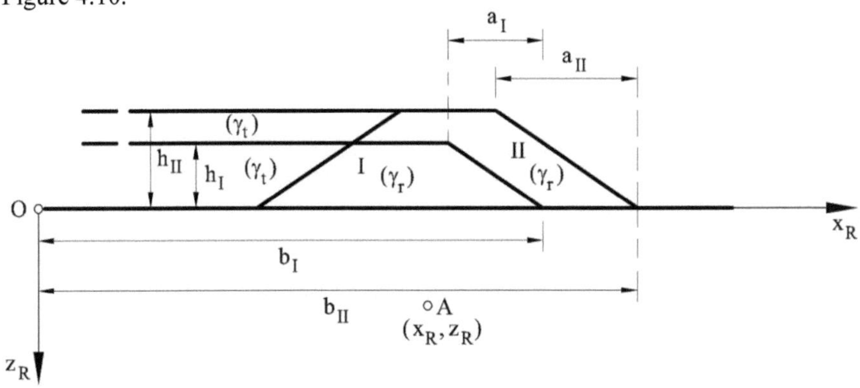

Figure 4.13 Geometry of two successive positions of the dam.

The jump in stress when geometry I changes (instantaneously) into II is given by

$$\Delta\sigma_z = \sigma_z^{\mathrm{II}} - \sigma_z^{\mathrm{I}}, \tag{4.4a}$$

$$\Delta\sigma_x = \sigma_x^{\mathrm{II}} - \sigma_x^{\mathrm{I}}, \tag{4.4b}$$

$$\Delta\tau_{xz} = \tau_{xz}^{\mathrm{II}} - \tau_{xz}^{\mathrm{I}}, \tag{4.4c}$$

where stresses are given in Equations (4.3) and (4.1) and the superscripts (II) and (I) refer to the two geometries indicated in Figure 4.13.

b) Pore water pressures

It will be assumed that the instantaneous increase in pore water pressure when the dam "II" is applied over dam "I" (Fig. 4.13) is given by the increment of mean or "octaedral" stress:

$$\Delta u_0(x,z) = \Delta\sigma_{oct} = \frac{\Delta\sigma_x + \Delta\sigma_z + \Delta\sigma_y}{3}$$
$$= \frac{\Delta\sigma_x + \Delta\sigma_z + \nu(\Delta\sigma_x + \Delta\sigma_z)}{3} = \frac{(1+\nu)(\Delta\sigma_x + \Delta\sigma_z)}{3}, \tag{4.5}$$

where the condition of plane strain ($\varepsilon_y = 0$) is introduced, assuming elastic conditions. In fact, if the soil is assumed elastic, shear induced dilatancy is not present and the initial pore pressure is given by the increase in mean stress.

Once generated at each of the loading jumps, pore pressures will dissipate in time. Consider again Point A (x_R, z_R) and the problem of finding its pore pressure at the time of failure, t_f, when it has experienced the stress history associated with embankment and tailing's pond construction. If the set of time instants at which loading jumps are applied is $t_1, \ldots, t_N$, the pore pressure at t_f will be obtained by superposition of a set of N consolidation records,

$$u_A(t_f) = \sum_{i=1}^{N} \Delta u_A\left(t_f - t_i\right), \tag{4.6}$$

where $(t_f - t_i)$ is the dissipation time for each of the instantaneous increments of loading in which the history of dam construction has been divided.

We are interested in changes in shear and effective stresses in the proximity of the dam; let us say the 20 upper meters of the clay foundation. The lower boundary of the deep blue clay, being 60 m apart, has no effect on the dissipation conditions of the upper levels. In addition, horizontal flow is probably of minor importance, given the high vertical gradients induced by the top pervious boundary. Therefore, a one-dimensional vertical dissipation of pore pressures will be assumed. Furthermore, given the thickness of the blue clay layer, the clay stratum may be considered to be semi-infinite.

The solution of the one-dimensional consolidation problem of a semi-infinite stratum subjected to a uniform increase in pore water pressure, u_0, is given by the expression (Scott, 1963),

$$W(Z,T) = \operatorname{erf}\left(\frac{Z}{2\sqrt{T}}\right), \tag{4.7}$$

where W is a dimensionless pore water pressure ($W = u/u_0$), Z is a dimensionless depth ($Z = z/H$) and T is the time factor ($T = c_v t/H^2$); u_0 is the initial increment of pore pressure, H is any reference distance, which is interpreted as the thickness of an upper clay layer, and c_v is the consolidation coefficient. The error function is defined as

$$\operatorname{erf}(x) = \frac{2}{\sqrt{\pi}}\int_0^x e^{-t^2}\,\mathrm{d}t. \tag{4.8}$$

The dam and tailings reservoir loading increments induce profiles of initial pore pressure, which, in general, are not uniform in depth. However, the initial distribution of excess pore pressures is fairly uniform under the reservoir and part of the dam. The simple assumption made, however, is that the local evolution of excess pore water pressures is given by

$$\Delta u\left(x,z,t\right) = \Delta u_0(X,Z)W\left(Z,T\right). \tag{4.9}$$

These values are then used in Equation (4.6) to find pore pressures due to the history of loading. This approximation was shown in Gens and Alonso (2006) to provide very similar results to a finite element two-dimensional coupled flow-deformation analysis.

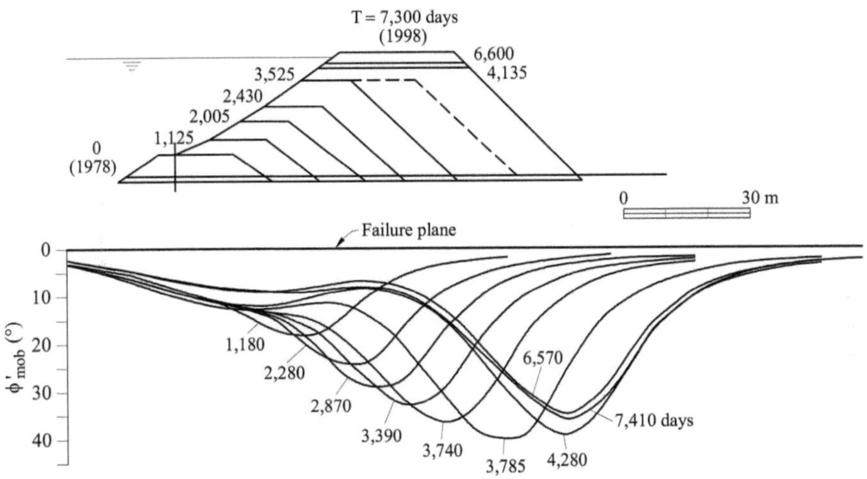

Figure 4.14 Mobilized friction angle on the position of the failure plane in the blue clay (Gens and Alonso, 2006).

Once pore pressures are calculated, total and effective stress can be found at any point of the foundation during the entire operation of the dam.

The procedure described above was validated against the recorded pore pressures in piezometers located in Profile 1 (Fig. 4.9b). Two calculations, for a horizontal plane joining piezometers P1, are shown in the figure. They correspond to two values of c_v, namely 10^{-3} cm^2/s and 2.3×10^{-3} cm^2/s.

Actual measurements fit well between the two plotted approximations. The curve for $c_v = 10^{-3}$ cm^2/s is closer to the measured pore pressures in the downstream side of the dam, a critical area to interpret stability conditions, as discussed later. Vertical profiles of calculated pore pressure (the shaded area) show the response associated with the two c_v values mentioned. Again a reasonably good agreement with measurements is found. A value $c_v = 10^{-3}$ cm^2/s was henceforth applied in the calculations.

c) Mobilized shear stresses in dam foundation

Consider now in Figure 4.14 the forward construction of the dam in the failed area. Attention is paid to the shear stress ($\tau \equiv \tau_{xy}$) development at the position of the failure plane (it was simplified as a horizontal plane, 14m below foundation level). Rather than directly plotting τ, it is more meaningful to plot the stress ratio (τ/σ'_n) where σ'_n is the effective normal stress on the plane, or, alternatively, the mobilized friction angle defined as $\phi'_{\text{mob}} = \text{atan}(\tau/\sigma'_n)$.

This mobilized friction angle, ϕ'_{mob}, in Figure 4.14, may be directly compared with the strength envelopes of the blue clay given in Figure 4.8. Each one of the curves plotted corresponds to the indicated time (in days) after the beginning of dam construction. The figure also shows the geometry of the dam for some particular times. The values of τ and $\sigma_n \equiv \sigma_z$ were calculated through Equations 4.3c and 4.3a. The pore pressure at each time instant was approximated as described in the previous section.

The mobilized friction angle advances with the dam, moving forward. The curves exhibit a significant maximum, which is located on the vertical of the downstream toe of the dam. Once a maximum is reached at a particular location, the forward motion of the dam implies a reduction in shear stress and therefore the mobilized friction is reduced. A maximum angle close to 40° is calculated 10.3 yrs after the beginning of construction (at around 1988). Later, the reduced rate of dam heightening beyond 1990 resulted in a sustained value of the mobilized friction on the potential failure plane at approximately $\phi'_{\text{mob}} = 35$–$37°$.

The calculated maximum values of ϕ'_{mob} are high values that are not resisted by the blue clay except if some effective cohesion is operating. As soon as some minor relative displacements (of tectonic origin, for instance) eliminates the apparent effective cohesion, the clay will not be able to resist the mobilized friction angle applied by the dam.

Therefore, in order to maintain equilibrium, some shear stress, which cannot be resisted by the blue clay under the toe of the downstream slope, will be transferred to points nearby, increasing in turn the mobilized friction under the entire dam foundation. This process is known as progressive failure. In view of Figure 4.14, this phenomenon could have started for a dam height of 18 m, at around 1986.

The value of ϕ'_{mob} was calculated for the cross-sectional area of the foundation and plots of interpolated contours of ϕ'_{mob} were drawn for several geometries of the advancing dam.

The results are given in Figure 4.15. The figure shows that the maximum value of (τ/σ'_n) is reached at a certain depth under the downstream toe of the dam. This depth changes slowly as the size of the dam increases. This critical depth is controlled not only by the dam cross-section but also by the rate of dissipation of excess pore pressures. It is expected that the progressive failure mechanism would develop around the maximum values of (τ/σ'_n) in a process of shedding shear stress towards the surrounding foundation soil. Therefore, the curve joining the position of maxima (shown in Fig. 4.15c) is a reasonable position of the sliding surface and indicates also the length of a "damaged" band. The position of this line is quite close to the actual elevation of the failure surface identified in the field.

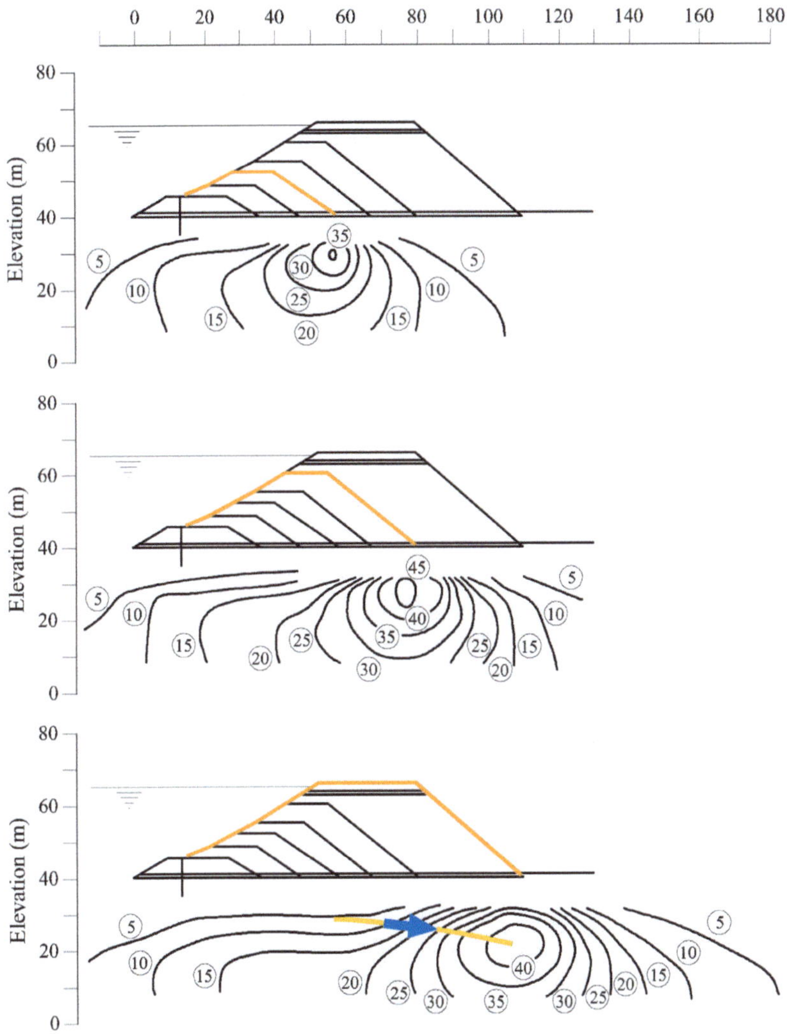

Figure 4.15 Mobilized friction angle in the foundation for three positions of the dam (Gens and Alonso, 2006).

This simple analysis, which makes use of the classical theories of elasticity and one-dimensional consolidation, is capable of providing an interesting insight into the nature of the shearing mechanism acting on the foundation during the forward construction of the dam.

4.4 Limit Equilibrium Analysis

Failure conditions in global terms can be investigated by means of a limit equilibrium analysis. The objective is now to find, by means of a back analysis of the failure, the average friction angle prevailing on the sliding surface. It will be assumed that effective cohesion was no longer available after the progressive failure mechanism mentioned above.

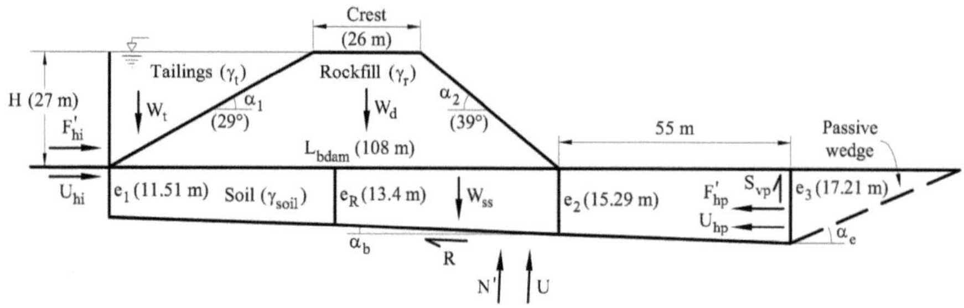

Figure 4.16 Model for the limit equilibrium analysis of the main wedge of Aznalcóllar slide.

Consider in Figure 4.16 a representative cross-section of the failed dam. The figure reproduces the reconstructed shape of the failure as revealed by field observations (Fig. 4.6b). It incorporates the following observations and interpretation of field data:

- The slide is limited in its upstream part by an essentially vertical wall at the position of the upstream toe of the dam. This vertical plane reflects the structure of the clay, described by a few families of vertical discontinuities and the sub-horizontal sedimentation planes. The cementation of tailings also explains that they were able to maintain a stable vertical cliff after the slide.
- The sliding surface follows a stratification plane. The inclination of the basal sliding plane is somewhat lower than the dip of the strata (estimated in the range 2–4° in field surveys). This geometry implies that the depth of the failure surface increases in the direction of motion.
- The failure mechanism ends, in the downstream direction, in a passive wedge (Fig. 4.17). This wedge does not start directly below the dam toe but at a distance of 55 m, following the interpretation of the initial shape of the failure surface in Figure 4.6a. The passive wedge is a simplified interpretation of the folded strata discovered by trenches performed after failure.
- Therefore, the mechanism is assumed to be described by a large wedge, which includes the dam, a mass of tailings overlying the upstream dam slope, and an exit passive wedge that offers a reaction against sliding. A vertical surface is assumed to be the interface between the two wedges

4.4.1 Backanalysis of failure

The upstream vertical surface receives the thrust exerted by the tailings in the upper part, by an intermediate granular layer 4 m thick, and by the lower clay. If a fissure was open in the clay, previous to sliding, as a consequence of the tailings thrust against the dam and the presence of vertical discontinuities, it would have been probably filled with saturated tailings. Then, under the hypothesis of active conditions, the following total horizontal force is calculated against the sliding body:

$$F_{hi} = F'_{hi} + U_{hi} = \frac{1}{2} K_a\, \gamma'_t (H + e_1)^2 + \frac{1}{2}\gamma_w (H + e_1)^2, \tag{4.10}$$

where $K_a = \left(1 - \sin\phi'_{\text{tailings}}\right) / \left(1 + \sin\phi'_{\text{tailings}}\right)$ is the active Rankine coefficient for the tailings, γ'_t is the effective unit weight of tailings, γ_w is the unit weight of water, H is the dam height and e_1 the thickness of the slab of moving soil directly under the toe of the upstream dam slope.

The weight of the main sliding wedge is made of three components:

- tailings: $$W_t = \frac{1}{2}\frac{\gamma_t H^2}{\tan\alpha_1}, \tag{4.11}$$

where α_1 is the slope angle of the upstream slope of the dam;

- dam: $$W_d = \frac{1}{2}\frac{\gamma_r H^2}{\tan\alpha_1} + Crest\ H\,\gamma_r + \frac{1}{2}\frac{\gamma_r H^2}{\tan\alpha_2}, \tag{4.12}$$

where α_2 is the slope angle of the downstream face of the dam, *Crest* is the width of dam crest and γ_r is the specific weight of the dam rockfill;

- soil slab: $$W_{ss} = \frac{1}{2}\gamma_{\text{soil}}\left(e_1 + e_3\right)\left(L_{b\text{dam}} + 55\right), \tag{4.13}$$

where γ_{soil} is the average effective unit weight of the foundation soil, $L_{b\text{dam}}$ the length of the base of the dam and e_1 and e_3 are conveniently defined in terms of the thickness of soil slab under the center of the dam base, e_R, as follows (Fig. 4.16):

$$e_1 = e_R + \frac{L_{b\text{dam}}}{2}\tan\alpha_b, \tag{4.14}$$

$$e_3 = e_R + \left(\frac{L_{b\text{dam}}}{2} + 55\right)\tan\alpha_b \tag{4.15}$$

where $L_{b\text{dam}}$ is the length of the base of the dam and α_b is the slope of the sliding plane. Lengths are expressed in meters.

Consider now the passive wedge resisting the slide in Figure 4.17. The water pressure forces against the wedge are assumed to be given by a hydrostatic condition associated with a water level at the surface. They are given by,

$$U_{hp} = \frac{1}{2}\gamma_w e_3^2, \tag{4.16}$$

$$U_p = \frac{1}{2}\frac{\gamma_w e_3^2}{\sin\alpha_e}, \tag{4.17}$$

where α_e is the exit inclination of the lower sliding surface of the passive wedge. Field observations indicated that α_e is close to 20°.

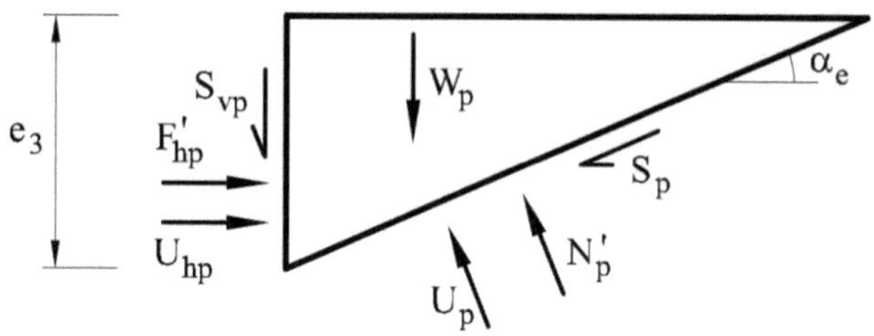

Figure 4.17 Passive wedge.

The weight of the wedge is given by

$$W_p = \frac{1}{2}\frac{\gamma_{soil} e_3^2}{\tan\alpha_e}. \tag{4.18}$$

The system of forces in Figure 4.17 must be in equilibrium. Shear S forces are related to effective N' forces through the Coulomb failure criterion. Since the effect of the upper granular layer is very small, the effective friction parameter is taken to be equal to the drained clay friction, ϕ_b'. Imposing the balance of forces in vertical and horizontal direction,

$$N_p' \cos\alpha_e + U_p \cos\alpha_e - W_p - S_{vp} - S_p \sin\alpha_e = 0, \tag{4.19}$$

$$F_{hp}' + U_{hp} - S_p \cos\alpha_e - N_p' \sin\alpha_e - U_p \sin\alpha_e = 0, \tag{4.20}$$

where $S_p = N_p' \tan\phi_b'$ and $S_{vp} = F_{hp}' \tan\phi_b'$.

These two equations will enter the equilibrium of the main wedge.

Pore pressures on the sliding plane have already been determined by the simplified procedure described above (see Fig. 4.9). They are plotted in Figure 4.18 in a more simple way in order to facilitate the calculations. They correspond to the date of the failure and are expressed in terms of water heads (in meters).

The integration of these pore pressures along the base sliding surface provides the water pressure force, U (in kN/m if lengths are in meters).

$$U = \frac{680+520}{2}\frac{H}{\tan\alpha_1}\frac{1}{\cos\alpha_b} + 520\,Crest\frac{1}{\cos\alpha_b} + \frac{520+260}{2}\frac{H}{\tan\alpha_2}\frac{1}{\cos\alpha_b} + \frac{260+172}{2}55\frac{1}{\cos\alpha_b}. \tag{4.21}$$

The balance of vertical and horizontal forces of the main wedge leads to the following expressions

$$W_t + W_d + W_{ss} - S_{vp} - U\cos\alpha_b - N'\cos\alpha_b - N'\tan\phi_b'\sin\alpha_b = 0, \tag{4.22}$$

$$F_{hi}' + U_{hi} - F_{hp}' - U_{hp} - N'\tan\phi_b'\cos\alpha_b + U\sin\alpha_b + N'\sin\alpha_b = 0, \tag{4.23}$$

which can be expressed in terms of quantities previously defined.

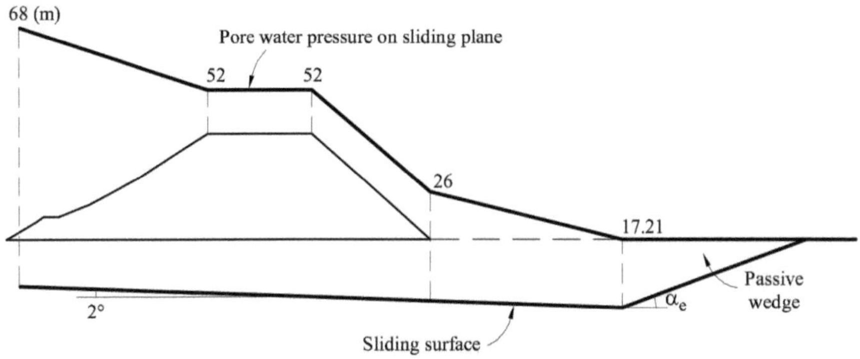

Figure 4.18 Pore pressures against the sliding surface.

Equations (4.22) and (4.23) were solved for the effective friction angle necessary to obtain strict equilibrium, ϕ_b'. The system of Equations (4.19), (4.20), (4.22), and (4.23) is nonlinear in ϕ_b' but it can be easily solved in an Excel sheet with the help of the command "solver".

The following parameters define the geometry of dam actually built and the failure surface as well as additional material parameters:

$\alpha_b = 2°$	$\alpha_e = 20°$	$\alpha_1 = 29°$	$\alpha_2 = 39°$
$\gamma_{soil} = 19.5$ kN/m^3	$\gamma_w = 10$ kN/m^3	$\gamma_r = 19$ kN/m^3	$\gamma_t = 31$ kN/m^3
$\phi'_{tailings} = 37°$	*Crest* = 26 m	$H = 27$ m	$e_R = 13.4$ m

When these values are introduced in the above equations, the solution is $\phi_b' = 18.09°$.

This value of the effective friction angle ϕ_b' is intermediate between the peak and residual friction angles. It is the average value between the two. However, one should imagine that, immediately before failure, some parts of the failure surface (namely those directly affected by the forward construction of the dam) would be close to residual conditions. Others would maintain strength properties closer to peak values: those parts not strained by the dam construction. The fact that the calculated friction for equilibrium is intermediate between peak and residual friction supports the mechanism of progressive failure. It also indicates that immediately after the initiation of the failure, there is a danger of progressive decay of friction towards the residual value (11°). The implications of this potential for further reduction of available strength at the sliding surface will be addressed in Chapter 6.

4.4.2 Undrained analysis

The main hypothesis so far is that the failure was a drained process. However, the low clay permeability also suggests that undrained conditions would also be a possibility. If the soil fails in an undrained manner, pore pressures are generated during the shearing process. The undrained strength, c_u, not only reflects the drained strength parameters but also the dilatancy conditions of the soil. The undrained analysis of the real geometry of the failure surface may be easily performed with the help of the set of preceding equations. Friction ϕ_b' is reduced to 0° and the shear forces at the sliding plane or at interfaces are simply the product of c_u by the involved length. The preceding set of Equations (4.19), (4.20), (4.22) and (4.23) was solved for c_u and strict equilibrium is found for $c_u = 70$ kPa. This value should be compared with the unconfined compression strength measured in specimens recovered in borings. Their q_u values increase slightly with depth. Average values at the elevation of the failure plane determined on recovered specimens fall in the range 250 to 300 kPa. Therefore, the safety factor against undrained failure is about 1.8–2.0 (undrained strength is half the value of the unconfined compression strength).

Undrained conditions in this overconsolidated foundation clay do not explain the failure. This is usually the case in overconsolidated clays, specially the plastic materials, which may exhibit a relatively high undrained strength (due to dilatancy and the associated increase in effective stress during undrained loading) and, at the same time, a reduced drained friction angle, prone to decrease towards residual conditions as shear straining accumulates.

Before discussing the lessons offered by this case, we will examine the failure from a different perspective, namely the three-dimensional nature of the problem.

4.4.3 Three-dimensional effects. The role of bedding planes

It was previously noted that the retaining dyke for the northern basin remained intact after the sliding failure of the southern dyke. However, the geotechnical conditions and dam geometry were essentially the same in both basins. A satisfactory explanations may be found if the three-dimensional nature of the sliding motion is analysed.

The situation is illustrated in Figure 4.1, which shows the direction and dip of sedimentation planes and its relationship with the expected direction of tailings thrust against the dykes. In the northern dyke, the thrust is sub-parallel to the direction of sedimentation planes.

However, in the southern dyke, the tailing's thrust is closer to the dip direction. It may be expected that this situation, other conditions being equal, will favour a slide of the southern dam. The problem is exceedingly difficult if considered in its entire complexity.

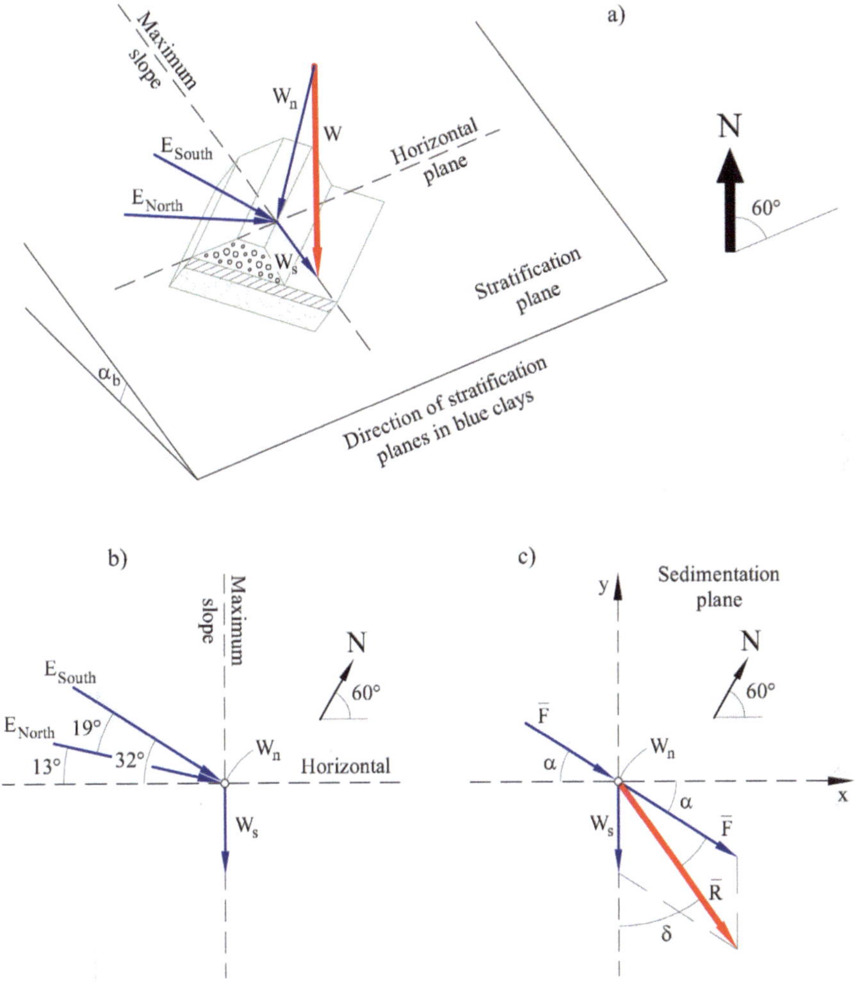

Figure 4.19 (a) Slice of the potential slide sitting on the sliding surface; (b) forces projected on the sliding plane; (c) vector composition of forces (Gens and Alonso, 2006).

Considerer, however, the simplified approach illustrated in Figure 4.19a. A slice of unit thickness of the actual slide, which includes the entire dam and a "slab" of foundation soils of approximately 14 m in thickness, is sitting on the sliding plane, which is actually a sedimentation plane. These planes are oriented in a direction N60°E and its true (maximum) dip is close to 3°.

The slice is now viewed as a rigid block, sliding on its base. This block receives the following external forces: its own weight, $\overline{W}$, and the thrust imposed by the tailings.

It will assumed that the latter is represented by a force vector $\overline{F}$ acting normal to the direction of the dam and parallel to the sliding plane (Fig. 4.19c). Now, the conditions of the northern and southern basins only differ in the direction of the force vector $\overline{F}$. In the northern basin, $\overline{F}$ is closer to the direction of the sedimentation planes and therefore the tailings thrust "sees" the sliding plane with an apparent direction close to 0°. By contrast, the change in direction of the southern dyke, implies that the motion of the slide is closer to the dip direction of sedimentation planes. The tailing thrust in the southern dyke "sees" an apparent inclination of the sliding plane close to 2°.

The problem now is to find the necessary forces, $\overline{F}$, to make the dam unstable, having in mind that vector $\overline{F}$ has different directions in northern and southern basins.

Figure 4.19a shows the force vectors acting on the unit slice. The weight, $\overline{W}$, will be divided into the normal component, $\overline{W}_n$, and the shear component $\overline{W}_s$ in the direction of the dip of the sedimentation planes. Also indicated are the thrust forces, E_{North} and E_{South}, of the two dyke sectors (northern and southern).

In Figure 4.19b, force vectors are represented in the plane of the clay strata. The two axes plotted coincide with the strata direction (horizontal axis) and the dip direction (vertical axis). For a purely frictional motion, the condition of force equilibrium reads

$$\left|\overline{R}\right| = \left|\overline{F} + \overline{W}_s\right| = W_n' \tan \phi_b', \tag{4.24}$$

where $\overline{R}$ is the resultant of frictional forces in the direction of sliding, $W_n' = W_n - U$, $W_n = W \cos \alpha_b$ and $W_s = W \sin \alpha_b$, U is the force resultant of pore pressure acting against the sliding surface and α_b is the dip angle. $\overline{R}$ has two components in the axis of Figure 4.19b:

$$\overline{R} \equiv \left(F \cos \alpha, F \sin \alpha + W \sin \alpha_b\right), \tag{4.25}$$

where α defines the direction of the force $\overline{F}$ with respect to the direction of sedimentation planes. Taking (4.25) into account, the equilibrium Equation (4.24) becomes,

$$\begin{aligned} &\left(W \cos \alpha_b \tan \phi_b'\right)^2 = \\ &= F^2 \left(\cos \alpha\right)^2 + F^2 \left(\sin \alpha\right)^2 + 2F\,W \sin \alpha_b \sin \alpha + W^2 \left(\sin \alpha_b\right) \end{aligned} \tag{4.26a}$$

or,

$$F^2 + 2F\,W \sin \alpha_b \sin \alpha + W^2 \left(\sin \alpha_b\right)^2 - \left(W \cos \alpha_b - U\right)^2 \left(\tan \phi_b'\right)^2 = 0. \tag{4.26b}$$

This equation provides the necessary force, F, to initiate the sliding motion as a function of the direction α. The remaining forces in Equation (4.26b) are known. It has been estimated that W = 100,428 kN/m and U = 70,682 kN/m for a slice one meter thick. Previous analysis indicate that ϕ_b' = 18.09°.

Equations (4.26) were represented in a polar diagram (Fig. 4.20) showing values of F for varying α angles. The direction of the tailing's thrust for the northern and southern basins is indicated in the plot (α = 13°, α = 32°). The corresponding forces to initiate sliding are 5.27 MN/m (northern basin) and 4.11 MN/m (southern basin).

It thus appears that the force needed to make unstable the dyke in the southern basin is substantially lower than the force necessary to initiate the slide in the northern dike. If the calculated force for the southern basin makes it unstable, the safety factor of the northern dyke against sliding would be 1.28 (= 5.27/4.11). This result explains that a failure of the northern dyke was unlikely. Note that this is an approximate calculation since there are other aspects not included (the resistance of the exit passive wedge in particular). They primarily serve to compare the effect of the dyke direction on stability conditions.

Equation (4.26b) is actually the equation of a circle. If plotted in the Cartesian axis ($x = F\cos\alpha$, $y = F\sin\alpha$), which represents the two components of the tailings thrust on the reference axis given by the horizontal direction of the sedimentation plane (x axis) and its dip direction (y axis), the equation recovers the more familiar form:

$$x^2 + (y-b)^2 = R^2, \tag{4.27}$$

where the radius R is given by

$$R = \left(W\cos\alpha_b - U\right)\tan\phi'_b \tag{4.28}$$

and the position of the center along the y axis,

$$b = W\sin\alpha_b. \tag{4.29}$$

The circle is shown in Figure 4.20. Its radius defines the available (effective) friction force offered by the slide weight. At the start of the motion, this force is exhausted by the combination of tailing's thrust and the slide weight component in the direction of the dip of the strata. Therefore, motions against the dip direction (for $\alpha = -90°$) require the largest applied thrust. Motions in the direction of the dip ($\alpha = 90°$) are the easiest. This is shown by the intensity of the thrust necessary to start the motion in each case. The "displacement" of the circle center (Eq. (4.29)) is, in fact, the weight component W_s in the direction of the dip of the sedimentation planes.

Also represented in Figure 4.19 are the directions of the displacement vectors for the two thrust forces represented. These vectors follow the direction of the resultant $\overline{R}$ in Figure 4.19c. It can be easily shown that they follow the direction of the radii of a circle (Eq. (4.26)). In other words, they are normal to the circle tangent.

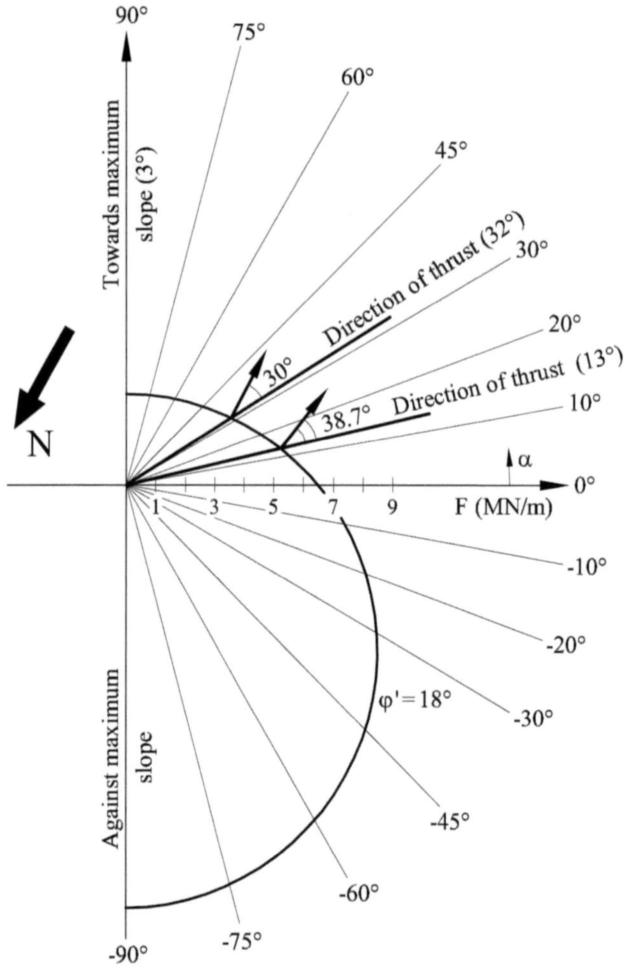

Figure 4.20 Polar diagram showing forces that induce dam sliding as a function of the direction of the thrust (α = 13°: northern dyke; α = 32°: southern dyke). Also indicated is the direction of the motion (arrows normal to the force function) (Gens and Alonso, 2006).

In plasticity theory, this is a condition for an "associated" plastic deformation. In fact, the sliding motion is entirely plastic in the sense that it does not have any elastic or reversible component. The circle drawn in Figure 4.20 is a yield locus for the thrust capable of setting the slide in motion. And the condition of associativity allows determining the direction of motion. Pure friction is therefore an "associated plastic mechanism", even if it is interpreted in an inclined plane. In plasticity theory, a displacement of the yield locus in the stress plane is described as a "kinematic hardening". In our parallel two-dimensional friction force plane, we interpret that the friction yield locus for a horizontal sedimentation plane (a circle centred at the origin of the force vectors) is displaced when the strata are inclined. The inclination of the strata results in a kinematic hardening in the direction opposing the maximum slope (dip direction). Note, finally, that the displacement vectors in Figure 4.20, if compared with the tailing's thrust vectors, are rotated towards the southern direction. This was a field observation which now receives an explanation.

4.5 Discussion

In the original design (Fig. 4.1) safety against dam sliding was checked by means of the Morgenstern – Price method of slices (Morgenstern and Price, 1965). The set of hypotheses made were particularly severe: Tailings were assumed to be liquefied and an earthquake of magnitude MSK = 7 was applied by means of pseudostatic accelerations a_h = 0.776g (horizontal) and a_v = 0.048g (vertical). Figure 4.21 shows the water level considered in calculations. It was assumed to reproduce steady state conditions of flow through the dam. Calculations were made under drained conditions and a zero effective cohesion and a drained friction angle of 25° were taken for the clay foundation. Tailings were characterized by a natural unit weight of 29.5 kN/m^3 (and zero strength).

Under this set of conditions, the critical failure slip shown in Figure 4.21 provided a safety factor of 1.3, which was considered satisfactory.

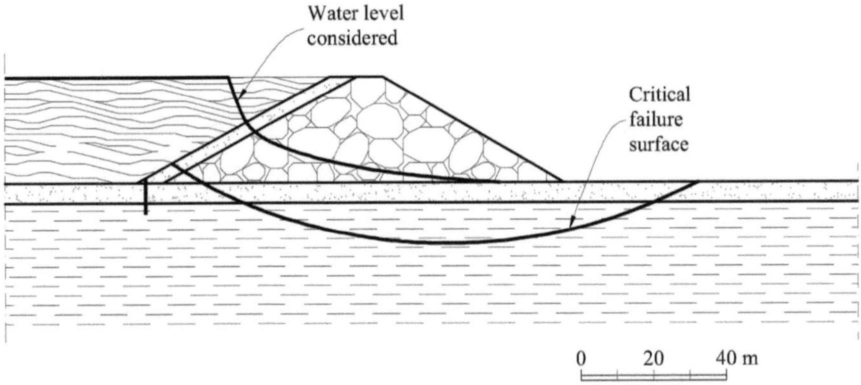

Figure 4.21 Stability calculations at the design stage.

Actual conditions at the time of failure are indicated in Figure 4.22. The downstream slope was increased from 30° to almost 39°. This change, although it did not have any significant effect on the stability of the rockfill slope itself, contributed to overstress and damage the clay foundation. Figures 4.14 and 4.15 show that the maximum stress ratio in the foundation soil occurs below the foot of the slope at a given depth. The intensity of this stress ratio increases with the embankment slope angle and, therefore, the change in downstream slope led to a more critical situation, if compared with the original design. The distribution of stress ratios on the foundation also depends on other aspects: the consolidation conditions and the evolution of dam geometry.

A major departure between assumed design hypotheses and field situation concerns the distribution of pore pressures in the foundation. Regarding the failure plane, Figure 4.22 shows the difference. No excess pore pressures due to tailings (three times heavier than water!) and the dam itself were considered. The design was based on some interpretation of steady-state flow conditions, from the perspective of a pure seepage through the dam.

The third significant departure refers to the clay strength. The assumed drained friction (25°) was very close to the average drained peak friction angle (24.1°; see Fig. 4.8) determined in direct shear tests. Assuming zero cohesion was also a wise design decision. However, no consideration was given to the reduction of friction on the potential failure plane, because of the brittle nature of the clay and the possible development of progressive failure mechanisms. The low residual friction angle (11°) explains that the drop in friction could be very substantial. On the other hand, the forward construction method selected for the dam favours conditions of progressive failure, as discussed before. These considerations were not present at the design stage.

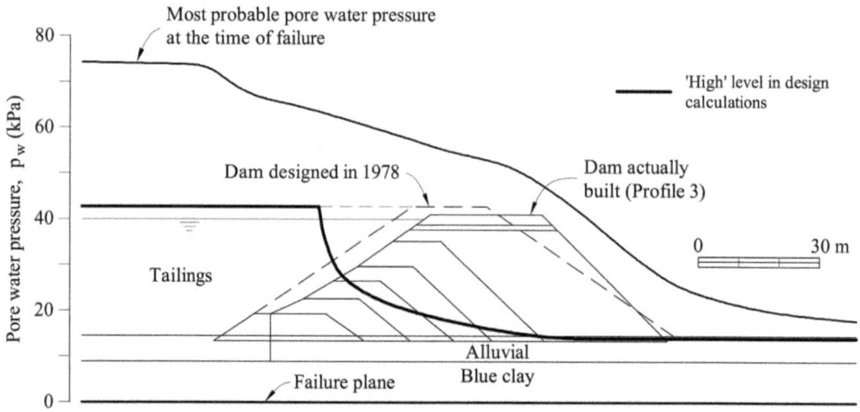

Figure 4.22 Dam actually built and the most probable water pressures acting on the failure plane. Also shown is the original design and the water saturation line assumed in stability calculations (Gens and Alonso, 2006).

These differences between assumed and field conditions were enough to offset the safety margin introduced in the design by some of the assumptions made and in particular, the occurrence of an earthquake of medium intensity and the liquefaction of tailings.

4.6 Mitigation Measures

The opening of the breach between the northern and southern dykes of the tailing's pond triggered a mud and water flood whose total volume was estimated in 5.5 Mm^3. In a flow gage station located in the Guadiamar river, 7 km downstream from the location of the failure, a maximum increase of water level of 3.6 m was measured 30 minutes after the beginning of the flood. This increase in level reduced to 20 cm, 12 hours later.

The mine company was able to seal the dam breach in 36 hours. The mud deposited on both margins of the rivers Agrio and Guadiamar (Fig. 4.23) occupied an estimated surface of 2,600 Ha. The thickness of the deposited layer ranged between 4 m and a few millimetres, 60 km away from the breach. The pyritic mud was thereafter exposed to the atmosphere in a very large area. Doñana National Park was protected by a wall rapidly built after the failure in the vicinity of Entremuros village (Fig. 4.23).

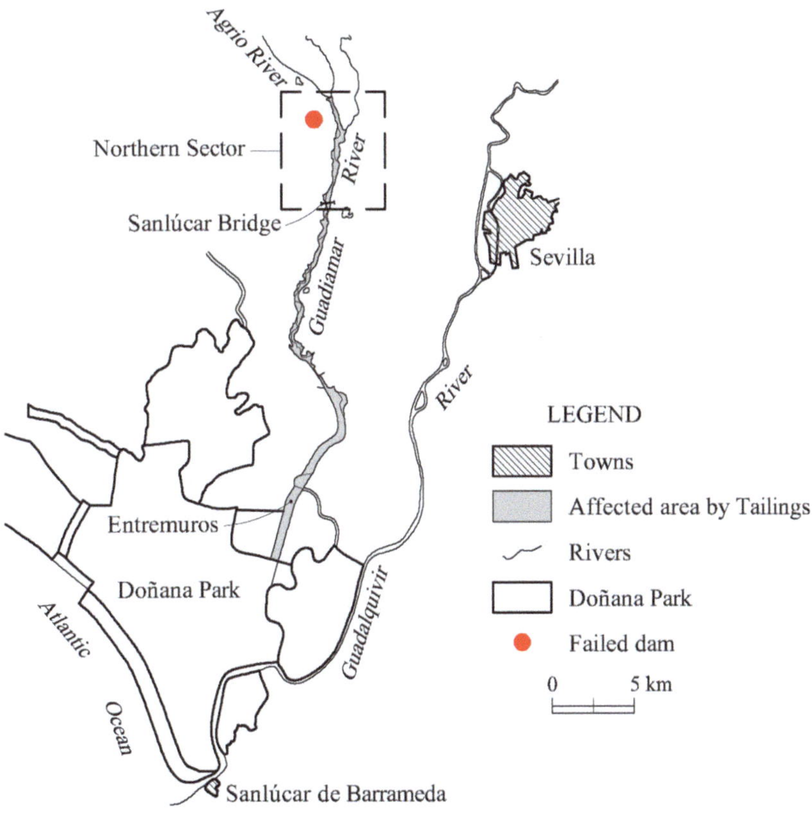

Figure 4.23 Map of affected area with reference points and rivers (Eriksson and Adamek, 2000).

When pyrite (FeS_2) is exposed to oxygen in the presence of water, an oxidation reaction begins. The oxidation reaction is represented as follows:

$$FeS_2 + 3.5O_2 + H_2O \rightarrow Fe^{2+} + 2SO_4^{2-} + 2H^+.$$

The resulting iron and the sulphates can dissolve in water. The free hydrogen creates a very acid environment. Sulphates are identified as white crusts which developed on the exposed surfaces of tailings. Within the pond, once the water was drained out, the eroded mud was covered by whitish sulphate crusts a few weeks after the failure.

The released water was extremely acid and it was responsible for the death of animal life in the Guadiamar river (37 tons of fish, crabs and shellfish were recovered). In addition, the mud had varying concentrations of many minerals: zinc, lead, arsenic copper, antimony, cobalt, thallium, bismuth, cadmium, and mercury, some of them highly toxic. The pyritic mud cover was therefore a major threat to the environment.

The decision to remove all the mud was immediately taken and an estimated weight of 10 million tons of mud and soil were removed from the river banks. This material was deposited in an existing open mine. This cleaning operation was essentially finished by December 1, 1998, seven months after the failure. A second cleaning operation was launched somewhat later, this time guided by the control of limiting concentrations of metallic minerals. An additional mass of one million tons of contaminated soil was removed and deposited in the open pit mine. A number of water wells in the area were also cleaned and controlled. Most of the vegetation on the riverbanks was totally removed.

The pond was also decommissioned (Eriksson and Adamek, 2000). It was first drained and a number of protective actions were taken: the entire pond area was covered by a protective sequence of layers (a geotextile in contact with the tailings, 0.5 m of waste rock, 0.1 m of binding layer, 0.5 m of compacted clay, and 0.5 m of a soil layer for the growth of vegetation); an impermeable cut-off wall was built on the northern and eastern sides of the pond; the slopes of the dam were reduced to 3:1 and its topography was modified to facilitate run-off and drainage.

In addition, a monitoring program was set out including inclinometers, piezometers, surface markers, and control wells. The recovery of the ecosystem was also closely followed by the authorities. Periodic controls included soil and water monitoring (from the surface and form irrigation wells). Health controls were also carried out on the human population and on the fauna. It has been reported that the recovery of the aquatic fauna was quite rapid.

The mining company resumed operations on June 1999. However, production was stopped on February 2001 and in September 2001, the mine was definitely closed. No indications of penal responsibility in the dam failure were declared by the judge in charge of the case and later by the regional Court of Andalusia.

4.7 Lessons Learned

4.7.1 Soft clay rocks, hard clay soils

These materials share properties of soils and rocks and present a challenge to geotechnical engineering because of a number of features, difficult to handle in practice: the presence of discontinuities, often at different scales, their brittleness, especially marked in the case of plastic clays, their very low permeability, which implies extremely long pore pressure dissipation times and their expansive nature, especially when montmorillonite is one of the clay minerals present.

4.7.2 Embankment loading

Embankments induce significant shear loading on the foundation soil. The attained stress ratio (ratio between the shear stress and the effective normal stress on any given plane and point) is a convenient parameter to investigate the safety conditions of a potential failure surface because it may be directly compared with the available strength. Maximum stress ratios are found directly below the toe of the embankment slope. The intensity of the stress ratio increases with the slope angle of the embankment. Drainage conditions of the foundation soil may lead to the occurrence of stress ratio maxima at some particular depth. This was the case of the foundation conditions of the Aznalcóllar dam.

4.7.3 Brittleness and progressive failure

Brittle materials are prone to progressive failure. This process results in a progressive reduction of available strength along a potential failure surface. Brittleness of clay soil increases with plasticity simply because the residual friction of clays decreases with plasticity. Clays containing significant proportions of montmorillonite may exhibit residual friction angles in the vicinity of 10°, a value significantly lower than the peak friction angle in most cases. Progressive failure mechanisms and, indeed, any shearing process, reduce fast the effective cohesion which may be measured in tests under peak conditions.

4.7.4 Bedding planes, discontinuities and tectonics

Sedimentation planes are particularly worrying because of their high lateral extent and the possibility of exhibiting a reduced strength if compared with the "bulk" or "matrix" clay strength. This strength reduction may be a consequence of tectonic motions but other mechanisms may also lead to strength degradation (unloading due to valley excavation, past sliding, soil expansion). Such mechanisms explain the presence of other systems of discontinuities, which may also show shear strengths significantly lower than bulk values. When a kinematically admissible failure mechanism integrates bedding planes and discontinuities, special attention should be paid to the actual strength conditions of the involved surfaces. In the case of Aznalcóllar, field evidence indicated the presence of striations and pre-shearing in some sedimentation planes. This implies a reduction of the available shear strength. In fact, under peak strength conditions ($c' = 64$ kPa, $\phi' = 24.1°$), no point in the foundation ever reaches plastic conditions and the process of progressive failure could not have started.

4.7.5 Operating strength

The problem of selecting an appropriate set of (drained) strength parameters in the presence of brittleness and progressive failure is not yet solved in practical terms. The case of Aznalcóllar suggests that the operative average effective friction at the initiation of failure (18°) was intermediate between the peak friction angle (24°) and the residual friction angle (11°). A few cases mentioned before suggest a similar result but it should be stressed that there is no fundamental reason behind such a simple rule.

4.7.6 Construction procedure

The forward construction procedure followed in Aznalcóllar favoured the development of progressive failure. It probably led to the formation of a damaged, low resistance clay surface under the dam which eventually propagated fast, upstream and downstream, at the time of failure. This event marked the initiation of the slide.

4.7.7 Pore pressures

Steady-state flow conditions are by no means the worst situation of the dam, when pore pressures are calculated. This comment refers, of course, to the joint consideration of the dam and its foundation. In general, when a drained safety analysis is performed, the prediction of pore pressures requires attention to the pore pressure generation due to dam construction and its subsequent dissipation through a consolidation process. In Aznalcóllar, the average degree of dissipation of pore pressures in the foundation was no more than 20% of the generated pore pressures. This is a result of the low clay permeability. The system of joints and bedding planes did not seem to have any effect on the field permeability in this case. Steady-state flow conditions were entirely irrelevant to explain the failure of the Aznalcóllar dam.

4.7.8 Undrained vs drained analysis

In hard clays, usually overconsolidated, the undrained strength determined in typical laboratory tests (triaxial, unconfined compression) often leads to an overestimation of safety against sliding. This is due to the dilatant nature of the clay mass. In general, the analysis should be made drained, providing special attention to discontinuities and to any evidence of previous shearing on them. The determination of the pore pressures therefore becomes a crucial aspect of the safety analysis. Aznalcóllar is a good example in this regard.

4.8 Advanced Topics

The reduction of available strength along the sliding plane due to progressive failure is a main reason for the catastrophic failure of Aznalcóllar. Progressive failure was identified as a mechanism leading to instability of overconsolidated clays (Skempton, 1964; Terzaghi and Peck, 1967; Bjerrum, 1967; Bishop, 1967, 1971). A review of the subject has been presented by Jardine *et al.* (2004). The propagation of rupture surfaces in idealized geometries is described in the work of Palmer and Rice (1973), Rice (1973), Chowdhury (1978) and Puzrin and Germanovich (2005). Early attempts to include progressive failure in limit equilibrium analysis are presented by Pariseau (1972), Gates (1973) and Lo and Lee (1973). Additional contributions using the finite element method are described in Yoshida *et al.* (1990), Potts *et al.* (1990), Dounias *et al.* (1996), and Potts *et al.* (1997). Stark and Eid (1994) reviewed a number of case histories involving first-time slides in stiff fissured clays, reported the value of the mobilized strength at failure, and compared it with two strength values: the "fully softened" and the residual value. In most instances, the mobilized strength at failures lies about midway between the two values. This was also the case of Aznalcóllar. The "fully softened strength" is defined as the peak drained strength of the reconstituted, normally consolidated material. It may easily be found in the laboratory by remoulding the intact material.

However, one should be cautious when trying to select an operational strength of hard clays because it is a function of the rate of degradation of post-peak strength and also on the strain mobilization along the sliding surface.

Well-documented case histories associated with progressive failure have been described by Cooper (1996), regarding the Selborne slope failure experiment and by Skempton (1985) and by Potts *et al.* (1990) regarding the Carsington dam failure.

References

Alonso, E.E. and Gens, A. (2006) Aznalcólar dam failure. Part 1: Field observations and material properties. *Géotechnique* 56 (3), 165–183.

Bishop, A.W. (1967) Progressive failure-with special reference to the mechanism causing it. *Proceedings of the Geotechnical Conference.* Oslo, Norway. Norwegian Geotechnical Institute. 2, 142–150.

Bishop, A.W. (1971) The influence of progressive failure on the method of stability analysis. *Géotechnique* 21, 168–172.

Bjerrum, L. (1967) Progressive failure in slopes of over-consolidated plastic clays and clay shales. *Journal of the Soil Mechanics and Foundations Division ASCE* 93, 3–49.

Chowdury, R.N. (1978) *Slope Analysis*. Elsevier, Amsterdam.

Cooper, M.R. (1996) The progressive development of a failure slip surface in over-consolidated clay at Selborne, UK. *Proceedings of the 7th International Symposium on Landslides*. Trondheim, Norway. Taylor & Francis. 2, 683–688.

Dounias, G.T., Potts, D.M. and Vaughan, P.R. (1996) Analysis of progressive failure and cracking in old British dams. *Géotechnique* 46 (4), 621–640.

Ericksson, N. and Adamek, P. (2000) The tailing pond failure at the Aznalcóllar mine. Spain. *Proceedings of the 6th International Symposium in Environmental Issues and Waste Management in Energy and Mineral Production.* Calgary, Canada. Singhal & Mehrota (eds). 8 pp.

Gens, A. and Alonso, E.E. (2006) Aznalcóllar dam failure. Part 2: Stability conditions and failure mechanism. *Géotechnique* 56 (3), 185–201.

Gates, R.H. (1973) Progressive failure model for clay shale. *Applications of the Finite Element Method in Geotechnical Engineering.* Vicksburg, USA. Desai (ed). 327–347.

Jardine, R.J., Gens, A., Hight, D.W. and Coop, W.R. (2004) Developments in understanding soil behaviour. Advances in geotechnical engineering. *The Skempton Conference.* London, UK. Jardine, Potts & Higgins (eds). Thomas Telford. 1, 103–206.

Lo, K.Y. and Lee, C.E. (1973) Stress analysis and slope stability in strain softening materials. *Géotechnique* 23 (1), 1–11.

Morgenstern, N.R. and Price, V.E. (1965) The analysis of the stability of general slip surfaces. *Géotechnique* 15 (1), 79–93.

Moya, J. (2004) Determination of the failure surface geometry in quick slides using balanced cross section techniques–Application to Aznalcóllar tailings dam failure. *Lecture Notes in Earth Sciences: Engineering Geology for Infrastructure Planning in Europe*. A European Perspective. Hack, Azzam & Charlier (eds). 104, 414–421.

Palmer, A.C. and Rice, J.R. (1973) The growth of slip surfaces in the progressive failure of over-consolidated clay. *Proceedings of the Royal Society of London*, A 332, 527–548.

Pariseau, W.G. (1972) Elastic-plastic analysis of pit slope stability. *Applications of the Finite Element Method in Geotechnical Engineering*. Vicksburg, USA. Desai (ed). 349–384.

Potts, D.M, Dounias, G.T. and Vaughan, P.R. (1990) Finite element analysis of progressive failure of Carsington embankment. *Géotechnique* 40 (1), 79–101.

Potts, D.M., Kovacevic, N. and Vaughan, P.R. (1997) Delayed collapse of cut slopes in stiff clay. *Géotechnique* 47 (5), 953–982.

Poulos, H.G. and Davis, E.H. (1974) *Elastic Solutions for Soil and Rock Mechanics*. Wiley. New York.

Puzrin, A.M. and Germanovich, L.N. (2005) The growth of shear bands in the catastrophic failure of soils. *Proceedings of the Royal Society* of *London*, A 461 (2056), 786–795.

Rice, J.R. (1973) The initiation and growth of shear bands. *Proceedings of the Symposium on Plasticity and Soil Mechanics*. Cambridge, UK. Palmer (ed). 263–274.

Scott, R.F. (1963) *Principles of Soil Mechanics*. Addison-Wesley. Massachussetts.

Skempton, A.W. (1964) Long term stability of clay slopes. *Géotechnique* 14 (2), 77–101.

Skempton, A.W. (1985) Geotechnical aspects of the Carsington Dam failure. *Proceedings of the 11th International Conference in Soil Mechanics and Foundation Engineering*. San Francisco, USA. 5: 2581–2591.

Stark, T.D. and Eid, H.T. (1994) Slope stability analyses in stiff fissured clays. *Journal of Geotechnical and Geoenvironmental Engineering* 123 (4), 335–343.

Terzaghi, K. (1943) *Theoretical Soil Mechanics.* Wiley. New York

Terzaghi, K. and Peck, R.B. (1967) *Soil Mechanics in Engineering Practice* (2nd edition). John Wiley. New York.

Yoshida, N, Morgenstern, N.R. and Chan, D.H. (1990) Finite element analysis of softening effects in fissured, overconsolidated clays and mudstones. *Canadian Geotechnical Journal* 28, 51–61.

PART III

DYNAMICS OF FAILURES

Chapter 5

Thermo-Hydro-Mechanics of a Rapid Slide: Vaiont Landslide, Italy

TABLE OF CONTENTS

E.E. Alonso et al., *Geomechanics of Failures. Advanced Topics*,
DOI 10.1007/978-90-481-3538-7_5, © Springer Science+Business Media B.V. 2010

Chapter 5
Thermo-Hydro-Mechanics of a Rapid Slide: Vaiont Landslide, Italy

5.1 Introduction

On October 9, 1963, a huge mass of rock, on the left bank of Vaiont reservoir, broke loose, accelerated and invaded the bottom of the valley at high speed. In seconds, the reservoir water was projected against the slopes of the opposite margin of the valley, where it reached a height of 250 m over the original level of the reservoir. Then the reservoir water turned downstream, over the crest of Vaiont arch dam, without breaking it, and flew along the river valley. The flood destroyed the village of Longarone and caused an estimated death toll of more than 2,000 lives.

The small village of Caso, located on the right margin of the Vaiont valley, 260 m above the reservoir level, barely escaped destruction. A Caso villager provided a vivid account of the failure:

> *"Rain fell heavily. At 22.15 hours a strong noise, as of rolling rocks, awoke me up. At 22.40 hours an extremely strong wind shook the house and broke the windows; suddenly the house roof was lifted and water and rocks invaded the house. The noise was frightful. In a few seconds the wind stopped and the valley remained calm"* (Valdés Díaz-Caneja, 1964).

This landslide, one of the largest known in historic times, has attracted continuous attention of geotechnical engineers, mainly because of the unexpected high velocity reached by the moving mass. In fact, the tragic consequences of the failure are directly attributed to this velocity. The main question is: why did Vaiont slide reached an estimated velocity of 30 m/s? Such a velocity can only be explained if a total loss of strength occurs at the sliding surface.

Unstable slopes around reservoirs are a common occurrence. They raise concern to dam designers and public authorities because of the risk associated with a rapid slide motion, just as in Vaiont. One is led to think that unless the dynamics of Vaiont motion are understood with some degree of confidence, limited progress will be achieved in predicting the risk of similar potentially dangerous situations.

In Chapter 2, an attempt was made to determine the run-out of the Vaiont landslide taking, as a reference model, the two-wedge representation of Cross-section 5 (Fig. 2.15). Starting at a condition of near equilibrium at $t = 0$, it was assumed that the strength of the plane separating the two wedges could degrade as shearing displacements developed along this plane during the motion. It was found that even in the extreme case of a fast and complete loss of cohesion acting on this plane (an unlikely event), the slide maximum velocity did not exceed 4.5 m/s. In order to explain the estimated high velocities of the slide (30 m/s), a consistent mechanism or physical process, leading to a total loss of basal shear strength has to be found.

The favourite explanation in a number of published contributions on the subject is associated with the development of frictional heat at the sliding surface. In some papers (Uriel and Molina, 1977; Nonveiller, 1987), the frictional heat is assumed to take pore water to the equilibrium state between liquid and vapour phases. In Uriel and Molina (1977), the phase diagram of water provides a criterion to find the water/vapour pressure. Nonveiller (1987) assumes a linear decrease of rock strength with temperature in the shear zone. In other approaches (Hendron and Patton, 1985; Voigt and Faust, 1982; Vardoulakis, 2002), the increase in pore pressure is related to the dilation of pore water as temperature increases and to temperature-induced plastic collapse of the shearing band (in the case of Vardoulakis).

In all cases, the fluid pressure developed at the sliding surface reduces the effective normal stress and, hence, the available strength. Before proceeding further, consider the results of two experiments: a simple one, which could be performed in any laboratory, and a complex "in situ" test.

5.1.1 A simple laboratory heating experiment

The idea is simply to heat a piece of saturated clayey rock in a microwave oven (Fig. 5.1). In the experiment performed, a thermocouple temperature sensor was inserted into a specimen of Opalinus clay, which had been maintained in a humid chamber to ensure saturation. Opalinus clay is a low permeability soft clayey rock of marine origin. Clay minerals (illite, illite-smectite mixed layers, chlorite and kaolinite) dominate its mineralogical composition (40 to 80%). Quartz, calcite, siderite, pyrite, feldspar, and organic carbon are also present. Natural porosity varies between 4 and 12% (Bossart *et al.*, 2002). Pore water has a concentration of 20 g/l of sodium chloride.

Permeability coefficients (Darcy) varying between 0.8×10^{-13} m/s and 7.3×10^{-13} m/s, Young's modulus ranging between 1,000 and 7,000 MPa and uniaxial compressive strength varying between 9 and 18 MPa have been reported for this clay shale by several authors (Thury and Bossart, 1999; Bock, 2001; Muñoz, 2007) on the basis of "in situ" and laboratory tests.

Figure 5.1a shows the piece of rock before heating. A thermal pulse having a nominal power of 1,400 watts was applied during 40 s. The recorded temperature is shown in Figure 5.2. The specimen broke, accompanied by a clearly audible cracking noise, shortly before the end of the application of the heating pulse. At that time, the temperature reached values in excess of 170°C (Fig. 5.2). The shale specimen cracked in an explosive manner and was reduced to small fragments.

The following explanation can be given for this phenomenon. When the temperature of a saturated porous material increases, the solid matter, as well as the water in pores, dilates. Probably, local equilibrium of temperature is achieved soon and therefore the temperatures of water and solid skeleton will be essentially equal.

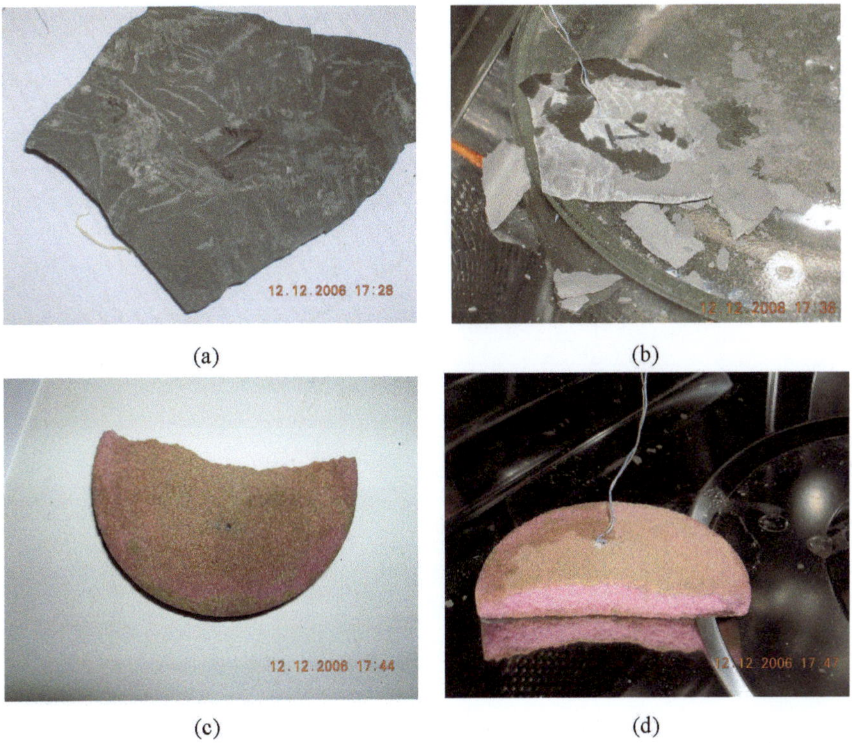

Figure 5.1 Heating experiment: (a) saturated fragment of Opalinus clay before heating; (b) the fragment, highly fissured and partially broken after heating in a microwave (cables indicate the position of the inserted thermocouple); (c) saturated porous stone before heating; (d) porous stone after heating.

The volume of pore water and solid skeleton will increase in direct proportion to their thermal dilation coefficients, β_w and β_s, respectively. The associated volumetric strains, for a common change in temperature, $d\theta$, can be written

$$d\varepsilon_{vw} = -\frac{dV_w}{V_w} = -\beta_w d\theta, \tag{5.1a}$$

$$d\varepsilon_{vs} = -\frac{dV_s}{V_s} = -\beta_s d\theta, \tag{5.1b}$$

where V_w and V_s are the volumes occupied by water and solid particles, respectively. β_w is substantially higher than β_s. Typical values for β_w and β_s are $3.4\times10^{-4}\,(^{\circ}\mathrm{C})^{-1}$ and $3.0\times10^{-5}\,(^{\circ}\mathrm{C})^{-1}$. Volumetric strains calculated through Equations (5.1a,b) for the range of temperatures 0–100°C were plotted in Figure 5.3, assuming a reference situation (zero strain) for θ = 4°C. Water dilates almost one order of magnitude more than solid particles. The thermal dilation of water and solid will result in an internal volumetric expansion. The soil expansion is explained by a decrease in effective stress. Therefore, in a saturated porous medium, if the total stress does not change, pore water pressure has to increase in order to reduce the effective stress. The increase in pore pressure will be proportional to the soil or rock stiffness. In the absence of external stresses, tensile effective stresses will develop. They may be able to overcome the tensile strength of the soil/rock and lead to a failure in tension, as observed in the photograph in Figure 5.1b.

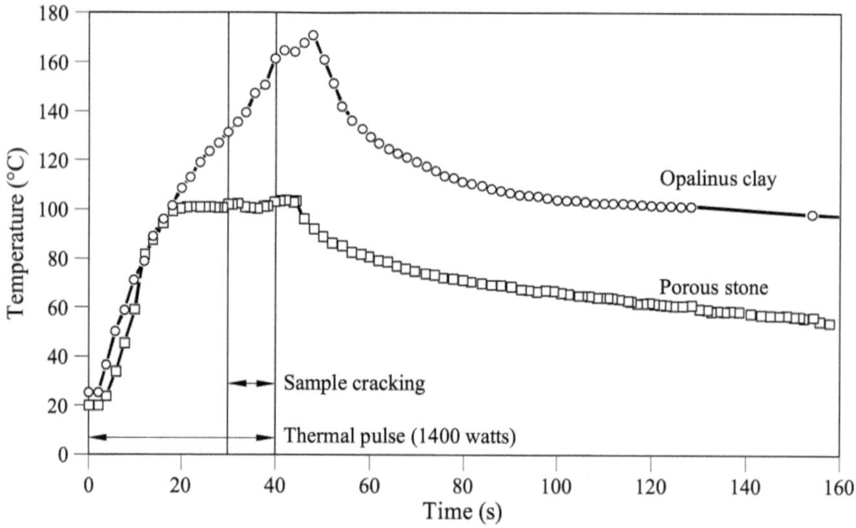

Figure 5.2 Recorded temperature during the two experiments performed in the microwave oven. A thermal pulse with a nominal intensity of 1,400 watts was applied during 40 s.

The volumetric strains plotted in Figure 5.3 are far from being negligible. For instance, for an increase of temperature from 4 to 50°C, a water volumetric strain of 1.5% is derived from Figure 5.3. It is concluded that the heat-induced expansive strain may cause a substantial increase in water pressure in an impervious stiff rock.

A simple explanation[1] for this increase in pore pressure can be given with the help of Figure 5.4, which shows a saturated pore. The rock or soil skeleton around the pore is represented by a thick spherical elastic shell. Holes in radial directions connect the inner pore water with neighbouring pores. In this representation, the skeleton stiffness is controlled by the thickness and the modulus of the shell material. The number and diameter of radial holes define the material's permeability. As a result of heating, pore water pressure will increase. In parallel

[1] A precise formulation of the effects of thermal dilation on the volumetric deformation and, eventually, on the development of pore water pressures in a saturated soil will be given later.

with the development of water pressure, a dissipation process will start as water begins to flow through radial holes. Therefore, for a given rate of increase of temperature, the attained pore water pressure will be the result of two competing mechanisms: the rate of increase of water volume, directly related to the rate of increase of temperature and the rate of dissipation governed by the permeability of the porous material (and also by the rock stiffness, in a process similar to the more familiar consolidation phenomenon).

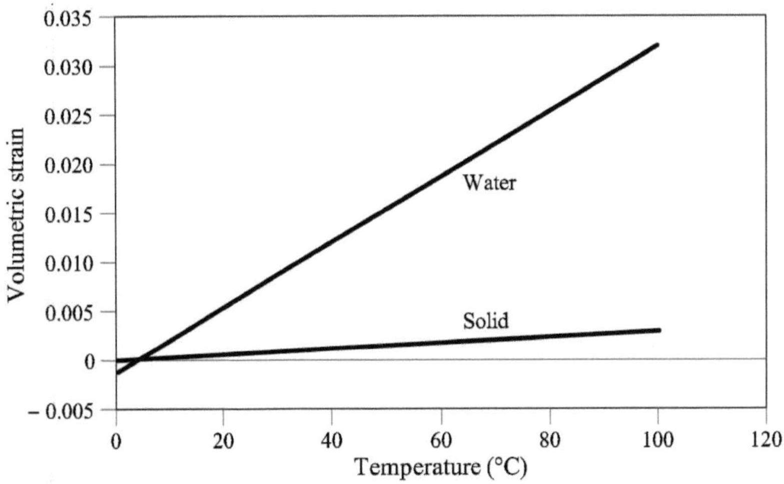

Figure 5.3 Volumetric strains of water and solid particles induced by temperature changes. $\beta_w = 3.4\times10^{-4}\,(°C)^{-1}$; $\beta_s = 3.0\times10^{-5}\,(°C)^{-1}$. Expansion is considered positive in plotting this graph.

For a given rate of temperature increase, the lower the soil or rock permeability and the stiffer the soil or rock, the higher the pore water pressure developed. Stiff clays and, particularly, clayey rocks are therefore prone to develop significant temperature-induced pore water pressures.

Note that the simple model of Figure 5.4 predicts that the pore pressure induced by the application of an external load decreases as soil or rock skeleton stiffness increases. In classical one-dimensional soil consolidation theory, the implicit assumption is that the soil skeleton has very low stiffness compared with the stiffness of water, and this implies that the external load is fully resisted by pore water: the skeleton spheres in Figure 5.4 are made of a very soft material.

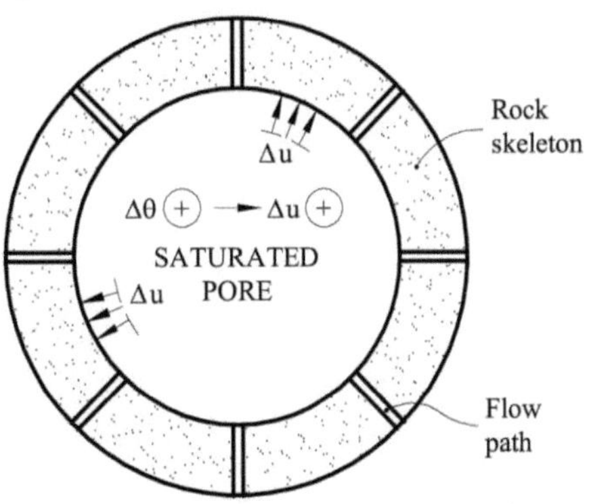

Figure 5.4 A saturated pore develops a positive pressure when temperature increases.

The pore water pressure was not measured in the simple experiment described but, interestingly, a small amount of liquid water – presumably escaped from the specimen – was also observed on the floor of the oven after the broken rock fragments were removed.

A second experiment, with a totally different material, a discarded highly pervious porous stone (Fig. 5.1c), was also run. The measured temperature is shown in Figure 5.2. No cracking noise was heard during heating and the specimen remained intact. Some water was also seen to escape from the stone. Unlike the previous experiment, the temperature record in this case showed an interesting behaviour: when the temperature measured by the thermocouple sensor reached 100°C, it remained constant at this temperature during the application of the power pulse. The water behaved as is to be expected in a free volume of water at atmospheric pressure: when the vapourization (boiling) temperature is reached, water evapourates in the pores and the boiling temperature remains constant, at 100°C, because the heat input is "spent" in vapourizing the remaining liquid water.

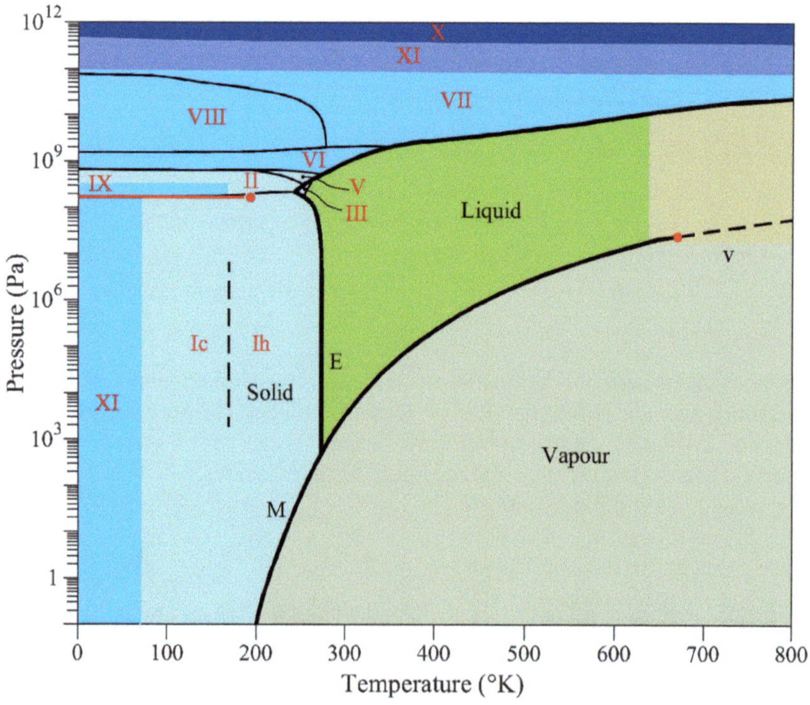

Figure 5.5 Phase diagram of water. Roman numerals indicate different types of ice. M, E and V stand for the average atmospheric conditions at the surface of Mars, Earth, and Venus, respectively (from London Southbank University website).

Another interesting information of these experiments is that the pore water in the shale specimen increased its temperature well beyond 100°C (it reached a peak value of 171°C (!) with no symptoms of decreasing during the power input phase).

Pore water in the claystone is adsorbed in a significant proportion by clay minerals and this prevents its vapourization.

The phase diagram of water provides additional information on the conditions leading to the vapourization of water (Fig. 5.5). At increasing pressure, the temperature for vapourization also increases. For instance, at a pressure of 120 m of water (1.2 MPa), which is the pore pressure likely to be acting at the lower horizontal sliding surface of Section 5 of Vaiont at the beginning of the failure (see Chapter 4) the boiling water temperature raises to 200 °C approximately (remember that °K = °C + 273°). The combination of the two effects, water adsorption by the clay minerals of the rock and the initial prevailing pore water pressure, implies that the sliding surface may reach fairly large temperatures before water is able to vapourize.

Pore pressures were not measured in the simple experiments reported and the explanation advanced for the failure of the clay shale may not be accepted by the reader. Consider, however, the following large-scale experiment.

5.1.2 An expensive field experiment

Deep geological disposal is an option favoured by several countries to store high level nuclear waste. A typical design is to locate the heat-emitting nuclear canisters in excavated galleries of massive and impervious rock, such as Opalinus clay, a clay shale common in northwest Switzerland. A ring of impervious bentonite is placed around the canister to improve isolation. One of the issues in this design is to investigate the long-term performance of natural rock, exposed to an increase in temperature as a result of the heat generated by the nuclear waste. The large-scale Heating Experiment (HE), performed in the Monterri underground research laboratory (Switzerland), addresses this aspect of nuclear waste disposal research. The experiment is described in detail in EUR (2006) and in Muñoz (2007).

The scheme given in Figure 5.6a summarizes the concept of the experiment. A cylindrical heater – which simulates the waste – is located in a centred position in a vertical borehole (30 cm of diameter) excavated in Opalinus clay from the floor of a tunnel. A ring of compacted bentonite blocks was placed around the heater. Piezometers and temperature sensors were located at different radial distances and depths below the floor of the niche where the experiment was located ($z = 0$). The temperature response of sensors located at increasing radial distances is shown in Figure 5.6b. Maximum temperature at the bentonite-borehole wall contact ($r = 0.05$m) was limited to 100°C.

Pore pressure sensors were installed at points A_1 and A_2 (Fig. 5.6a), located at a radial distance of 0.65 m from the axis of the borehole, at two different elevations ($z = -5$ m and $z = -6.5$ m). As temperature increased (at a rate of 0.25°C/day) until it reached a value of 40°C in sensor A_1, pore water pressures also increased at measured rates of 0.012 MPa/day and 0.007 MPa/day in the two sensors, until they reached maximum values of 1.1 and 0.65 MPa respectively. Note that a substantial pressure peak developed before pore pressure began to decrease, when the rate of temperature increase slowed down. The low permeability of Opalinus clay explains the continuous accumulation of pore pressure due to the relatively slow rate of increase of temperature. When the (permeability controlled) dissipation rate of excess water pressure dominated the process, the pore water pressure began to drop, at an essentially constant temperature.

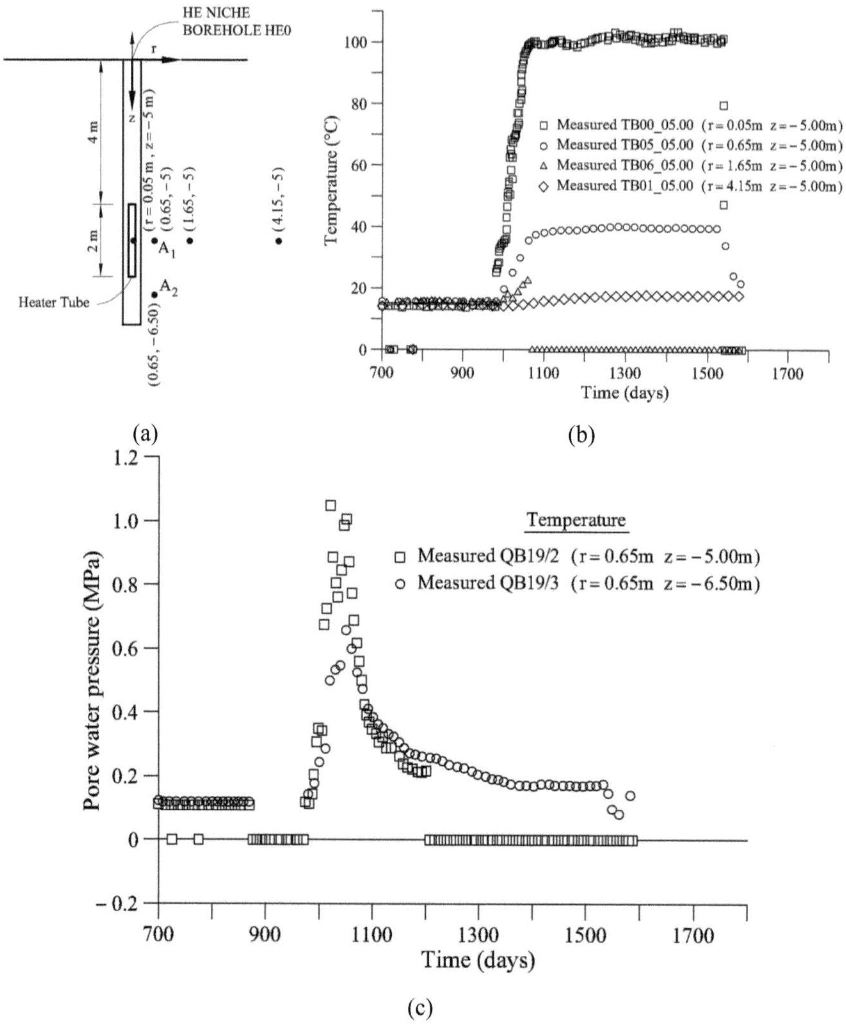

Figure 5.6 Field heating experiment of Opalinus clay: (a) schematic representation of the borehole, heater and instrumented points; (b) recorded temperature; (c) pore water pressures at Point A_1 and A_2. Heating HE Experiment, Monterri, Switzerland (Muñoz, 2007).

The maximum excess water pressure recorded in this experiment (0.9 MPa) is relatively large in absolute terms. Such water pressure is equivalent to the weight per unit area of a column of rock with a height of 40 m (if the rock had a bulk specific weight of 22.5 kN/m^3). The base of such a column of Opalinus clay, if heated in the location of Piezometer QB19/3 in Figure 5.6c, will reach a zero vertical effective stress and, therefore, it will not be able to develop any frictional shear strength.

This chapter discusses the conditions leading to this situation in the case of Vaiont and their consequences in practice.

5.1.3 Summary of main points

a) When temperature increases in a relatively impervious and saturated porous material, pore water pressures will develop. They will reduce the prevailing effective stress. Negative effective stresses may develop if the porous rock exhibits a tensile strength and, eventually, a fragile splitting type of failure may occur in unloaded specimens.
b) In saturated pervious granular materials liquid pore water under atmospheric pressure will not reach, if heated, temperatures in excess of 100°C (because of local equilibrium, this temperature will also be the temperature of the entire rock, provided that some free water remains in the specimen). Under similar conditions, the pore water of a low porosity clayey rock may reach significantly higher temperatures, with no clear evidence of liquid-vapour phase transition of the pore water.
c) The temperature for the liquid-vapour change of phase increases for free water as water pressure increases. For a water pressure of 1.2 MPa (a column of 120 m of water), water boils at approximately 200°C.
d) Points b) and c) suggest that the shear surface of Vaiont, located in clayey materials of high plasticity, may undergo relatively high temperatures, in excess of 200 °C, without reaching a vapourization state.

5.2 The Problem

A common observation in translational and rotational slides is that deformations are confined to sliding surfaces of negligible thickness. Direct observations of sliding surfaces in clayey materials indicate that their thickness is very small, typically in the range of a few millimetres. One example is given in Figure 5.7, which shows a portion of the sliding surface of Cortes landslide (Alonso *et al.*, 1992). The sliding surface was easily identified when it was exposed after a large excavation, because of its greenish-gray colour in contrast with the brown tonalities of the marl layer, 2 m thick, where it was embedded. Massive limestone strata, which essentially slid as a rigid body, covered the marl layer. The thickness of the striated layer ranged between 3 and 5 mm.

The thickness of shear bands has been reported by several authors (Morgenstern and Tchalenko, 1967; Roscoe, 1970; Vardoulakis, 1980; Scarpelli and Wood, 1982; Desrues, 1984). An important conclusion of the basic research is that shear band thickness is related to a characteristic grain size. For instance, Vardoulakis (2002) proposes a value $e \approx 200d_{50\%}$ for clays. The grain size analysis of specimens recovered from the Vaiont sliding surface (Hendron and Patton,

1985; Tika and Hutchinson, 1999) indicates that $d_{50\%} \approx 0.01$ mm. The reported direct observation at the Cortes slide is not far from the thickness suggested by the preceding relationship.

(a)

(b)

Figure 5.7 (a) Sliding surface of Cortes landslide showing motion grooves; (b) view of the sliding surface in cross-section. The upper layer of gray clay, overlying the brownish lower marl, was identified as the sliding surface.

Vaiont landslide was significantly bigger than the Cortes slide, but they had some similarities. In both cases, rigid and massive limestone and marl banks slid on a fairly continuous layer of clay. It is expected, however, that the sliding surface itself had a reduced thickness, probably a few millimetres, as in Cortes.

Consider now in Figure 5.8a, a representative cross-section of Vaiont. The clay stratum at the base of the slide was reported (Hendron and Patton, 1985) to have a thickness in the order of 1 m (Fig. 5.8b). The shear band proper will be located within the clay layer (Fig. 5.8c). Its thickness is many orders of magnitude smaller than the horizontal and vertical dimensions of the slide. If the slide moves as a rigid body with a velocity v_{max}, shear straining, which will be concentrated on the shear band, will induce an average shearing strain rate of

$$\dot{\gamma} = \frac{v_{max}}{2e}, \tag{5.2}$$

where $2e$ is the thickness of the shear band. Therefore, during the sliding motion, all the straining work will be concentrated inside the band. The volumetric deformation of the clay material, which constitutes the band, will be very small compared with the extremely large shear deformations induced by sliding on a thin clay band. Thus, the rate of work input per unit volume of band material will be essentially given by

$$\dot{W} = \tau_f \dot{\gamma} = \frac{\tau_f v_{max}}{2e}, \tag{5.3}$$

where τ_f is the shear strength offered by the shear band. This work input will be transformed entirely into heat. Therefore, the band will increase its temperature and, in view of the tests previously discussed, a pore water pressure in excess of that initially existing will develop.

It has been argued before that excess pore pressure is essentially caused by the thermal dilation of the water. Therefore, despite its potential large effect in modifying effective stresses, the absolute amount of increment of water volume in the band will be very small. Its dissipation will take place in the immediate vicinity of the band. In other words, the band and its "zone of influence" will have a small thickness (Fig. 5.8d) similar to the thickness of the band itself.

It then becomes reasonable to assume that, for the purposes of investigating the behaviour of pore pressures in the band and its vicinity, the band is essentially a planar feature located within an infinite domain. The lateral extent of this band is very large compared with its thickness and, in addition, points within the band are similar to each other. Water and energy transfer out of the band will take place in the direction normal to the band. The problem of the interaction of the band and its surroundings becomes a one-dimensional problem in which the spatial coordinate (z) is directed normal to the band plane (Fig. 5.8d).

The equation of the slide motion is given in Chapter 2 for the two-wedge model (Eq. (2.23)). The slide velocity depends on the water pressures existing on the sliding surface (terms P_{w1}, Pw_2 and P_{wf} in Eqs. (2.27)). But now pore water pressures will not depend only on the hydrostatic water conditions assumed in

Chapter 2 but also on the additional pore pressures developed in the band as a result of its heating. These pore pressures depend on the work input into the shear band and therefore on the slide velocity (Eq. (5.3)) which is the unknown variable of the problem.

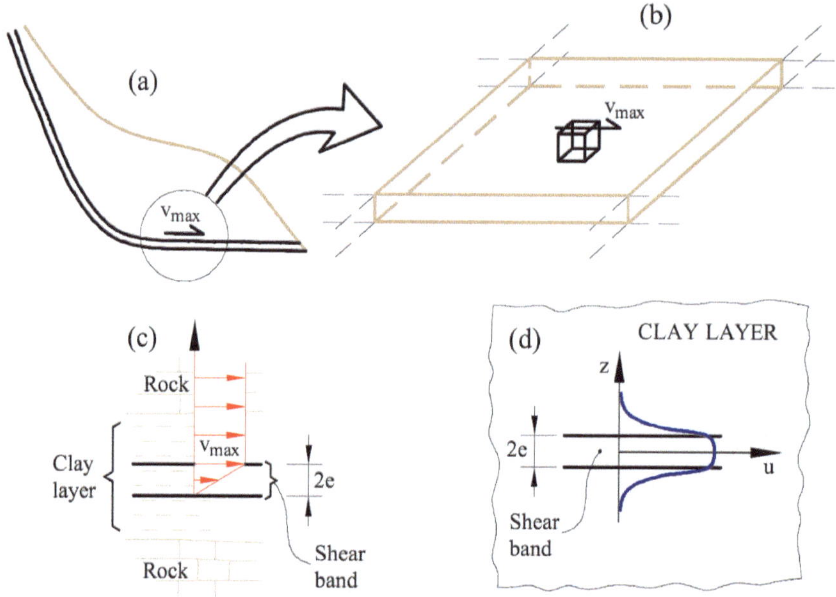

Figure 5.8 The sliding surface: (a) "in situ" conditions; (b) representative element of the sliding surface; (c) shear band; (d) local axis in the shear band.

A procedure to find excess pore pressures from a given heat (or strain work) input has first to be found. Then the calculated pore pressures will be used to solve the equation of the slide motion.

In classical consolidation theory, pore pressures induced by applied stresses are the solution of a partial differential equation which expresses the condition of mass balance of water flowing in a deformable soil. In our heat-driving process we will need an additional balance (or conservation) equation, namely the condition of energy conservation.

When consolidation theory is derived in most soil mechanics textbooks, only the balance equation of water is used as a starting point despite the fact that solid matter (soil particles) also moves. The approximation is perfectly justified in most applications but in our case, for the reasons explained later, it will be convenient to also add the mass balance of solids to the remaining conservation laws.

Therefore, the problem of finding pore water pressures in the band when it is sheared by means of the application of a boundary velocity will be approached by formulating the three conservation equations just mentioned (solid, water, and energy). They will be written for a general (three-dimensional) case but the solution will be found for the one-dimensional case previously described. The solution of this problem will also enable the calculation of temperatures in the shear band. The final step will be the solution of the equation of the slide motion.

Summarizing the main assumptions, deformation and heat generation will take place within the shear band. The band material is a saturated porous deformable material and the remaining sliding masses will move as rigid bodies.

5.3 Balance Equations in the Shear Band

5.3.1 Solid and water

Consider a shear band of indefinite length (L) and thickness ($2e$) (Fig. 5.9). Since $L >> e$, the excess pore pressure, $u_w(z,t)$, temperature, $\theta(z,t)$, and velocity, $v(z,t)$ are assumed to be exclusively a function of the position normal to the band direction (z) and time (t). A common temperature is considered for solid particles and pore fluid. This is a result of the assumption of local thermal equilibrium between both species (solid and water).

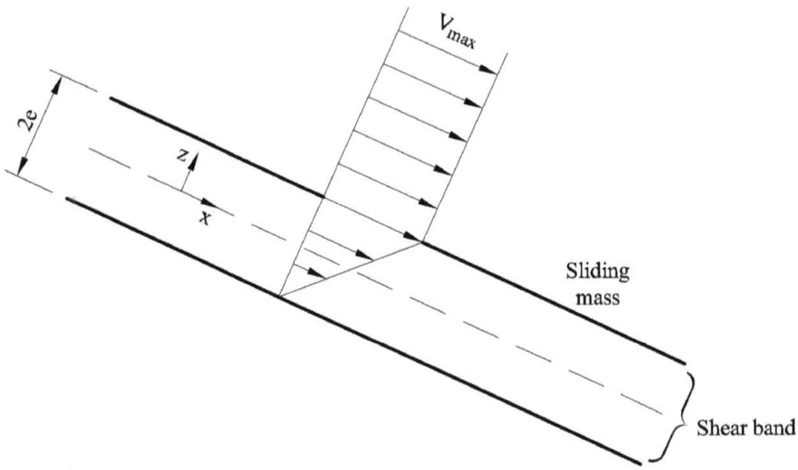

Figure 5.9 Geometry of the planar shear band.

Solid mass balance was already derived in Chapter 3 for general three-dimensional conditions (Eq. 3.22). The water mass balance was found (Eq. 3.25) for a partially saturated soil. Here pores will be assumed to be full of water. Therefore, Equation (3.25) becomes

$$n\frac{\mathrm{D}\rho_w}{\mathrm{D}t}+\rho_w\frac{\mathrm{D}n}{\mathrm{D}t}+\rho_w n\,\mathrm{div}(\mathbf{v})+\mathrm{div}(\rho_w\mathbf{q})=0. \tag{5.4}$$

The rate of change in porosity, $\mathrm{D}n/\mathrm{D}t$, can be expressed in terms of changes in solid density and in terms of the skeleton deformation through the mass balance of solids (Eq. (3.22)). Substitution of $\mathrm{D}n/\mathrm{D}t$ from Equation (3.22) into the water mass balance Equation (5.4) leads to

$$n\frac{\mathrm{D}\rho_w}{\mathrm{D}t}+\rho_w\left[\frac{(1-n)}{\rho_s}\frac{\mathrm{D}\rho_s}{\mathrm{D}t}+(1-n)\mathrm{div}(\mathbf{v})\right]+ \\ \rho_w n\,\mathrm{div}(\mathbf{v})+\mathrm{div}(\rho_w\mathbf{q})=0 \tag{5.5}$$

and, finally, to

$$\frac{n}{\rho_w}\frac{\mathrm{D}\rho_w}{\mathrm{D}t}+\frac{(1-n)}{\rho_s}\frac{\mathrm{D}\rho_s}{\mathrm{D}t}+\mathrm{div}(\mathbf{v})+\frac{1}{\rho_w}\mathrm{div}(\rho_w\mathbf{q})=0, \tag{5.6}$$

which provides the mass conservation condition for water and solid.

Constitutive equations

In order to proceed, constitutive equations should now be considered. Let us start with the (material) rate of change of solid and water densities. Solid grains will be assumed to be incompressible against stress changes but not against temperature (θ) changes. Thermal dilation implies the increase of volume, V_s, of a given constant mass. Therefore, the change in density, ρ_s, associated with a change in volume is written as

$$\mathrm{d}\rho_s=-\rho_s\frac{\mathrm{d}V_s}{V_s}=-\rho_s\beta_s\mathrm{d}\theta, \tag{5.7}$$

where the linear dilation volumetric strain laws given in Equation (5.1) have been used. β_s is the thermal expansion coefficient for solid particles, which will be accepted as a material parameter independent of temperature. The negative sign indicates that temperature increments induce a reduction in density.

Equation (5.7) provides, by simple integration, the following state equation for the solid:

$$\rho_s=-\rho_s^0\exp\left[\beta_s\left(\theta-\theta_0\right)\right], \tag{5.8}$$

where ρ_s^0 is the density of solid particles at reference temperature, θ_0.

Differentiating of Equation (5.8) leads to

$$\frac{\mathrm{D}\rho_s}{\mathrm{D}t}=-\beta_s\rho_s\frac{\mathrm{D}\theta}{\mathrm{D}t}. \tag{5.9}$$

Water density depends on its current pressure, p_w, and temperature, θ. None of these effects are usually found in common geotechnical applications. However, in our particular problem, pressures and temperatures may reach unexpectedly high values. Already discussed, by means of introductory tests, is the fundamental effect of temperature to control pore water pressure of impervious porous materials.

The following state equation is assumed for water:

$$\rho_w = \rho_w^0 \exp\left[\alpha_w\left(p_w - p_w^0\right) - \beta_w\left(\theta_w - \theta_0\right)\right]. \tag{5.10}$$

This expression is formally equivalent to the state equation of solid (Eq. (5.8)). ρ_w^0 is the reference water density at the reference temperature (θ_0) and reference liquid pressure (p_w^0). α_w and β_w are the coefficients of compressibility and thermal expansion, respectively. These coefficients have been accepted to be constant. Differentiating of Equation (5.10) leads to

$$\frac{D\rho_w}{Dt} = \alpha_w \rho_w \frac{Dp_w}{Dt} - \beta_w \rho_w \frac{D\theta}{Dt}. \tag{5.11}$$

If Equations (5.9) and (5.11) are substituted in Equation (5.6), the following expression is obtained for the mass conservation of solid and water:

$$-\left[n\beta_w + (1-n)\beta_s\right]\frac{D\theta}{Dt} + n\alpha_w \frac{Dp_w}{Dt} + \mathrm{div}(\mathbf{v}) + \frac{1}{\rho_w}\mathrm{div}(\rho_w \mathbf{q}) = 0 \tag{5.12}$$

The first term is a “source” term due to the thermal expansion of liquid and solid; the second term describes the volume change of water associated with change in water pressure; the third term represents the volume change of the skeleton; and the fourth term provides the volume change associated with the flow of water.

The hydrostatic component of pore pressure, which depends on the position of the water table, will not change within the short interval of the slide and therefore the time derivative of p_w will depend only on the excess pore pressure u_w

$$\frac{Dp_w}{Dt} = \frac{Du_w}{Dt}. \tag{5.13}$$

The mass balance equations derived above will be applied to the shear band where “oedometric” conditions can be assumed, as explained before. Therefore, under elastic conditions, the volumetric strain can be estimated from the one-dimensional compressibility coefficient, m_v, and the increment of (normal to the band) effective stress,

$$\mathrm{div}(\mathbf{v}) = -\frac{\partial \varepsilon_{vol}}{\partial t} = -m_v\left(\frac{\partial \sigma_n}{\partial t} - \frac{\partial p_w}{\partial t}\right) = -m_v\left(\frac{\partial \sigma_n}{\partial t} - \frac{\partial u_w}{\partial t}\right), \tag{5.14}$$

where Equation (3.21) was used. In Equation (5.14), σ_v is the total stress acting in a direction normal to the shear band. This stress will change somewhat during motion due to changes in slide geometry (see Fig. 2.15 in Chapter 2). Time variation of hydrostatic pressure can be neglected with respect to changes of excess pore pressures. Therefore, p_w can be replaced by u_w in Equation (5.14).

The final term in Equation (5.12) refers to flow through pores due to the head gradient (Darcy’s law). A generalized Darcy law for compressible fluid describes the relative flow velocity $\mathbf{q}$ in terms of gradients of pore water pressure and the gradient of elevation as follows:

$$\mathbf{q} = -\frac{k}{\rho_w g}\left[\mathbf{grad}(p_w) + \rho_w g\ \mathbf{grad}(z_g)\right] \approx -\frac{k}{\rho_w g}\frac{\partial u_w}{\partial z}, \tag{5.15}$$

where k is the hydraulic conductivity (the term permeability will also be used), which will be assumed to be constant, and z_g is the vertical coordinate.

Since the analysis is one-dimensional in a direction normal to the shear band (z-direction), the gradient is simply the derivative with respect to z. The flow due to gradients of hydrostatic pressure and gradients of level (z_g term) can be neglected with respect to changes of pore water pressure. In addition, the spatial variation of hydrostatic pore water pressure can be neglected during the slide and p_w can be replaced by u_w. Therefore, Darcy's flux depends only on the excess pore pressure (u_w).

Introducing Equations (5.13), (5.14) and (5.15) into Equation (5.12), the water and solid mass balance equation results in

$$\begin{aligned} -\left[n\beta_w + (1-n)\beta_s\right]\frac{\mathrm{D}\theta}{\mathrm{D}t} + n\alpha_w \frac{\mathrm{D}u_w}{\mathrm{D}t} + m_v\left(\frac{\partial u_w}{\partial t} - \frac{\partial \sigma_n}{\partial t}\right) - \\ \frac{1}{\rho_w}\frac{\partial}{\partial z}\left(\rho_w \frac{k}{\rho_w g}\frac{\partial u_w}{\partial z}\right) = 0. \end{aligned} \tag{5.16}$$

At this point, a further simplification will be introduced. Since the expected velocity of solids, $\mathbf{v}$, will be small compared with the (Eulerian) rates of change of the variables of the problem (θ, u_w) total and partial derivatives are equivalent and Equation (5.16) becomes

$$\begin{aligned} -\left[n\beta_w + (1-n)\beta_s\right]\frac{\partial\theta}{\partial t} + (m_v + n\alpha_w)\frac{\partial u_w}{\partial t} - \\ m_v\frac{\partial \sigma_n}{\partial t} - \frac{k}{\gamma_w}\frac{\partial^2 u_w}{\partial z^2} = 0, \end{aligned} \tag{5.17}$$

valid for the shear band, $z \in [-e, e]$. In Equation (5.17), the water specific weight $\gamma_w = \rho_w g$ is introduced.

Equation (5.17) synthesizes the mass balance equations of solid (grains) and water. It is a parabolic second-order differential equation with two unknowns: the temperature and the excess pore water pressure. The contribution of the solid mass balance equation was to provide an expression for the change in porosity when the soil is heated and loaded under one-dimensional conditions. This result was then used in the water mass balance equation, which led to Equation (5.6). This equation will allow the calculation of excess pore water pressures.

Equation (5.17) also provides an explanation to the phenomenon of pore water pressure development during heating, a topic discussed in Section 5.1.1 with the help of Figure 5.4. If no change in total external stress occurs ($\partial\sigma_n/\partial t = 0$) and the soil is very impervious ($k \rightarrow 0$), the rate of increase of pore water pressure due to changes in temperature will be given by

$$\frac{\partial u_w}{\partial t} = \frac{n\beta_w + (1-n)\beta_s}{n\alpha_w + m_v} \frac{\partial \theta}{\partial t}. \tag{5.18}$$

This expression provides the theoretical background for the qualitative discussion on heat-induced development of pore water pressure in Section 5.1.1. In view of (5.18), stiff soils or rocks (low values of m_v) will develop high pore water pressures upon heating. Also, since the dilation coefficient of water is one order of magnitude higher than the dilation coefficient of solids (Fig. 5.3), the higher the porosity, the stronger the development of heat-induced water pressures. Note that Equation (5.18) is a direct consequence of the principle of effective stress, which was introduced in Equation (5.14) to describe the volumetric deformation of the soil. In other words, temperature changes lead to volumetric deformations of the soil skeleton. These deformations are explained by a change in effective stress. If the total stress does not change, the pore water pressure will increase to reduce effective stress, which will lead to soil expansion: the imposed thermal deformation. In the limit, if the soil skeleton is rigid ($m_v \rightarrow 0$) (the spherical shells in Figure 5.4 are made of steel), the pore water pressure developed when heating will be controlled by water compressibility (α_w).

A final remark at this point concerns the use of material derivatives. The reader may wonder why they were used at all if, at the end, the simplified mass balance Equation (5.17) will be used in calculations. One reason for doing it was to provide a general balance expression (5.6) which may be useful in other applications. The use of total derivatives in this case simplifies the notation. The joint consideration of the solid and water mass balances are also properly handled in terms of material derivatives. Note also that the solid mass balance provided a general expression, Equation (3.22) in Chapter 3, for the change in porosity, which included the effect of solid density changes. Other processes leading to porosity changes may be found in nature (solid mass dissolution or precipitation, for instance), which may be relevant in geotechnical engineering. In all these cases the formal derivation of the solid mass balance relationship is a first step towards finding the field equations of the problem.

5.3.2 Energy

The rate of work input into the shear band (Eq. (5.2)) dissipates as heat (H) and results in a temperature increase (θ) of the material in the band. The rate of work is homogeneous in the band because a uniform distribution of shear strain across the band was assumed. The shear strength in Equation (5.3) is given by Coulomb's frictional law:

$$\tau_f = \sigma'_n \tan(\varphi') = (\sigma_n - p_w)\tan\varphi', \tag{5.19}$$

where σ'_n is the effective stress acting in a direction normal to the shear band, p_w is the total pore water pressure (hydrostatic plus excess pore pressure) and φ' is the effective frictional angle of the material in the shear band, which in our case corresponds to residual conditions.

The shear strength τ_f can be calculated if normal stresses on the band are known. Normal stresses will be derived from the conditions of mechanical equilibrium of the slide described later. But first the temperature rise in the shear band will be investigated. To do so the energy balance equation in the band has to be derived. The procedure is already known because it follows precisely the same steps already given in Chapter 3 to derive balance equations for solid and water mass. Mass is now substituted by heat. In fact, heat is proportional to mass and temperature. The proportionality constant is the “specific heat” (c) which characterizes different species (in our case, solid mass (c_s) and water (c_w)). Therefore, the products $\rho_w c_w \theta$ or $\rho_s c_s \theta$ identify the heat stored in a unit volume of water and solid (grains) respectively. Constants “c” have the units of Joule/(kg·°C) = Newton·m/(kg·°C).

The heat stored in a unit volume of saturated soil, having a porosity n, is the sum of two terms

$$\rho c_m = (1-n)\rho_s c_s + n\rho_w c_w, \tag{5.20}$$

where c_m is the specific heat of the mixture (the saturated soil) and ρ is the saturated soil density.

Heat will flow whenever mass flows. This type of heat transfer is the advective component. However, heat is also transferred across bodies fixed in space. This common experience is a “conductive” phenomenon described by means of Fourier law

$$\mathbf{q}_c = -\Gamma \mathbf{grad}(\theta), \tag{5.21}$$

which states that the flow rate of heat follows the gradient of temperature. Γ is the conductivity constant.

The balance of heat can now be directly written if one of the preceding balance equations (for instance, Eq. (5.6) for water balance) is taken as a “model”:

$$\begin{aligned} H = {} & \frac{\mathrm{D}(\rho c_m \theta)}{\mathrm{D}t} + \mathrm{div}\left[-\Gamma \mathbf{grad}(\theta)\right] + \\ & \mathrm{div}\left(\rho_w c_w \theta(\mathbf{q} + n\frac{\partial \mathbf{u}}{\partial t}) + (1-n)\rho_s c_s \theta \frac{\partial \mathbf{u}}{\partial t}\right). \end{aligned} \tag{5.22}$$

Unlike the balance equations for solid and water, there is now a source term, the heat input into the band (H), which should be included into the total balance. Conduction and advective terms can now be identified in Equation (5.22).

In a general situation, heat will flow through the band boundaries. However, the failure of Vaiont was very fast (a few seconds) and conductive, as well as advective heat transfer will not be relevant. In fact, only the fast generation of heat in the band will essentially control the development of excess pore water pressures. Indeed, the error introduced by this simplification was investigated and a comparison between the results obtained with the full formulation (Eq. (5.22)) and the simplified one (5.23) was performed. Minimum discrepancies could be found. Therefore, the problem of finding the temperature increase in the band is greatly simplified if the heat balance Equation (5.22) reduces to

$$H = \rho c_m \frac{\partial \theta}{\partial t}, \tag{5.23}$$

where the source term $H(t)$ is, for the time being, an unknown function of time.

Before examining the stability conditions of the slide, let us consider briefly the balance conditions outside the shear band.

5.4 Balance Equations Outside the Shear Band

The soil in contact with the shear band will be affected by pore pressures and temperature developed in the shear band. In general, shear band boundaries are permeable to fluid and heat flow. Excess pore pressure induced by frictional heating in the shear band will tend to dissipate in a process essentially controlled by the permeability and compressibility of the band and surrounding material.

The assumption made before is that heat flow out of the band is negligible during fast sliding. Therefore, it is not necessary to perform a temperature analysis outside the band. The temperature outside the shear band will remain constant and equal to its initial value.

Water flow outside the band will be governed by an equation similar to Equation (5.17), which was derived for general one-dimensional conditions. The situation is now simplified because no temperature gradients exist and the mass conservation equation becomes

$$\left(m_v^r + n_r \alpha_w\right) \frac{\partial u_w}{\partial t} - m_v^r \frac{\partial \sigma_n}{\partial t} - \frac{k_r}{\gamma_w} \frac{\partial^2 u_w}{\partial z^2} = 0, \tag{5.24}$$

valid outside the shear band, $z \in (-\infty, -e] \cup [e, \infty)$. Index r indicates that a different material ("rock") is now considered, although clay material will typically exist on both sides of the band.

5.5 Dynamics of an Infinite Planar Slide

5.5.1 Introduction

The development of pore pressures in the sliding surface is the fundamental information required to perform the dynamic analysis of the slide. Therefore, Equation (5.17) has to be solved. However, the "thermal" term in this equation depends (Eq. (5.23)) on the work input $H = \dot{W}$ which is proportional to the sliding velocity and shear strength (Eq. (5.3)), which, in turn, depend on pore pressures developed in the sliding surface. This is a fully coupled problem that requires the simultaneous solution of the balance equations just developed for the sliding band and the sliding mass and the equilibrium equations of the whole slope.

Before tackling Vaiont, a simpler case will be analyzed: the behaviour of an infinite planar slide. The geometry of this case is defined by a constant base inclination of angle β and a constant thickness D (Fig. 5.10). The x-axis is parallel to the slope surface and indicates the motion direction, while z-axis is the normal direction. The origin of axis is located in the mid plane of the shear band.

The broken line in Figure 5.10 indicates water level (height h_w over the sliding surface) which, for simplicity, was accepted parallel to the surface of sliding. In this way the problem becomes one-dimensional. The sliding mass is assumed rigid and deformation is localized in the shear band. If the slide is unstable, it will reach an increasing velocity. The sliding mass translates as a rigid solid. Its velocity is the maximum velocity acting on one boundary of the shear band (v_{max} in Fig. 5.9).

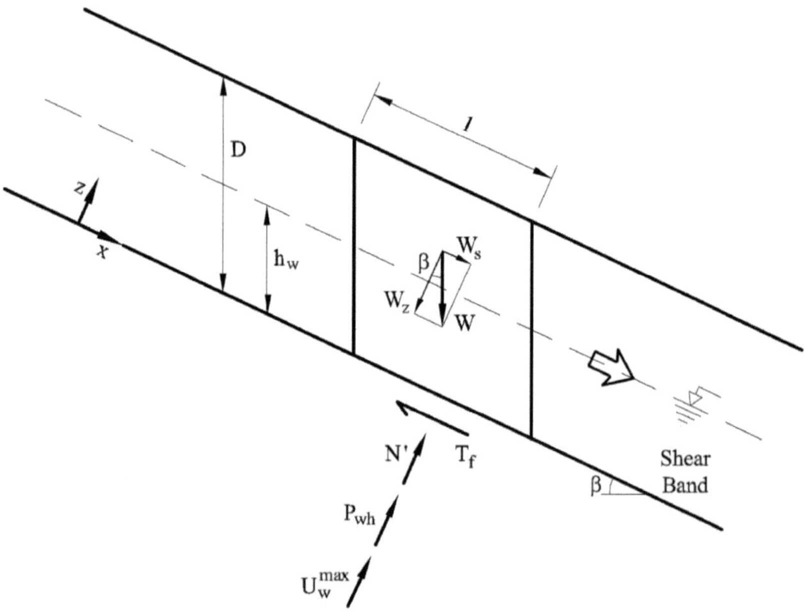

Figure 5.10 Geometry of the infinite planar slide.

At any time, the motion of the unstable mass is accelerated by the action of gravity and is resisted by the shear strength acting on the sliding surface. Shear strength is proportional to the effective normal stress and therefore it depends on the weight component normal to the motion direction and on the pore water pressure on the sliding surface. Note that the pore water pressure that acts in the sliding surface and controls the landslide motion is the maximum pore pressure that develops in the shear band. Since the excess pore water pressure generated in the shear band can dissipate through the boundaries, the maximum excess of pore pressure (u_w^{max}) will be developed in the central plane of the shear band.

Dynamic equilibrium equations are written, in z- and x-directions, for a slice of unit width of moving slide:

$$N' = W_z - P_{wh} - U_w^{max} \tag{5.25}$$

and

$$W_x - T_f = M\frac{dv_{max}}{dt}. \tag{5.26}$$

Note that the Newton's Second Law applied in Equation (5.26), which states that the resultant forces will be equal to the time variation of momentum, has been simplified since the involved mass is constant.

In view of Figure 5.10, $W_x = W\sin\beta$ and $W_z = W\cos\beta$ are the x-component and z-component of the total weight of the sliding mass, respectively; P_{wh} is the force due to hydrostatic pressure on the failure surfaces; U_w^{max} is the force due to the excess of water pressure evaluated at $z = 0$; N' is the normal effective resultant force developed in the sliding surface; T_f is the frictional shear force that acts on the base of the slide and is calculated by means of Coulomb's frictional law ($T_f = N'\tan\varphi'$); v_{max} is the slide velocity evaluated at $z = e$; and M is the total mass of the slide.

Since the analysis is made per unit of length of moving slide, forces T_f, U_w^{max} and P_{wh} can be replaced by frictional shear stress, τ_f, excess water pressure, u_w^{max}, and the hydrostatic pore water pressure p_{wh}, respectively.

Therefore, the rigid-body motion is described by Newton's Second Law,

$$M\frac{dv_{max}}{dt} = F_{sliding} - F_{resisting}, \tag{5.27}$$

where $F_{sliding} = W\sin(\beta)$ and

$$F_{resisting} = \tau_f \cdot 1 = \left[W\cos(\beta) - p_{wh}\cdot 1 - u_w^{max}\cdot 1\right]\tan\varphi'. \tag{5.28}$$

The total slice weight is $W = \gamma_r D\cos(\beta)$ and its mass $M = \rho_r D\cos\beta$, where $\gamma_r = \rho_r g$ is the unit weight of the sliding material and ρ_r is the bulk density ($\rho_r = (1-n_r)\rho_s + n_r\rho_w$). It has been assumed that $D \approx D + e$.

5.5.2 Formulation

Summarizing previous results, the set of equations governing the motion of an infinite planar slide are:

a) Equilibrium conditions and the Mohr-Coulomb strength law

$$\tau_f(t) = \left[W\cos(\beta) - p_{wh} - u_w^{max}(t)\right]\tan\varphi'; \tag{5.29a}$$

b) First Law of Thermodynamics (shear band)

$$H(t) = \tau_f(t)\frac{v_{max}(t)}{2e} \quad \text{for } z \in [-e, e]; \tag{5.29b}$$

c) Mass balance (water and solid) and heat balance in the shear band

$$-\left[(1-n)\beta_s+\beta_w n\right]\frac{H(t)}{\rho c_m}+(m_v+n\alpha_w)\frac{\partial u_w(z,t)}{\partial t}= \\ =\frac{k}{\gamma_w}\frac{\partial^2 u_w(z,t)}{\partial z^2} \quad \text{for } z\in[-e,e]; \tag{5.29c}$$

d) Mass balance (water and solid) in the sliding mass outside of the shear band

$$\left[m_v^r+n_r\alpha_w\right]\frac{\partial u_w(z,t)}{\partial t}=\frac{k_r}{\gamma_w}\frac{\partial^2 u_w(z,t)}{\partial z^2} \quad \text{for } z\in(-\infty,-e]\cup[e,\infty); \tag{5.29d}$$

e) Dynamic equilibrium

$$\frac{\mathrm{d}v_{\max}(t)}{\mathrm{d}t}=\frac{1}{M}\left[W\sin(\beta)-\tau_f(t)\right]. \tag{5.29e}$$

Notice that in Equations (5.29c,d) the term of time variation of external stress has not been included. In the case of an infinite planar slide there are no changes in total stress during sliding because of the simple geometry.

This system of equations can be immediately reduced to three equations if the strength (τ_f) and heat rate (H) expressions are replaced in Equations (5.29c,d,e). A single equation for the dependent variable u_w could eventually be found, but the hope of solving it in closed form is remote.

To solve these equations it is also necessary to define the appropriate initial and boundary conditions. A natural initial condition for the dynamic problem is a situation in which static equilibrium has been slightly exceeded. It would imply the initiation of motion. In such a situation, the initial excess pore pressure and slide velocity would be zero and no heat would be generated. Therefore,

$$u_w(z,t_0)=0, \tag{5.30a}$$

$$v(z,t_0)=0, \tag{5.30b}$$

$$\theta(t_0)=\theta_0, \tag{5.30c}$$

where θ_0 is the reference initial temperature at the beginning of the slide motion.

It was mentioned before that frictional heat is generated at a constant rate within the shear band, between $z=-e$ and $z=e$. No heat is generated, at any time, outside of the shear band. Therefore, the heat generated excess pore pressure is constant in the shear band and zero in the remaining of the domain. However, the unbalance of water pressures between points inside and outside of the shear band induces its dissipation. It will be also accepted that the soil outside the two boundaries of the shear band is described by a common set of material properties.

Since the gradient of hydrostatic pressure may be neglected in the band, given its small thickness, it follows that the axis $z = 0$ (Fig. 5.8d) is a symmetry axis.

Therefore, the solution of the problem will be sought for $z \geq 0$ and symmetry conditions will be forced at $z = 0$. This condition implies a zero flow through $z = 0$:

$$\mathbf{q}\big|_{z=0} = -\frac{k}{\gamma_w}\frac{\partial u_w}{\partial z} = 0 \;\Rightarrow\; \left.\frac{\partial u_w}{\partial z}\right|_{z=0} = 0. \tag{5.31}$$

At the other boundary, $z = e$, continuity of excess pore pressure and flow rate has to be satisfied on both sides of the shear band-rock interface:

$$u_w\big|_{z=e^-} = u_w\big|_{z=e^+}, \tag{5.32a}$$

$$\mathbf{q}\big|_{z=e^-} = \mathbf{q}\big|_{z=e^+} \;\Rightarrow\; k\left.\frac{\partial u_w}{\partial z}\right|_{z=e^-} = k_r\left.\frac{\partial u_w}{\partial z}\right|_{z=e^+} \tag{5.32b}$$

Changes in water pressure outside the band will extend to relatively small distances because the volume of water expelled by the band is very small. Small changes in porosity within a limited distance outside the band will be able to absorb the transient flow of water. Therefore, no effect on the calculated pore pressures outside the band will be noticed if a zero excess pore water pressure is specified at an infinite distance:

$$u_w\big|_{z=\infty} = 0. \tag{5.33}$$

The problem, summarized in Equations (5.29) to (5.33), was solved by means of a finite difference approximation developed in Appendix 5.1.

5.5.3 Results and discussion

The accelerated motion of a deep planar slide (D = 240 m; see Fig. 5.10) will be investigated. The depth of the sliding surface is taken from the average thickness of the lower wedge of Vaiont (Section 2). For a residual friction angle of 12° the infinite slope becomes strictly unstable for an inclination $\beta = 9.5°$ and a height of the water table over the sliding surface of h_w = 119.1 m.

The thermal and compressibility parameters for water and the solid mineral constituent of the shear band were taken from Olivella *et al.* (1996). They are given in Table 5.1.

A relevant parameter of the analysis is the thickness of the shear band. It will be assumed that the band is embedded in a much thicker clay layer. As a reference, in the case of Vaiont, Hendron and Patton (1985, page 20) mention that the clay layer at the base of the sliding mass had a thickness varying between 1 and 3 m. In the Cortes landslide (Alonso *et al.*, 1992), the shear band was located within a 2 m thick marl layer. Most probably, the clay material in the immediate vicinity of the band will have essentially the same properties as the band itself. It is also expected (and the computations reported below demonstrate it) that the transient changes in water pressure around the shear band will affect a thickness of the encasing material, which will be of the same order of magnitude as the thickness of the band. Given the expected dimensions of the band (a few millimeters), its effect will only extend a small distance at both sides of the band and, for the purposes of the pore pressure analysis reported here, the entire domain of the material outside the band will be a clay material having the properties of the band.

Table 5.1 Material properties.

Parameter	Symbol	Value	Unit
Water			
Density	ρ_w	1,000	kg/m^3
Coefficient of compressibility	α_w	5×10^{-10}	1/Pa
Thermal expansion coefficient	β_w	3.42×10^{-4}	1/°C
Specific heat	c_w	4.186×10^{3} 1.0	J/kg·°C cal/ kg·°C
Solid particles			
Density	ρ_s	2,700	kg/m^3
Thermal expansion coefficient	β_s	3×10^{-5}	1/°C
Specific heat	c_s	8.372×10^{-2} 0.20	J/kg·°C cal/ kg·°C
Shear band material			
Porosity	n	0.2	–
Permeability	k	10^{-11}	m/s
Compressibility coefficient	m_v	1.5×10^{-9}	1/Pa
Friction angle (residual)	φ'	12	°
Sliding mass material			
Density	ρ_r	2,350	kg/m^3

However, shear band and sliding mass have to be differentiated in the formulation of the problem simply because heat is generated in the shear band and not outside. A shear band thickness of 5 mm was selected to perform the dynamic analysis of an infinite slide reported here.

The remaining material properties of the shear band (and the clay material on both sides) are given in Table 5.1. Porosity and residual friction angle approximate the actual values of the Vaiont sliding clay surface and were taken from Hendron and Patton (1985). An average rock density ρ_r = 2,350 kg/m^3 was assumed to calculate the weight of the sliding mass. The initial temperature for the analysis is 10°C.

No precise laboratory information on the permeability of the clay sliding surface seems to be available. Hendron and Patton (1985) use the value k =

1.6×10^{-10} m/s in their analysis. Vardoulakis (2002) uses k = 1.1×10^{-11} m/s. The high plasticity values consistently measured (see Chapter 2) and the presence of montmorillonite probably favours low clay permeability. A value k = 1.0×10^{-11} m/s was selected here as a base case. Shear band permeability is one of the key parameters of the model and it is subjected to high uncertainty. A sensitivity analysis, discussed later, was performed to analyze the effect of changing clay permeability.

Similar difficulties were found to select a value for the clay stiffness. Hendron and Patton (1985) report an elastic modulus of 1,000 MPa, which is equivalent to an edometric deformability coefficient m_v = 5×10^{-10} Pa^{-1} (for ν = 0.3). Vardoulakis (2002) selects a much softer value, m_v = 1.5×10^{-8} Pa^{-1}, which is perhaps a high compressibility for the geologically old and indurated Jurassic clay levels at the base of the landslide. An intermediate number, m_v = 1.5×10^{-9} Pa^{-1}, was selected here for the base case.

The geometry and material properties described above, lead to an infinite slide in strict equilibrium. To activate the slide, the water level was increased by a small amount: 10 cm. The calculated response of the slide is shown in Figures 5.11 to 5.14.

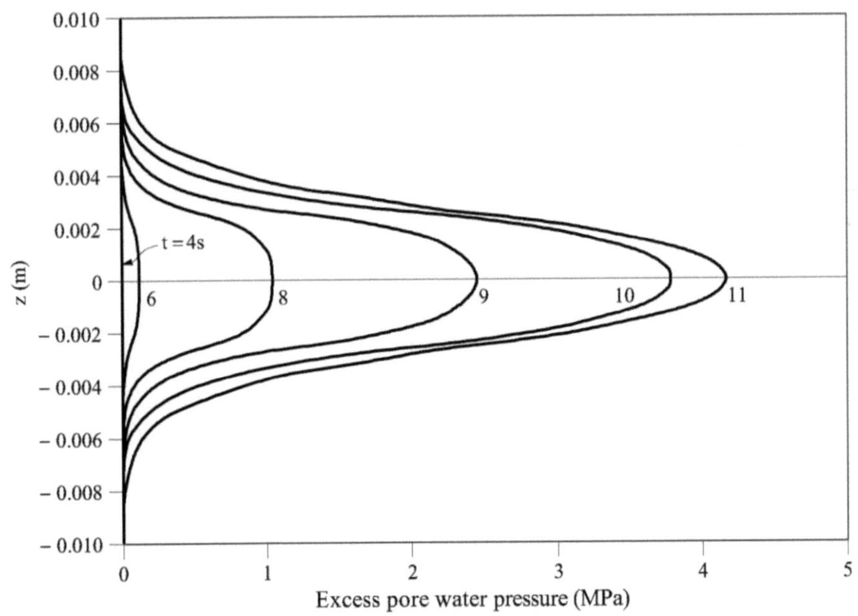

Figure 5.11 Excess of water pressure isochrones in a section normal to the slide direction. Infinite slide (z = 0 is in the center of the shear band).

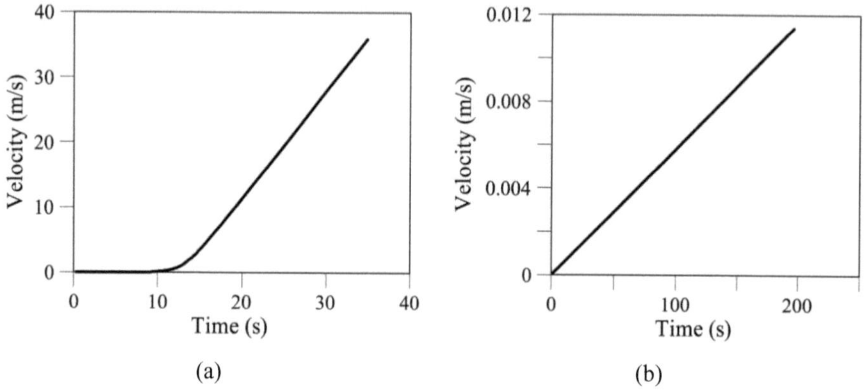

Figure 5.12 Dynamic analysis of infinite slide. Base case (shear band permeability, $k = 10^{-11}$ m/s). Time evolution of slide velocity: (a) heat generation considered; (b) no heat generation considered.

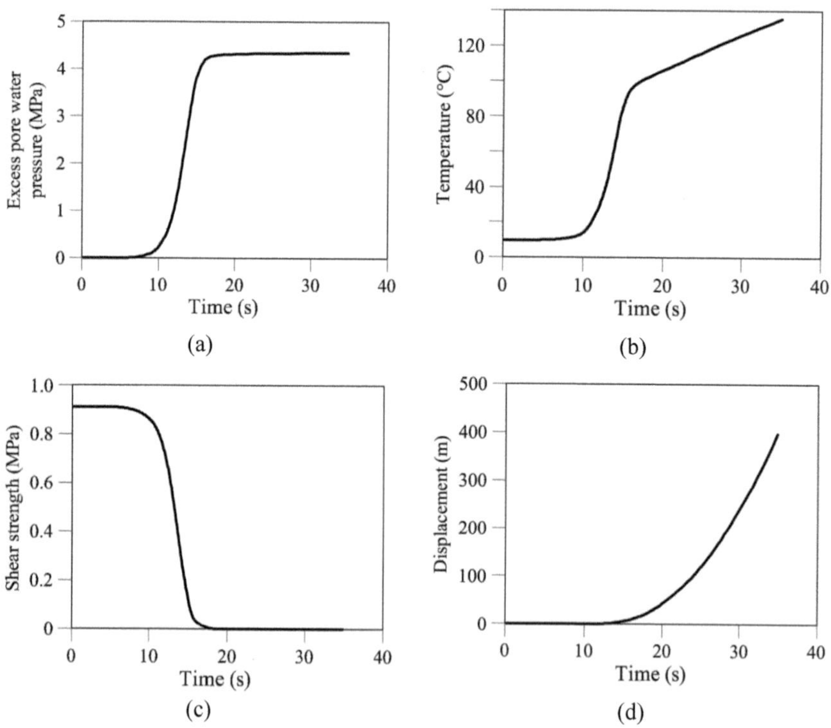

Figure 5.13 Dynamic analysis of infinite slide. Base case (shear band permeability, $k = 10^{-11}$ m/s). Evolution in time of (a) excess pore water pressure in the middle of the band; (b) temperature; (c) shear strength of shear band; (d) slide displacement.

Isochrones of excess pore pressures in the shear band and adjacent clay material are plotted in Figure 5.11. As frictional heat accumulates during the slide displacement, pore pressure increases. The dissipation towards the boundaries of the band is slow due to the low permeability of the clay material. Note that changes in pore pressure only extend to a reduced thickness outside the band. Maximum pore pressures are always calculated at the center of the band (z = 0). This is the point where effective normal stresses are calculated when establishing the dynamic equilibrium of the slope.

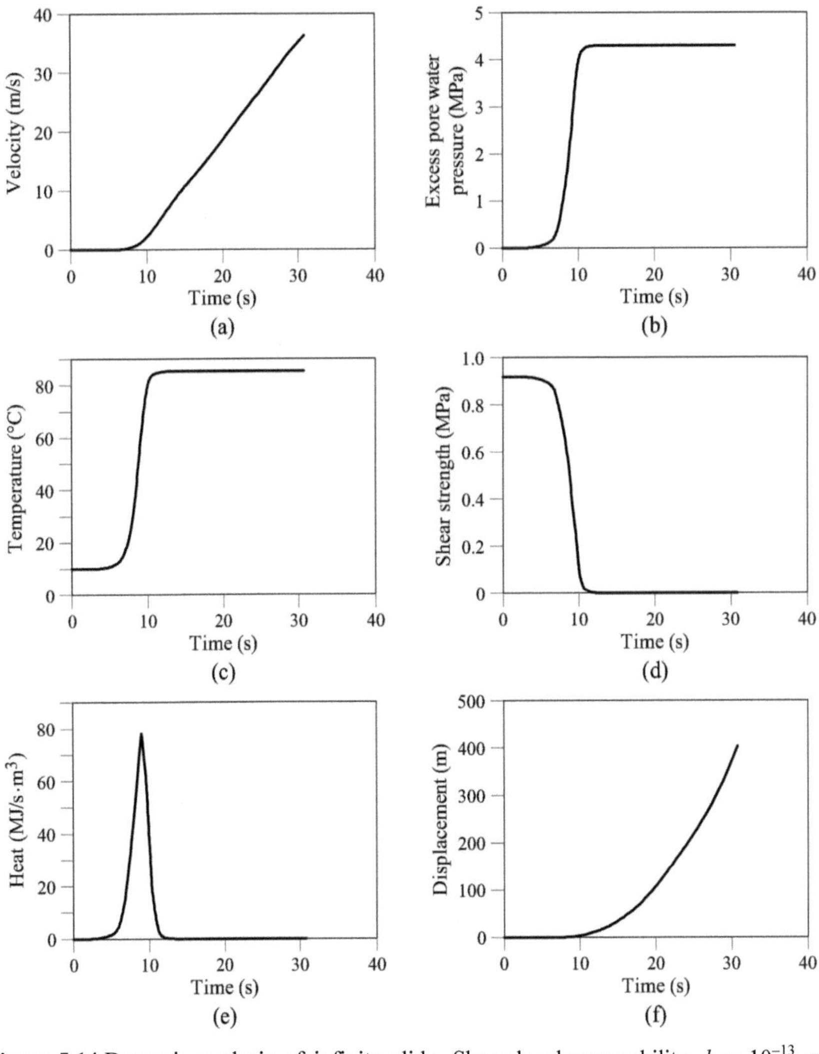

Figure 5.14 Dynamic analysis of infinite slide. Shear band permeability, k = 10^{-13} m/s. Evolution in time of (a) velocity; (b) excess pore water pressure in the middle of the band; (c) temperature; (d) shear strength of shear band; (e) heat generated in the band; (f) slide displacement.

The fundamental effect of heat-induced pore water pressure generation is shown in Figure 5.12 where the development of slide velocity with or without heat generation is compared. An unstable infinite slope will eventually reach an infinite velocity, irrespective of the heat generation at the shearing band. However, adding the heat effect results in a much faster acceleration. In fact, after a common sliding time of 30 s, the "standard" slope, which was made (slightly) unstable, reached a velocity of 2 mm/s. If the heat phenomenon is considered, the calculated velocity is 30 m/s for $t = 27$ s.

Results for the base case ($k = 10^{-11}$ m/s) are summarized in Figure 5.13. The following variables have been plotted along time: maximum excess pore pressure and temperature generated in the band, shear strength, and slide displacement.

During the first 10 s, the generated heat does not have any relevant effect. The excess of pore pressure remains negligible because the frictional work generated is very small and the heat released is not enough to sufficiently increase the pore pressure. Eventually, as time increases, pore pressure build-up is capable of reducing the resisting shear strength. Then the driving force increases, the slide accelerates, the work input and the temperature in the shear band increase, and additional pore pressures are generated. The shear strength reduces to a very small value at $t = 16$ s. However, the increasing velocity still provides a heat input into the shear band and the temperature continues to rise. The pore pressure generation is now almost compensated with flow-induced dissipation. The calculation was stopped when total displacement was 400 m.

The effect of changing the permeability of the shear band (and surrounding clay) is presented in Figures 5.14 ($k = 10^{-13}$ m/s) and 5.15 ($k = 10^{-9}$ m/s). When the band is more impervious, the results are essentially the same as in the base case. Now the shear strength becomes essentially zero beyond $t = 10$ s and the band temperature remains constant at $T = 85$°C. The work input into the band reaches a maximum and then decreases to a small value, which is enough to compensate for the slow pore pressure dissipation. As a result, the excess pore pressure remains constant and reaches a value close to 4.3 MPa.

When the band is more pervious ($k = 10^{-9}$ m/s), Figure 5.15b, the pore pressure does not increase as fast as in the previous two cases, because dissipation is enhanced. Therefore, the shear strength does not fall so fast. The combined effect of increasing sliding velocity and non-negligible residual strength in the band leads to an increase of the work input (Fig. 5.15e) and to a significant elevation of temperature in the band (Fig. 5.15c). The calculated temperatures far in excess of 1,000°C implies that some of the assumptions made in the derivation of the governing equations for the mass and heat balance of the band may not be satisfied. In particular, if water vapourizes, pore water pressure may be different from the value associated solely with water dilation and (liquid) flow. The behaviour of the solid phase will also be affected by high temperatures. The problem now will require modification of the formulation, which is not attempted here.

The effect of band permeability on slide velocity and pore pressure generation, during the interval necessary for the slide to reach a displacement of 400 m is illustrated in Figure 5.16. Increasing the permeability leads to a slower response of the pore pressure build-up and a delayed acceleration of the slide. The results seem, at a first sight, to be consistent with the physics of the problem, but the high temperatures developed in the band for permeabilities in excess of 10^{-9} m/s force to be cautious in the high permeability range.

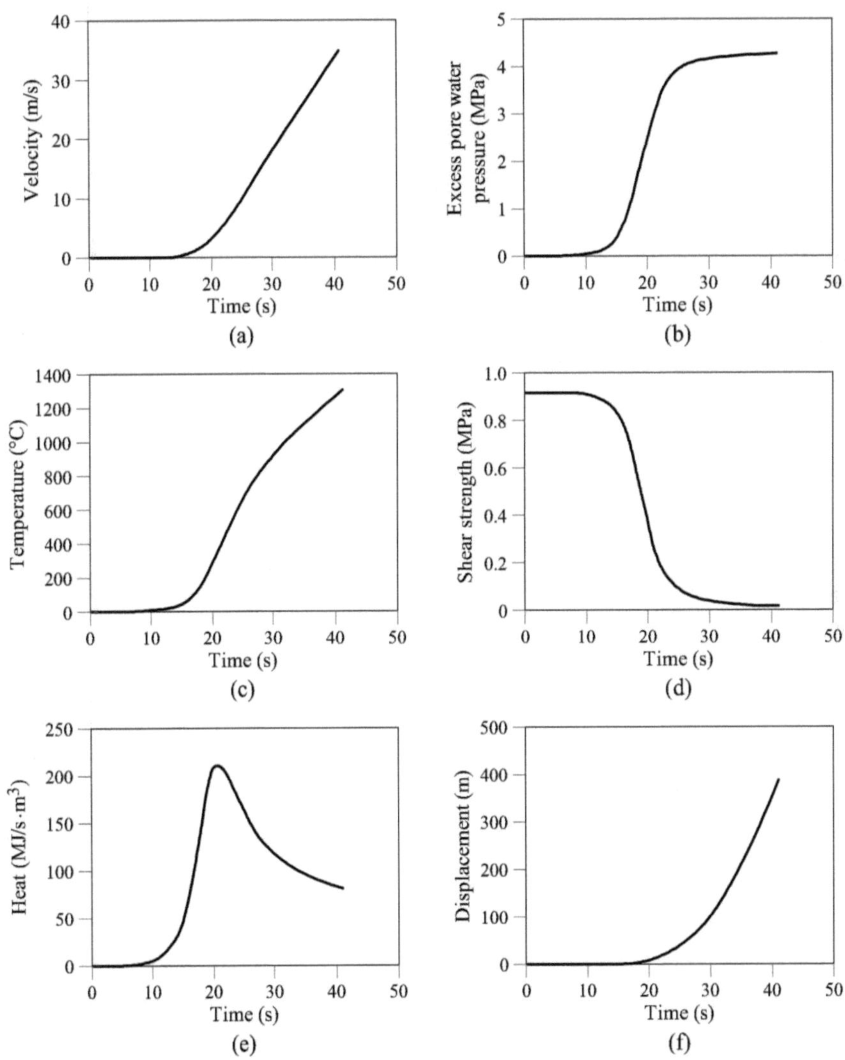

Figure 5.15 Dynamic analysis of infinite slide. Shear band permeability, $k = 10^{-9}$ m/s. Evolution in time of (a) velocity; (b) excess pore water pressure in the middle of the band; (c) temperature; (d) shear strength of shear band; e) heat generated in the band; (f) slide displacement.

The infinite slide is a crude approximation of reality. The cross-section of Vaiont can be approximated by two interacting wedges, the lower one resting on an essentially horizontal sliding surface. The sliding mechanism in this case is substantially different because, in the absence of phenomena leading to strength reduction, the motion of an initially stable geometric configuration tends to decelerate, as discussed in Chapter 2. The following section explores the dynamics of Vaiont.

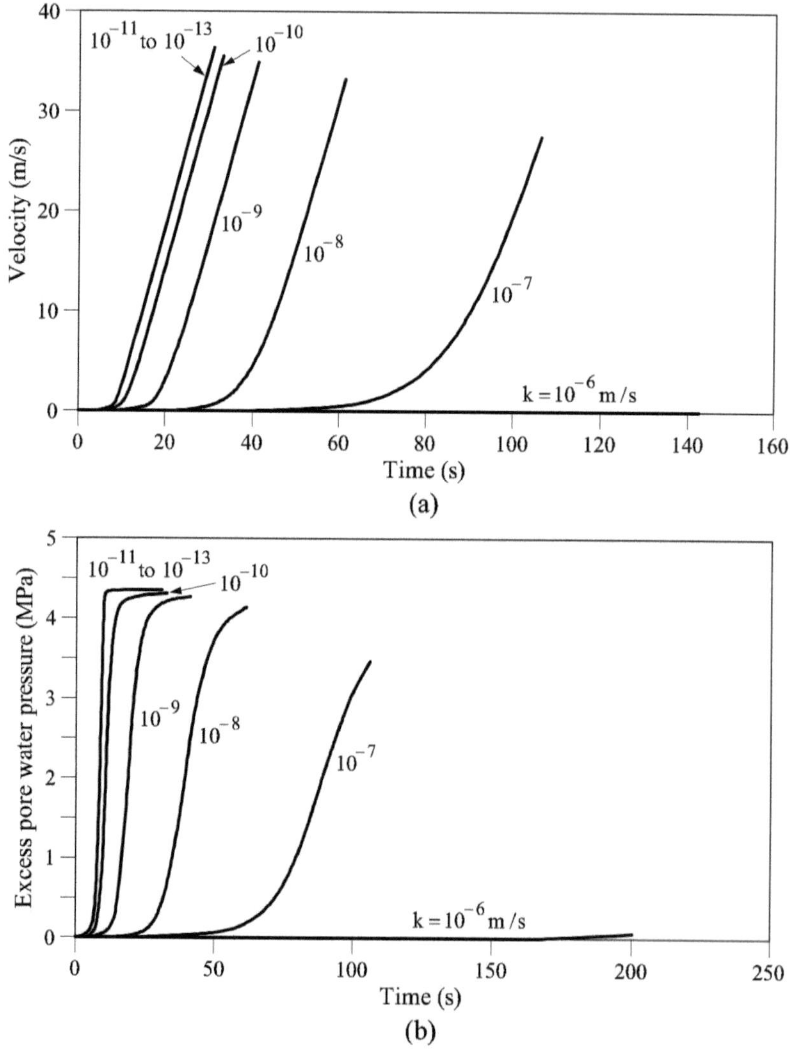

Figure 5.16 Dynamic analysis of infinite slide. Effect of shear band permeability (k) on the development of (a) velocity and (b) excess pore water pressure in the middle of the band during the interval necessary to reach a displacement of 400 m.

5.6 Two Interacting Wedges

5.6.1 Geometry

The analysis of the infinite slide presented above is useful in understanding the thermo-hydraulic process that takes place in a shear band and its effect on the overall slide motion. However, the geometry of the slide introduces significant changes, which will be presented here. The slide is now divided into two wedges (1 and 2), following the discussion presented in Chapter 2. Section 5 of Vaiont is schematized in Figure 5.17.

The analysis follows the calculation procedure developed for the infinite slide: mass and energy balance have to be written for the shear bands limiting the two wedges and the overall dynamic equilibrium of the two wedges has to be satisfied.

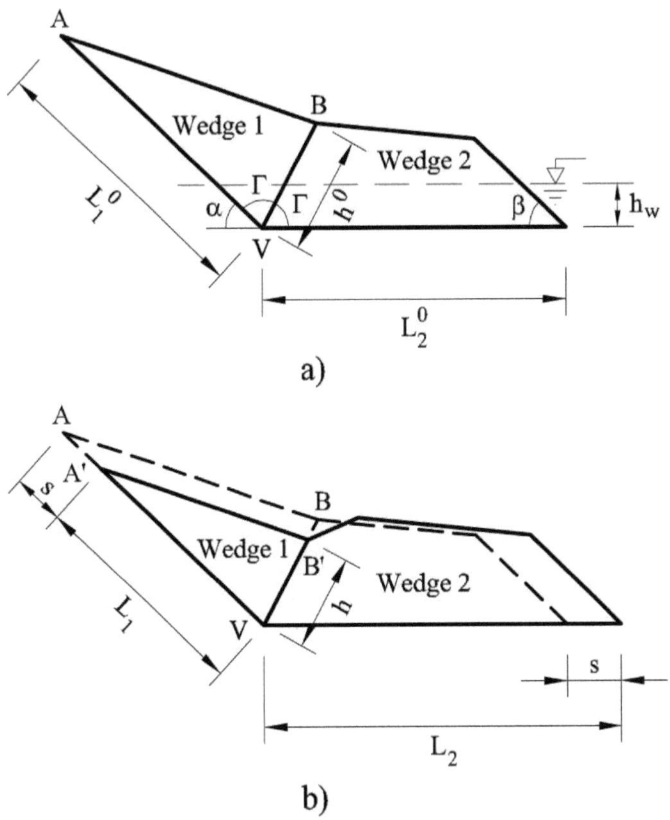

Figure 5.17 Cross-section 5 of Vaiont: (a) initial geometry; (b) geometry after a displacement *s*.

The lower wedge (Wedge 2), resting on a horizontal plane, supports (passively) the unstable upper wedge (Wedge 1), which slides on a sloping plane.

This geometry was used in Chapter 2 for the analysis of static and dynamic equilibrium of the slope without considering the effect of water dilation due to the heat generated on the basal shear band.

Changes in geometry have to be considered in a dynamic analysis. Figure 5.17 indicates the evolving geometry of the slide when a displacement s is considered. The initial basal length of Wedge 1 (L_1^0) is reduced to

$$L_1 = L_1^0 - s. \tag{5.34}$$

Displacement increases the initial basal length of Wedge 2 (L_2^0) to

$$L_2 = L_2^0 + s. \tag{5.35}$$

Length h (see Fig. 5.17b) can be obtained, for a given displacement s, knowing that

$$\frac{h}{L_1} = \frac{h^0}{L_1^0}, \tag{5.36}$$

because triangles AVB and A′VB′ are similar.

Now, the volume of Wedge 1 for a given displacement can be obtained as

$$V_1 = \frac{1}{2} L_1 h \cos\left(\frac{\alpha}{2}\right). \tag{5.37}$$

Volume reduction of Wedge 1 contributes towards increasing the volume of Wedge 2 by the same amount and therefore its current volume becomes

$$V_2 = V_2^0 + \left(V_1^0 - V_1\right). \tag{5.38}$$

Wedge weights (W_1 and W_2) and masses ($M_1 = W_1/g$ and $M_2 = W_2/g$; where g is the gravity acceleration) can be computed from these volumes. A specific weight of the rock ($\gamma_r = 23.5$ kN/m^3) was used in calculations.

5.6.2 Balance equations

Mass and energy balance (of the lower shear band) and equilibrium conditions (for the entire moving mass) will be written separately for each wedge. By forcing the slide to move as a single unit, the governing equations of the movement of the landslide will be obtained.

The effective interaction forces across the common plane (VB′; see Fig. 5.17 and 5.18) between the two wedges have two components, N'_{int} and Q_{int}, normal and tangential to the plane. Forces due to pore water pressures P_{w_1}, $P_{w\text{int}}$ and P_{wf} will be considered as constant during the landslide.

Since the shear resistant forces of each wedge (T_1 and T_2) are different (although a unique frictional angle is considered, normal effective resultant forces on the basal planes, N_1' and N_2', are not equal), the work input into the bounding shear bands of the two wedges will be different. Therefore, two different values for the shear band temperature (θ_1 and θ_2) and for the excess pore water pressures (U_{w1} and U_{w2}) will be developed in the two wedges. Specific balance equations should be written for each one of the two wedges. To avoid confusions, each part of the shear band will be denoted by shear band 1 or 2 according to the wedge involved. Equal thickness and material properties will be assumed in the two bands (they are taken from Table 5.1).

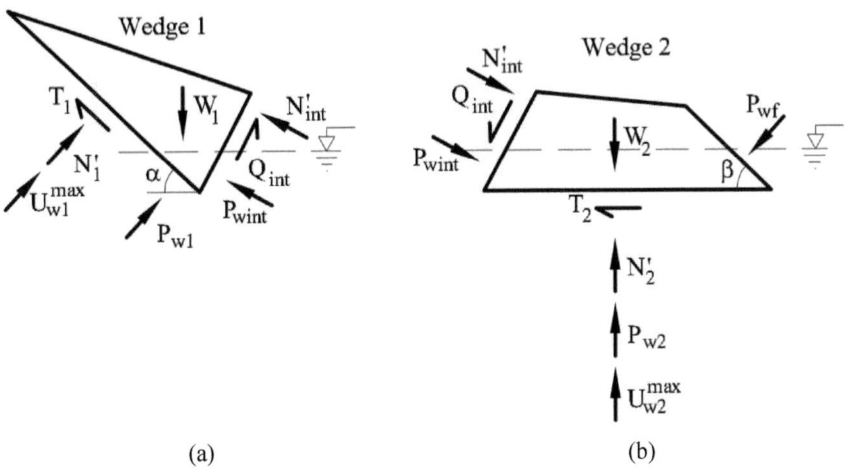

Figure 5.18 Geometry and forces on wedges: (a) Wedge 1; (b) Wedge 2.

Consider first the one-dimensional balance equations already developed for the infinite slide in Section 5.5. They will now be directly applied to Wedge 1. The z_1-direction corresponds to the normal direction of Shear Band 1. From the First Law of Thermodynamics, the generated heat (H_1) in the Shear Band 1 is expressed as

$$H_1(t) = \tau_{f1}(t)\frac{v_{\max}(t)}{2e} \quad \text{for } z_1 \in [-e, e]. \tag{5.39}$$

The frictional strength (τ_{f1}) can be derived from equilibrium conditions, as done previously for the infinite slide.

Neglecting conduction and diffusion of heat, heat balance in the Shear Band 1 reads

$$H_1(t) = \rho c_m \frac{\partial \theta_1(t)}{\partial t} \quad \text{for } z_1 \in [-e, e]. \tag{5.40}$$

Mass balance of water and solid inside and outside of Shear Band 1 results in

$$-\left[(1-n)\beta_s+\beta_w n\right]\frac{\partial\theta_1(t)}{\partial t}+\left[m_v+n\alpha_w\right]\frac{\partial u_{w1}(z_1,t)}{\partial t}-$$
$$m_v\frac{\partial\sigma_{n1}(t)}{\partial t}=\frac{k}{\gamma_w}\frac{\partial^2 u_{w1}(z_1,t)}{\partial z_1^2}\quad \text{for } z_1\in[-e,e], \tag{5.41a}$$

$$\left[m_v^r+n_r\alpha_w\right]\frac{\partial u_{w1}(z_1,t)}{\partial t}-m_v\frac{\partial\sigma_{n1}(t)}{\partial t}=\frac{k_r}{\gamma_w}\frac{\partial^2 u_{w1}(z_1,t)}{\partial z_1^2}$$
$$\text{for } z_1\in(-\infty,-e]\cup[e,\infty) \tag{5.41b}$$

Regarding Wedge 2, the generated heat can be expressed as

$$H_2(t)=\tau_{f2}(t)\frac{v_{\max}(t)}{2e}\quad \text{for } z_2\in[-e,e], \tag{5.42}$$

valid in the normal direction (z_2) to Shear Band 2. The heat balance will be given by

$$H_2(t)=\rho c_m\frac{\partial\theta_2(t)}{\partial t}\quad \text{for } z_2\in[-e,e]. \tag{5.43}$$

Likewise, mass balance of water and solid inside and outside of the Shear Band 2 is written as

$$-\left[(1-n)\beta_s+\beta_w n\right]\frac{\partial\theta_2(t)}{\partial t}+\left[m_v+n\alpha_w\right]\frac{\partial u_{w2}(z_2,t)}{\partial t}-$$
$$m_v\frac{\partial\sigma_{n2}(t)}{\partial t}=\frac{k}{\gamma_w}\frac{\partial^2 u_{w2}(z_2,t)}{\partial z_2^2}\quad \text{for } z\in[-e,e], \tag{5.44a}$$

$$\left[m_v^r+n_r\alpha_w\right]\frac{\partial u_{w2}(z_2,t)}{\partial t}-m_v\frac{\partial\sigma_{n2}(t)}{\partial t}=\frac{k_r}{\gamma_w}\frac{\partial^2 u_{w2}(z_2,t)}{\partial z_2^2}$$
$$\text{for } z_2\in(-\infty,e]\cup[-e,\infty). \tag{5.44b}$$

These expressions complete the balance equations for the two shear bands.

5.6.3 Dynamic equilibrium of the two wedges

At this point the reader may wish to read first Chapter 2, where a detailed presentation of the equilibrium equations of the two wedges is made. Reference is made here to Figure 5.18. Regarding the previous analysis for the infinite landslide, the main difference now is that masses and weights depend on the displacement and then they are not constant in time.

For Wedge 1, the dynamic equilibrium equations for directions parallel and normal to the basal sliding plane are:

$$W_1(t)\sin(\alpha)-T_1(t)-N'_{\text{int}}(t)\cos\left(\frac{\alpha}{2}\right)-$$
$$Q_{\text{int}}(t)\sin\left(\frac{\alpha}{2}\right)-P_{w\text{int}}\cos\left(\frac{\alpha}{2}\right)=\frac{\mathrm{d}\left(M_1(t)v_{\max}(t)\right)}{\mathrm{d}t}, \tag{5.45a}$$

$$W_1(t)\cos(\alpha)-N'_1(t)+N'_{\text{int}}(t)\sin\left(\frac{\alpha}{2}\right)-Q_{\text{int}}(t)\cos\left(\frac{\alpha}{2}\right)+$$
$$P_{w\text{int}}\sin\left(\frac{\alpha}{2}\right)-P_{w1}-u_{w1}^{\max}(t)L_1(t)=0. \tag{5.45b}$$

The right-hand term of Equation (5.45a) can be developed as:

$$\frac{\mathrm{d}\left(M_1(t)v_{\max}(t)\right)}{\mathrm{d}t}=M_1(t)\frac{\mathrm{d}v_{\max}(t)}{\mathrm{d}t}+v_{\max}(t)\frac{\mathrm{d}M_1(t)}{\mathrm{d}t} \tag{5.46}$$

and the time variation of mass of the wedge can be expressed as a function of the time variation of the displacement ($\mathrm{d}s/\mathrm{d}t$), which is equal to the velocity (v):

$$\frac{\mathrm{d}M_1}{\mathrm{d}t}=\delta_r\frac{dV_{\text{Wedge 1}}}{\mathrm{d}t}=-\delta_r\left(L_{10}-s\right)\frac{h^0}{L_{10}}\cos\left(\frac{\alpha}{2}\right)\frac{\mathrm{d}s}{\mathrm{d}t}. \tag{5.47}$$

The shear resistance force on the base of Wedge 1 (T_1) is expressed, following the Mohr – Coulomb strength criterion, as

$$T_1(t)=N'_1(t)\tan\left(\varphi'_b\right), \tag{5.48}$$

where φ'_b is the effective residual friction angle of the sliding surface.

The mobilized shear force on the common plane between wedges is given by

$$Q_{\text{int}}(t)=c'_r h(t)+N'_{\text{int}}(t)\tan\left(\varphi'_r\right), \tag{5.49}$$

where c'_r is the effective cohesion of the rock, and φ'_r, the effective friction angle of the rock. The values of these strength parameters are indicated in Table 5.2. These values are accepted and justified in Chapter 2.

Table 5.2 Strength parameters of the sliding rock mass.

Sliding mass material		
Cohesion	c'_r	762.2 MPa
Friction angle	φ'_r	38°

The water pressure force due to the presence of a water table of height h_w acting against Wedge 1 is

$$P_{w1} = \frac{h_w^2 \gamma_w}{2\sin\alpha}. \tag{5.50}$$

The water pressure force acting against the right boundary of Wedge 1 (Fig. 5.18) is calculated as

$$P_{w\text{int}} = \frac{h_w^2 \gamma_w}{2\cos\alpha} \tag{5.51}$$

Dynamic equilibrium expressions for Wedge 2 (parallel and normal to the slide direction, respectively) are

$$N'_{\text{int}}(t)\cos\left(\frac{\alpha}{2}\right) - Q_{\text{int}}(t)\sin\left(\frac{\alpha}{2}\right) - T_2(t) = \frac{\mathrm{d}\left(M_2(t)v_{\max}(t)\right)}{\mathrm{d}t}, \tag{5.52a}$$

$$\begin{gathered} W_2(t) - N'_2(t) + N'_{\text{int}}(t)\sin\left(\tfrac{\alpha}{2}\right) + Q_{\text{int}}(t)\cos\left(\tfrac{\alpha}{2}\right) + \\ P_{w\text{int}}\sin\left(\tfrac{\alpha}{2}\right) + P_{wf}\cos(\beta) - P_{w2}(t) - u_{w2}^{\max}(t)L_2(t) = 0. \end{gathered} \tag{5.52b}$$

The shear resistance on the base of Wedge 2 (T_2) is given by

$$T_2(t) = N'_2(t)\tan(\varphi'_b) \tag{5.53}$$

and the value of P_{w2} is given by

$$P_{w2}(t) = L_2(t)h_w\gamma_w. \tag{5.54}$$

Note that these equations also depend on displacement, s, travelled by the wedges.

If Equations (5.45) to (5.54) are properly combined, a single motion equation for the total sliding mass is obtained as follows:

$$\begin{gathered} t_{W_1}W_1(t) + t_{W_2}W_2(t) + t_{P_{w\text{int}}}P_{w\text{int}} + t_{P_{wf}}P_{wf} + t_{P_{w1}}P_{w1} + \\ t_{P_{w2}}P_{w2}(t) + t_{u_{w1}}u_{w1}^{\max}(t)L_1(t) + t_{u_{w2}}u_{w2}^{\max}(t)L_2(t) + \\ t_{c_r}c_r h(t) + t_{dM_1dt}\frac{dM_1}{dt}v_{\max}(t) + t_{dM_2dt}\frac{dM_2}{dt}v_{\max}(t) = \\ = \left(t_{M_1}M_1(t) + t_{M_2}M_2(t)\right)\frac{\mathrm{d}v_{\max}(t)}{\mathrm{d}t}, \end{gathered} \tag{5.55}$$

where t coefficients depend on the section geometry and on the cohesive and frictional parameters of the materials involved, as indicated in Appendix 5.3.

The strength acting on the basal sliding surface of the two wedges is found as the ratio of total resistance forces T_1 or T_2 and current base lengths L_1 or L_2. T_1 and T_2 are given by

$$T_1(t) = \Big[r_{W_1} W_1(t) + r_{W_2} W_2(t) + r_{P_{w\,\mathrm{int}}} P_{w\,\mathrm{int}} + r_{P_{wf}} P_{wf} + r_{P_{w1}} P_{w1} + r_{P_{w2}} P_{w2}(t) + r_{u_{w1}} u_{w1}^{\max}(t) L_1(t) + r_{u_{w2}} u_{w2}^{\max}(t) L_2(t) + r_{c_r} c_r h(t) + r_{dM_1dt} \frac{\mathrm{d}M_1}{\mathrm{d}t} v_{\max}(t) + r_{dM_2dt} \frac{\mathrm{d}M_2}{\mathrm{d}t} v_{\max}(t) \Big] \Big/ \big[r_{M_1} M_1(t) + r_{M_2} M_2(t) \big], \tag{5.56}$$

$$T_2(t) = \Big[s_{W_1} W_1(t) + s_{W_2} W_2(t) + s_{P_{w\,\mathrm{int}}} P_{w\,\mathrm{int}} + s_{P_{wf}} P_{wf} + s_{P_{w1}} P_{w1} + s_{P_{w2}} P_{w2}(t) + s_{u_{w1}} u_{w1}^{\max}(t) L_1(t) + s_{u_{w2}} u_{w2}^{\max}(t) L_2(t) + s_{c_r} c_r h(t) + s_{dM_1dt} \frac{\mathrm{d}M_1}{\mathrm{d}t} v_{\max}(t) + s_{dM_2dt} \frac{\mathrm{d}M_2}{\mathrm{d}t} v_{\max}(t) \Big] \Big/ \big[s_{M_1} M_1(t) + s_{M_2} M_2(t) \big], \tag{5.57}$$

where coefficients r and s are function of geometry and of wedge masses. They are collected in Appendix 5.3. These expressions allow the calculation of heat generation through Equations (5.39) and (5.42).

Summarizing the preceding results, the system of equations to be solved includes the balance equations for the two shear bands (2+2 equations) and the equation for the dynamic equilibrium of the entire landslide (one equation):

$$-\big[(1-n)\beta_s + \beta_w n\big] \frac{H_1(t)}{(\rho c)_m} + \big[m_v + n\alpha_w\big] \frac{\partial u_{w1}(z_1,t)}{\partial t} - m_v \frac{\partial \sigma_{n1}(t)}{\partial t} = \frac{k}{\gamma_w} \frac{\partial^2 u_{w1}(z_1,t)}{\partial z_1^2} \quad \text{for } z_1 \in [-e,e], \tag{5.58a}$$

$$\big[m_v^r + n_r \alpha_w\big] \frac{\partial u_{w1}(z_1,t)}{\partial t} - m_v^r \frac{\partial \sigma_{n1}(t)}{\partial t} = \frac{k_r}{\gamma_w} \frac{\partial^2 u_{w1}(z_1,t)}{\partial z_1^2} \quad \text{for } z_1 \in (-\infty,-e] \cup [e,\infty), \tag{5.58b}$$

$$-\big[(1-n)\beta_s + \beta_w n\big] \frac{H_2(t)}{\rho c_m} + \big[m_v + n\alpha_w\big] \frac{\partial u_{w2}(z_2,t)}{\partial t} - m_v \frac{\partial \sigma_{n2}(t)}{\partial t} = \frac{k}{\gamma_w} \frac{\partial^2 u_{w2}(z_2,t)}{\partial z_2^3} \quad \text{for } z_2 \in [-e,e], \tag{5.58c}$$

$$\big[m_v^r + n_r \alpha_w\big] \frac{\partial u_{w2}(z_2,t)}{\partial t} - m_v^r \frac{\partial \sigma_{n2}(t)}{\partial t} = \frac{k_r}{\gamma_W} \frac{\partial^2 u_{w2}(z_2,t)}{\partial z_2^2} \quad \text{for } z_2 \in (-\infty,-e] \cup [e,\infty), \tag{5.58d}$$

$$\begin{aligned} & t_{W_1} W_1(t) + t_{W_2} W_2(t) + t_{P_{w\mathrm{int}}} P_{w\mathrm{int}} + t_{P_{wf}} P_{wf} + t_{P_{w1}} P_{w1} + \\ & t_{P_{w2}} P_{w2}(t) + t_{u_{w1}} u_{w1}^{\max}(t) L_1(t) + t_{u_{w2}} u_{w2}^{\max}(t) L_2(t) + \\ & t_{c_r} c_r h(t) + t_{dM_1 dt} \frac{dM_1}{dt} v_{\max}(t) + t_{dM_2 dt} \frac{dM_2}{dt} v_{\max}(t) = \\ & = \left(t_{M_1} M_1(t) + t_{M_2} M_2(t) \right) \frac{\mathrm{d}v_{\max}(t)}{\mathrm{d}t}, \end{aligned} \tag{5.58e}$$

where heat generation rates H_1 and H_2 are given by Equations (5.39) and (5.42).

5.6.4 Results and discussion

The system of Equations (5.58) was solved by finite differences following the same calculation procedure explained in Appendix 5.1 for an infinite slope. Note that for the case of two interactive wedges, the geometry of each wedge, their weights and masses, the total normal stress under each wedge and the hydrostatic pore water pressure under Wedge 2 depend on the displacement and should be updated at each time interval of the calculation. Initial and boundary conditions for each one of the shear bands are identical to the conditions described for the infinite slope. The computer program developed is included in Appendix 5.2 to facilitate calculations for other cases not covered here and to show all the details of the calculation procedure.

Results for the base case ($k = 10^{-11}$ m/s) are given in Figures 5.19 and 5.20. A shear band thickness $2e = 5$ mm was assumed. Calculation ended when the slide reached a displacement of 400 m. The physical explanation of phenomena taking place in the shear band and the response of the slide were already given when discussing the results for the infinite slope. Isochrones of excess pore water pressure in the shear band below Wedge 1 are given in Figure 5.19a for the first 12 s of motion, when the slide velocity was 10 m/s. The excess pore water pressure reached a maximum value of 1.7 MPa for $t = 10$ s. At this time the available shear strength at the center of the shear band was already very small and the heat generated (and the associated pore pressure build-up) decreased sharply.

As a result, pore pressure dissipation under Wedge 1 towards the surrounding soil dominated the following time steps ($t > 10$ s). Note also in Figure 5.19a that excess pore pressures become negative outside the shear band. This is a consequence of the unloading associated with the loss of weight of Wedge 1 as the slide moves forward. This effect is of minor importance within the shear band itself where the excess pore pressure is dominated by heating effects. In Wedge 2 (Fig. 5.19b), excess pore pressures reach higher values due to the higher weight of the wedge. The average normal total stress on the horizontal sliding surface under Wedge 2 decreases also during the slide displacement and a decrease in pore water pressure outside the band is also calculated (Fig. 5.19b).

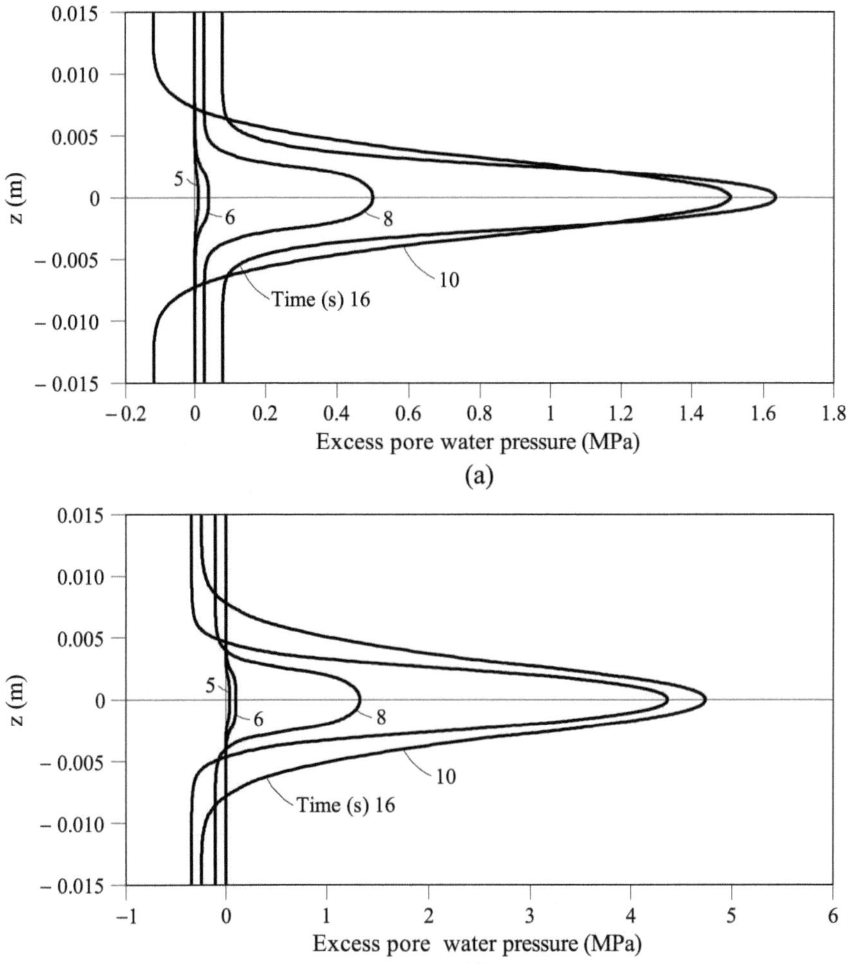

Figure 5.19 Dynamic analysis of Section 5 of Vaiont. Base case (shear band permeability, $k = 10^{-11}$ m/s). Excess pore water pressure isochrones in the shear band of (a) Wedge 1 and (b) Wedge 2 and adjacent soil. Shear band extends from z = 0.0025 m to z = –0.0025 m.

Figure 5.20 provides additional details. Global performance variables for Wedge 1 have been plotted against time. The slide reaches a displacement of 400 m, 30 s after the initiation of the motion. At this time the velocity is 27 m/s (close to 100 km/h). These are values consistent with field observations (see Chapter 2).

The development of excess pore pressure at the center of the shear band is shown in Figure 5.20b and has already been explained. Further insight is provided by the evolution of temperature, the drop in strength and the work (or heat) input into the shear band. The maximum temperature calculated in this case is somewhat higher than 100°C. The drop of shear strength is rapid from t = 8 to 10

s. The work performed increases fast during this period due to the rapid increase in velocity, but it later decays because of the very low value of shear strength. The entire behaviour of the band and, hence, of the landslide, depends in a fully coupled manner on the mass and heat transfer phenomena in the thin shear band and its immediate vicinity.

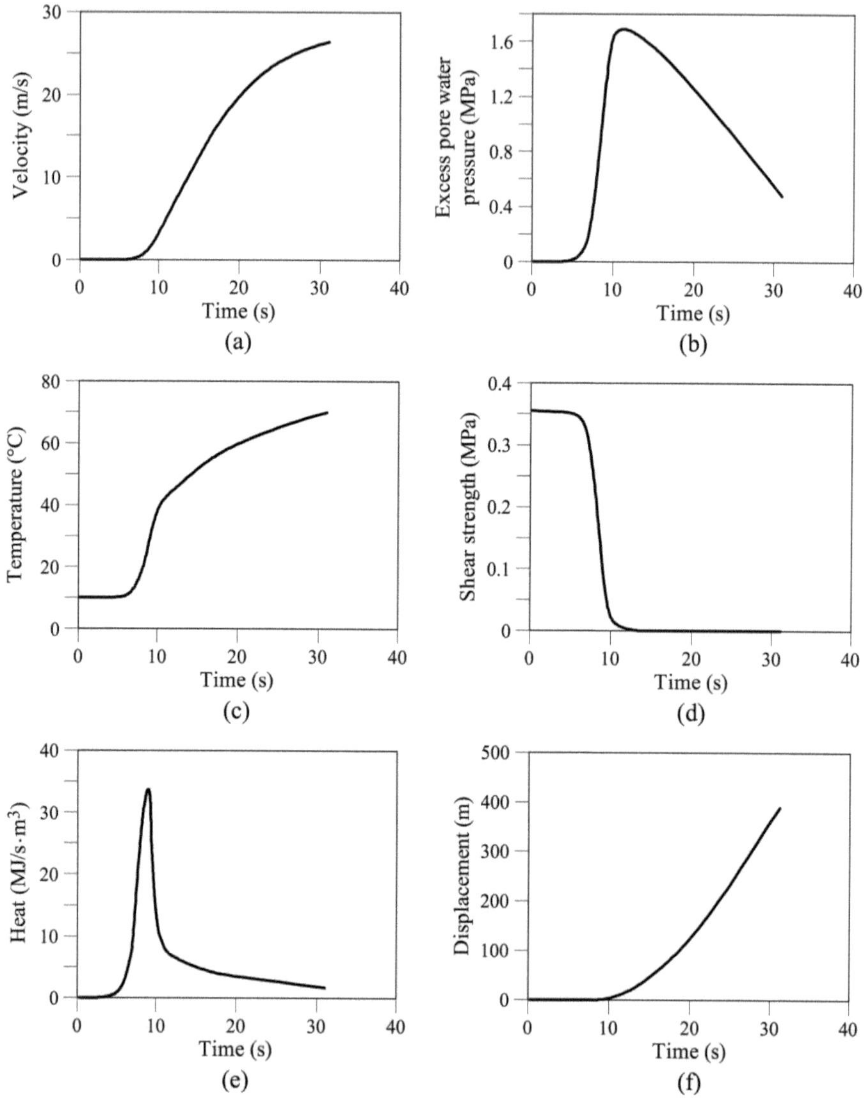

Figure 5.20 Dynamic analysis of Section 5 of Vaiont. Base case (shear band permeability, $k = 10^{-11}$ m/s). Wedge 1. Evolution in time of: (a) velocity; (b) excess pore water pressure in the middle of the band; (c) temperature; (d) shear strength of shear band; (e) heat generated in the band; (f) slide displacement.

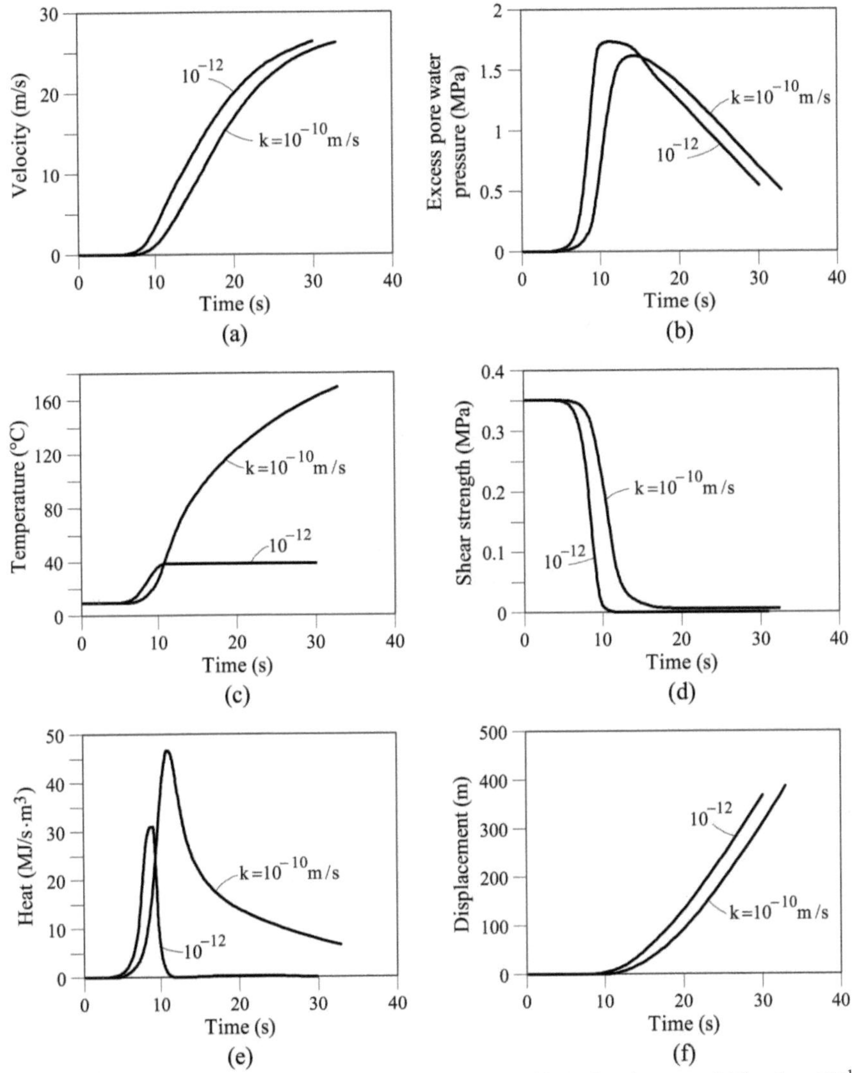

Figure 5.21 Dynamic analysis of Section 5 of Vaiont. Shear band permeability k = 10^{-10} and 10^{-12} m/s (remaining properties as in Base Case). Wedge 1. Evolution in time of: (a) velocity; (b) excess pore water pressure in the middle of the band; (c) temperature; (d) shear strength of shear band; (e) heat generated in the band; (f) slide displacement.

Changing the permeability of the shear band leads to significant changes in behaviour. A more impervious band leads to minor changes, when compared with the base case. When it is made more pervious, pore water pressure dissipation becomes more significant and the effective normal stress (and shear strength) maintains higher values. The slide also accelerates fast and high velocities coupled with relatively higher shear strengths lead to larger heat inputs into the band and to higher temperatures. These effects can be followed in Figure 5.21, where the cases for $k = 10^{-12}$ and $k = 10^{-10}$ m/s, which are also accepted values for the clay band material of Vaiont, have been represented. The three calculated cases lead to the same basic result: a fast accelerated motion is predicted and the final velocities (for $s = 400$ m) are similar in all of the three cases.

Vaiont was an extreme case. Landslides are commonly of a much lower volume. Therefore is it interesting to explore the effect of reducing the size of the slide. In other words, to raise the question of scale effects on the dynamic behaviour of slides. This practical issue will be examined in the next section.

It was also interesting to perform a more complete sensitivity analysis than just varying the band permeability. In this way, a relevant practical issue, namely establishing a general criteria for landslide acceleration, could be analyzed. Shear band permeability is only one of the parameters controlling the development of pore pressures. Relevant parameters are also band thickness and stiffness. To some extent permeability and band thickness provide the same information: both are related to grain size distribution. Narrow or, alternatively, thick shear bands are expected in impervious or pervious materials, respectively. Stiffness is a different type of property and rock-like or soil-like materials may be found for the same mineralogy and grain size distribution. An analysis of the combined effect of permeability, band thickness and stiffness will be presented.

5.7 Scale Effects

Vaiont was a very large landslide (a mobilized volume close to 260 million m^3 was estimated). A slide 100 times smaller is still a very large landslide. For instance, the 5 million m^3 Cortes landslide, described in Alonso *et al.* (1992), posed a significant threat to the 100 m high Cortes concrete arch dam. Its overall dimensions (length, height) were roughly 1/10 of Vaiont dimensions. Moreover, many dangerous rock and soil slides described in the literature are one order of magnitude smaller than Cortes slide. Vaiont was an extreme case, of very rare occurrence, on a world basis. Therefore, a relevant question remains: is the velocity reached by Vaiont also a common occurrence or, at least, a real possibility in smaller and much more frequent landslides?

A comprehensive answer to this question would require a lengthy analysis of the dynamic behaviour of different types of landslides. But a simple answer can be given if the main characteristics of Vaiont (a displacement type of motion involving a mass of rigid rock, sliding on a clay layer) are maintained and the geometrical dimensions are reduced without any further change in material properties or geometrical arrangement. In fact, if all the dimensions of Vaiont are reduced by a factor of 10, a landslide very similar to the Cortes slide is obtained. If this slide becomes (sligthtly) unstable, how would it evolve if heat-induced water pressure develops at the sliding surface?

The program included in Appendix 5.2 is of direct application in this case. A new case has been run, modifying the scale of the Vaiont landslide. The new geometry is defined by reducing the dimensions (lengths and heights) of Wedges 1 and 2 (Fig. 5.17a) by a factor of 10. The water level was located at a position which brought the slope to a state of strict equilibrium. The remaining properties (including the shear band thickness) have not been modified and they are given in Table 5.1. The motion was triggered by a slight increase (10 cm) of the water level.

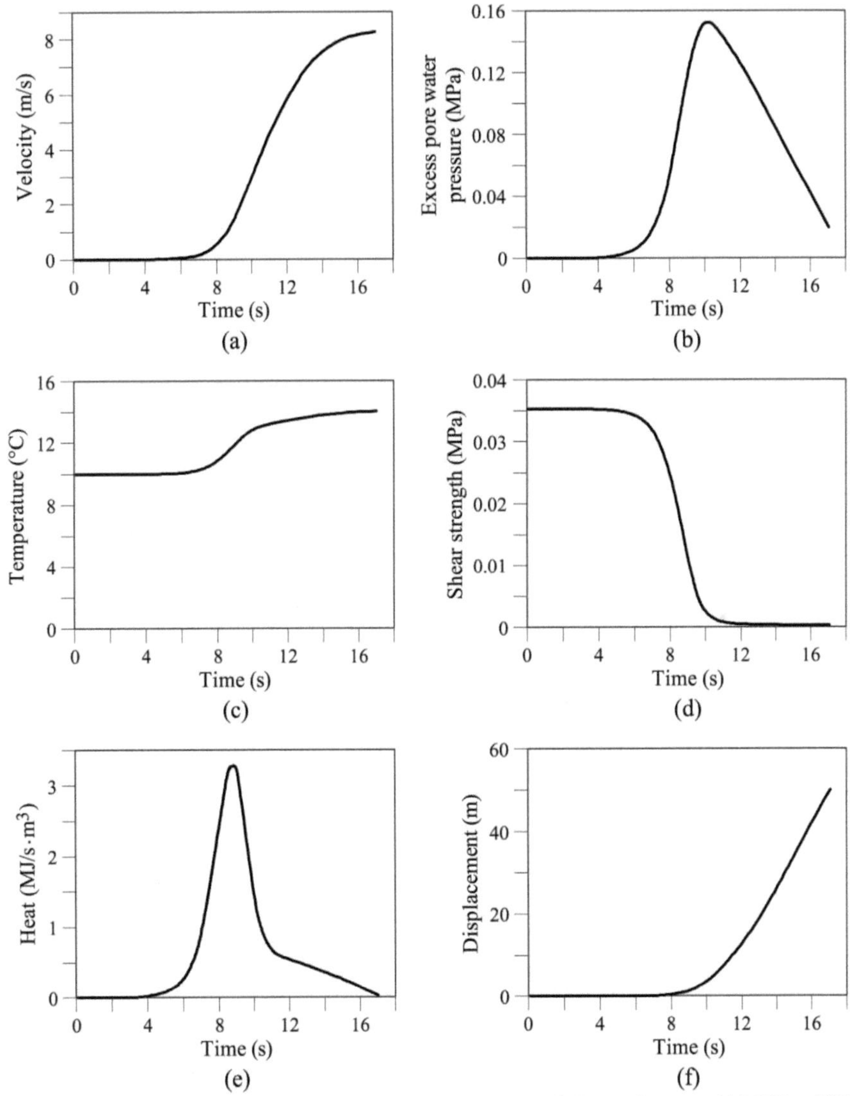

Figure 5.22 Reduced Vaiont landslide (dimensions × (1/10); volumes × (1/100)). Slide response for a base case (shear band permeability, $k = 10^{-11}$ m/s). Wedge 1. Evolution in time of: (a) velocity; (b) excess pore water pressure in the middle of the band; (c) temperature; (d) shear strength of shear band; (e) heat generated in the band; (f) slide displacement.

The calculated response of this slide is shown in Figure 5.22 for a base case ($k = 10^{-11}$ m/s). Calculations were run in time until the slide reached a displacement of 50 m. The calculated heat input into the shear band and the maximum excess pore pressures are now one order of magnitude smaller than in the previous case. As a result, the temperature increase of the band is very moderate (3.5 °C). The shear strength, however, is lost after a few seconds and the slide is able to reach a significant velocity. A maximum value around 9 m/s is obtained at the end of the calculation period. The implication is that this reduced slide may be also dangerous if the circumstances of the analysis are fulfilled in practice.

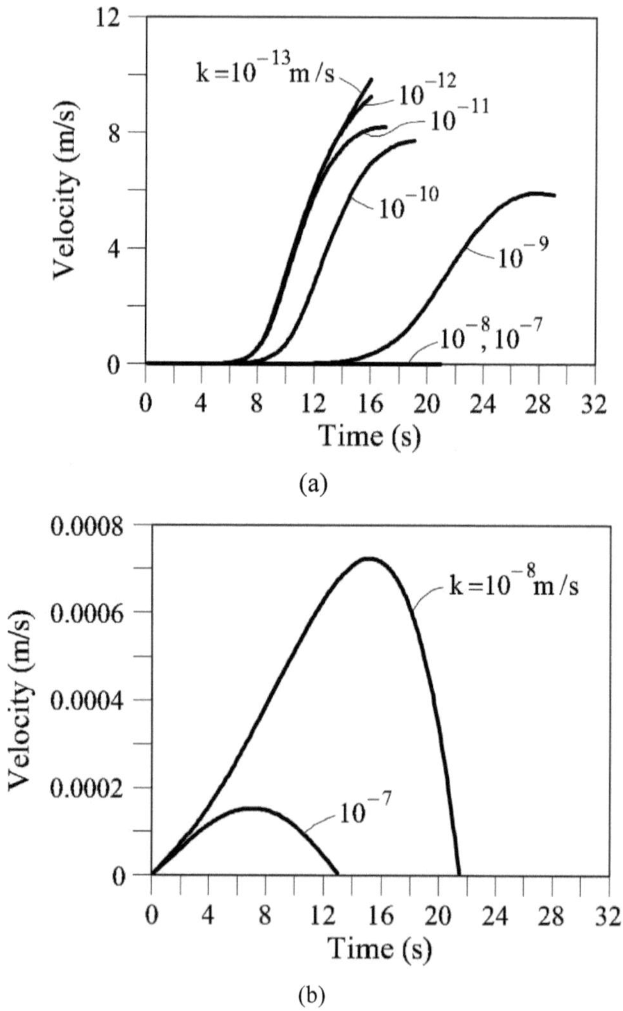

Figure 5.23 Reduced Vaiont landslide (dimensions × (1/10); volumes × (1/100)). Wedge 1. (a) Effect of the permeability on the landslide velocity; (b) detail for high band permeability.

Band permeability is a key parameter to control the response of the slide. This is shown in detail in Figures 5.23-5.26 which show the calculated velocity, excess pore pressure, temperature and displacement for varying band permeability (Wedge 1 in all cases). Band permeabilities of 10^{-8} m/s and larger do not trigger any heat-induced effect. This threshold is obviously associated with the band thickness used in calculations (5 mm), but a more consistent analysis is given below. Since the two-wedge mechanism analyzed has a self-equilibrating response, the small initial triggering effect (increasing water pressure in the shear band by 10 cm) is, in those cases, "absorbed" by the changing geometry and the slide comes to rest after a small increase in velocity (Fig. 5.23b). If permeability decreases below this threshold, the coupled thermo-hydro-mechanical processes taking place in the band result in a progressive accumulation of pore pressures (Fig. 5.24) and in an accelerated slide motion. The temperature increase in the band, when the slide accelerates ($k < 10^{-9}$ m/s) is now quite moderate in most cases. However, for the reasons already explained, there are some specific k values (around $k = 10^{-9}$ m/s) which result in a strong dissipation of energy at the band and, accordingly, in a significant temperature increase (30 °C are obtained – Fig. 5.25d – at the end of the calculation interval). The attained displacements for a given time (Fig. 5.26) reflect also the preceding comments.

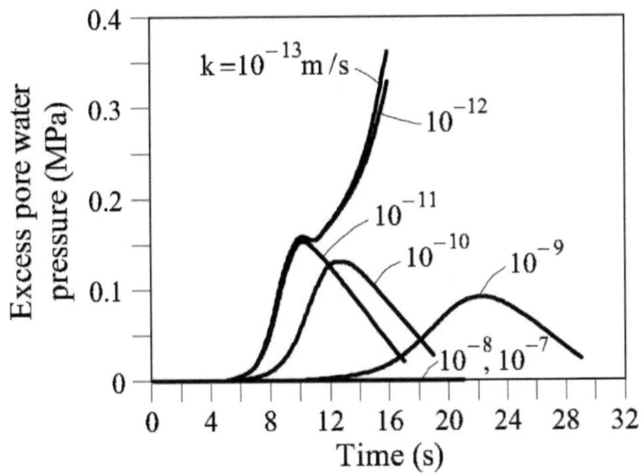

Figure 5.24 Reduced Vaiont landslide (dimensions × (1/10); volumes × (1/100)). Wedge 1. Effect of shear band permeability on the excess pore water pressure in the middle of the band.

Summarizing, smaller slides, similar in shape to the Vaiont case, may also reach significant velocities. It appears that band permeability is a key parameter controlling slide acceleration. Below a certain threshold value (around $k = 10^{-8}$ m/s for the geometry and parameters selected for the case analyzed), the slide may reach a high velocity. However, when the size of the slide decreases, the generated band excess pore pressures and temperatures reduce. In fact, it appears that for slides having the size of a "reduced Vaiont" by a factor of 10 in the scale of dimensions, maximum temperature increments in the shear band will be no more than a few degrees. In most cases, it turns out that the generated temperature depends also strongly on the thickness of the shear band. Before general conclusions are reached in this regard, it is convenient to perform a sensitivity analysis of the calculated solution when the thickness, permeability and stiffness of the band are varied between acceptable limits.

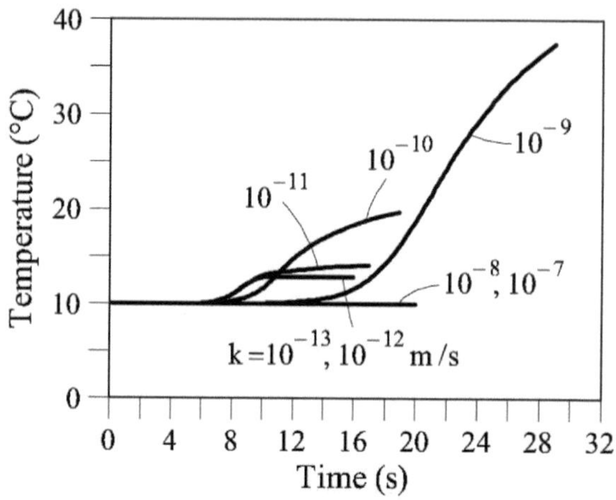

Figure 5.25 Reduced Vaiont landslide (dimensions × (1/10); volumes × (1/100)). Wedge 1. Effect of shear band permeability on the temperature in the band.

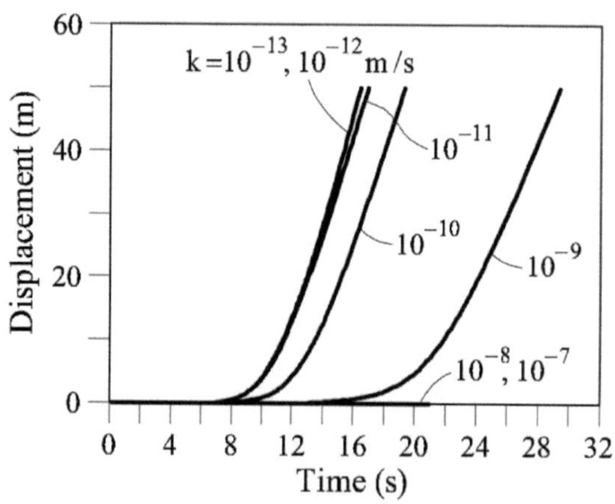

Figure 5.26 Reduced Vaiont landslide (dimensions × (1/10); volumes × (1/100)). Wedge 1. Effect of shear band permeability on the slide displacement.

5.8 Discussion

A better insight into the physics of the problem is gained if a sensitivity analysis of the main controlling factors is performed. In view of previous results, permeability values in the range 10^{-12} to 10^{-7} m/s and band thickness varying between 0.5 and 10 mm were chosen. Two moduli of confined stiffness were selected, having in mind that in most cases the shearing surfaces in landslides are located in soft clayey rocks: $m_v = 10^{-9}$ Pa^{-1} (a relatively stiff clay rock) and $m_v = 10^{-8}$ Pa^{-1} (a relatively soft clayey rock). Then, for each combination (k, $2e$, m_v), the program given in Appendix 5.2 was run for the geometry of the scaled Vaiont geometry and maximum velocities for a runout of 50 m and temperatures in the shear band were calculated.

Consider first the case of a stiff shearing band ($m_v = 10^{-9}$ Pa^{-1}) in Figure 5.27. The calculated velocities for varying band thickness remain in a narrow band. Velocities reach high values (8 – 9 m/s) when the permeability is low (10^{-12} to 10^{-10} m/s). For relatively large permeabilities (higher than 10^{-8} m/s) the velocity of the slide drops to zero. In these cases the initially unstable situation is soon counter-acted by the self-stabilizing mechanism of the slide (weight transfer from the upper to the lower wedge).

The transition from the "rapid regime" to the "slow" or self-stabilizing situation occurs for permeabilities in the range 10^{-9} to 10^{-8} m/s.

Calculated temperatures for varying band permeability and band thickness are represented in Figures 5.27b,c for Wedge 1 and in Figures 5.27d,e for Wedge 2. The normal effective stress in Wedge 1 against the sliding plane is significantly smaller than the value calculated for Wedge 2. Resisting shear stresses react in the same manner and the work input for Wedge 1 is smaller if compared with Wedge 2. The consequence is that temperatures in Wedge 1 remain at moderate values in the "fast" and "slow" ranges of permeabilities. Temperature increases in the intermediate "regime" because the combination of non-negligible shear strength and a substantial sliding velocity leads to a significant mechanical work input into the band.

Band thickness controls the temperature development. A maximum temperature of 259 °C for Wedge 1 is calculated for $k = 10^{-9}$ m/s and $2e = 0.5$ mm. Temperatures are higher in Wedge 2 for the reason given and they reach a peak value close to 800 °C for $k = 10^{-9}$ m/s and $2e = 0.5$ mm. These high temperatures would require a more precise formulation of the constitutive model of the band material and, possibly, the presence of additional physical phenomena (water vapourization) which are outside the limits of this chapter. But in most cases in practice the maximum temperature calculated is moderate and the analysis developed should represent reasonably well the relevant physical phenomena.

Similar qualitative results were obtained for the softer band material ($m_v = 10^{-8}$ Pa^{-1}) (Fig. 5.28). In order to explain the results, consider the balance equation for solid and water mass (Eq. (5.41a)), written now in the following form:

$$\frac{\partial u_w}{\partial t} = \frac{k}{\gamma_w\left[m_v + n\alpha_w\right]}\frac{\partial^2 u_w}{\partial z^2} + \frac{m_v}{\left[m_v + n\alpha_w\right]}\frac{\partial \sigma_n}{\partial t} + \frac{\left[(1-n)\beta_s + \beta_w n\right]}{\left[m_v + n\alpha_w\right]}\frac{\tau_f v_{\max}}{2e\rho c_m} \qquad (5.59)$$

The rate of pore pressure change has been isolated. Use has been made also of Equations (5.39) and (5.40) which provide the relationship between temperature and rate of work input into the band. Three phenomena contribute to change pore water pressures in the band: the dissipation due to flow of water (first term), the variation of total normal external stress (second term), and the generation of pore pressures due to water and solid dilation, controlled by heat (or rate of work input) (third term). The three terms are affected by the compressibility coefficient of the band material, m_v (in the denominator).

The term $m_v + n\alpha_w$ becomes:

$$10^{-9}+0.2\times 0.476\times 10^{-9}\ \text{Pa}^{-1} = 1.095\times 10^{-9}\ \text{Pa}^{-1}\ \text{for the stiffer band}$$

$$10^{-8}+0.2\times 0.476\times 10^{-9}\ \text{Pa}^{-1} = 10.095\times 10^{-9}\ \text{Pa}^{-1}\ \text{for the softer band}$$

Therefore, the rate of heat-induced generation of pore pressures is reduced ten times when the compressibility of the material increases ten times. The sketch in Figure 5.4 also explains qualitatively the effect of rock skeleton stiffness on pore pressure generation when temperature increases: a softer rock pore accommodates better an increase in water dilation, and leads to a lower pore pressure.

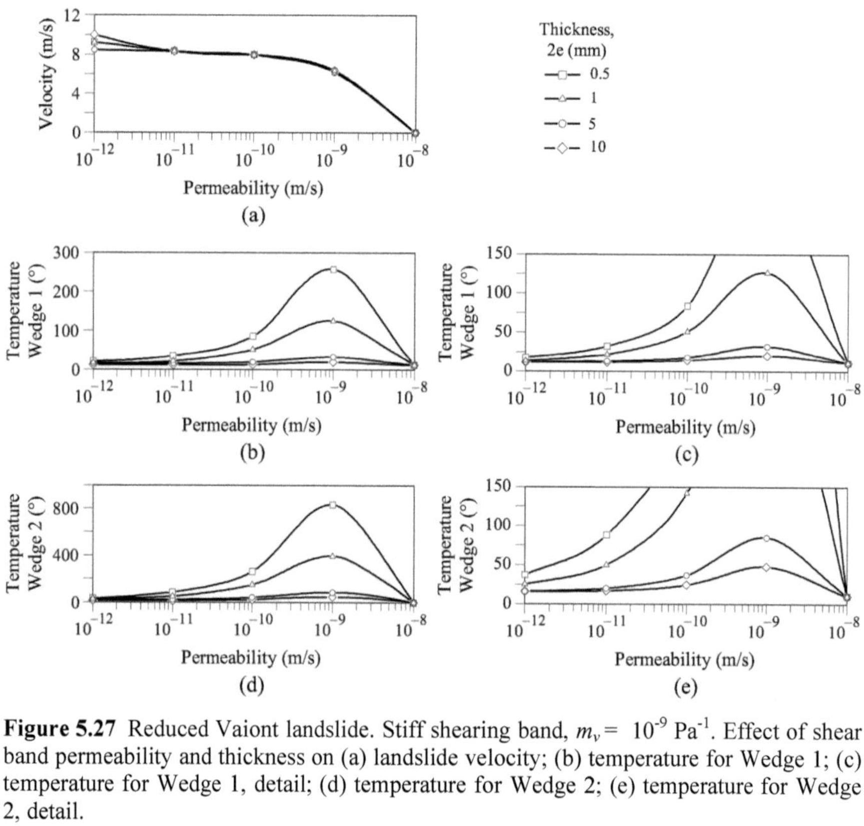

Figure 5.27 Reduced Vaiont landslide. Stiff shearing band, $m_v = 10^{-9}$ Pa^{-1}. Effect of shear band permeability and thickness on (a) landslide velocity; (b) temperature for Wedge 1; (c) temperature for Wedge 1, detail; (d) temperature for Wedge 2; (e) temperature for Wedge 2, detail.

The remaining terms in Equation (5.59) are also controlled by m_v (an increase in m_v also results in a decrease in the dissipation rate of pressures) and it is difficult to predict the final result in a particular case without actually performing the calculations. If the heat-associated term dominates, excess pore pressures, other conditions maintained, will decrease when m_v increases – softer material – and the normal effective force on the sliding surface will increase. This implies, in general terms, a higher resistance to sliding and a reduced velocity.

The plot in Figure 5.28a shows the final velocities when the slide has displaced 50 m. It indicates that the increase in m_v leads to a systematic reduction in calculated velocities for any value of permeability. A consequence of this reduction is that the transition permeability between the fast and slow regimes now ranges between 10^{-10} and 10^{-9} m/s. Another obvious consequence is that the time to reach a given displacement should increase when m_v increases.

Consider finally the effect of m_v in the development of temperatures in the shear band (compare plots in Figs. 5.27b,c,d, and Figs. 5.28b,c,d). The plotted temperatures correspond to the end of the calculation period, when the slide in all cases has reached a displacement of 50 m. Therefore, the plot provides an accumulated quantity which is proportional (in the absence of any dissipation by conduction and advection, given the fast phenomena analyzed) to the total work input during the sliding time.

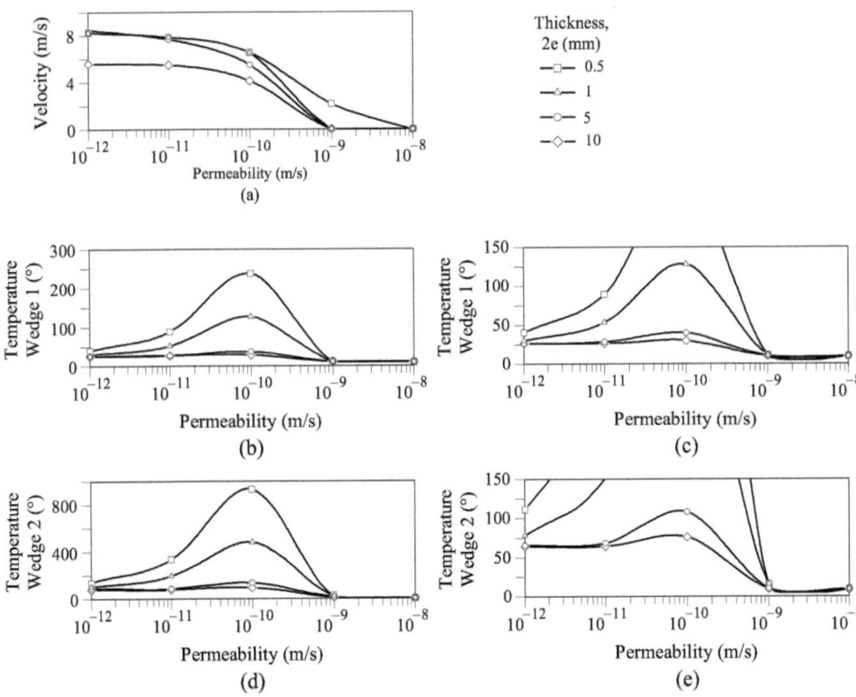

Figure 5.28 Reduced Vaiont landslide. $m_v = 10^{-8}$ Pa^{-1}. Effect of shear band permeability and thickness on (a) landslide velocity; (b) temperature for Wedge 1; (c) temperature for Wedge 1, detail; (d) temperature for Wedge 2; (e) temperature for Wedge 2, detail.

In the "fast" regime (low permeability) the pore pressure generation term for a softer soil (third term in Eq. (5.59)) is smaller than the same term for a rigid material. Therefore, in order to accelerate (pore pressures should increase until effective stresses reduce to vey low values), heat has to accumulate during a longer time for a softer material. The direct consequence is that the time to get the accelerated motion of the slide should increase for a softer material, a result already advanced. It turns out that the accumulated heat for the softer material (the integrated value of the slide velocity times the available shear strength) is higher than the calculated value for the stiffer material. As a result, temperatures increase in the case of softer materials. This is shown in the plots for the low range of permeabilities. When the permeability increases and enters into the "slow" regime, the comparison of the temperature results for the two m_v values cannot be discussed in the preceding terms because the slide stops soon after the initial instability because of the fast dissipation of excess pore pressures.

The preceding set of comments illustrates the intricate coupling among the different phenomena and the difficulty to make predictions on the basis of a simple reasoning.

This is a case in which predictions, even if they are qualitative, require the help of a computational tool.

5.9 Mitigation Measures

In 1960, engineers in charge of the dam were already aware of the fact that the left margin of the Vaiont River was a very large ancient slide whose mobilized volume was roughly estimated as 260×10^6 m^3. Previously, in March 1959, a big landslide, whose volume was estimated in 3×10^6 m^3, slid rapidly into the Pontesei reservoir, built on the Piave river, North of Longarone. It created a huge wave, 25 m high, which flew over the dam, although no damage was reported. Also, in 1960, a relatively large slide, close to 10^6 m^3, fell into the Vaiont reservoir and created a 2 m high wave.

These two events prompted the performance of a hydraulic impact test at the University of Padova. Reduced scale tests were performed. The landslide volume was simulated by means of gravel that fell into the reservoir water. They reported that a maximum wave height of 26 m could develop in Vaiont, in the case of full landslide. It was estimated that the consequences of this maximum wave could be managed in an acceptable way.

Predicting the wave generation induced by a moving mass entering a body of water requires information on the shape and velocity of the mass, as well as the bathymetry of the lake invaded.

In the case of Vaiont, the possibility of a major slide into the reservoir was well accepted. It was feared that the slide could effectively close the valley. In fact, a diversion tunnel was built on the right margin to allow the connection between the two resulting isolated volumes of the reservoir.

However, the landslide speed could not be predicted or even imagined by engineers. The hydraulic tests performed in Padova were inaccurate and this was attributed to the use of gravel to simulate the slide. In tests performed in 1974, when a similar risk was investigated for the reservoir of Libby dam, it was found that the height of the wave generated in the lake was very much dependent on the details of the shape of the material used to simulate the slide (Leonards, 1987). If a solid vertical wall is used instead of a mass of gravel, the wave height increases substantially. The fact is that the predicted wave was 26 m against the actual wave heights over the Vaiont dam (150 and 250 m on the left and right abutments, respectively).

Summarizing what has been discussed here and in Chapter 2, the critical path to adopt (or not) remedial measures was:

a) to decide how the initiation of the slide could be controlled. Straight use of static models does not offer much confidence and the observational method adopted seems a conceptually correct decision. But the observational approach also requires a model to interpret field measurements. A model was implicitly adopted and it is essentially derived from the plot in Figure 2.12. Observations were interpreted in the sense that a reduction in reservoir level results in an immediate reduction of sliding velocity. Unfortunately, this conclusion is not based on a precise mechanical analysis of the phenomenon, especially in the presence of progressive failure, not to mention the thermo-hydro-mechanical interactions developed in this chapter. The necessary knowledge and calculation tools were not available in 1960 and, even if some progress was made in the following decades, the fundamental aspects remain obscure.
b) by accepting the failure as a probable event, the next step is to ascertain its consequences. The main concern was the possibility of water spilling over the dam (and possibly the damage of the dam itself if it becomes directly hit by the slide). The efforts made (reduced-scale hydraulic tests) suffered two limitations: the lack of any basis to decide the sliding velocity, a subject currently under research, and the effect of some testing details (the actual representation of the sliding mass) on the results.

In neither of these two aspects there was reliable information at the time of failure. Forty-five years afterwards, many theoretical and practical aspects remain insufficiently known: the limitations to build reliable slope stability models, which include time effects and phenomena such as progressive failure when brittle materials are involved; the difficulty to know the state of stressing and available strength in ancient landsliding areas; the difficulties to perform accurate geotechnical investigations at great depths in a rugged terrain, and the lack of established and verified criteria to decide when an impending slide may accelerate and reach vey high velocities. This chapter provides some answers to the last question.

5.10 Lessons Learned

5.10.1 Heating a saturated soil or rock

Heating leads to a pore water pressure rise in saturated soils and rocks. It has been shown that this phenomenon is a consequence of the thermal dilation of water and solid skeleton and of the principle of effective stress. Excess pore pressures dissipate in a consolidation type of process. Therefore, the pore pressure reached at a particular time depends on the thermal dilation coefficients, as well as on the consolidation parameters, namely permeability and stiffness.

5.10.2 Consequences of strain localization and the First Law of Thermodynamics

Field observations indicate that translational and rotational slides move largely as solid bodies sliding on thin bands where shear strains are highly localized. Therefore, as the slide displaces, strain work is concentrated in a small volume. On the other hand, the First Law of Thermodynamics states that the change in internal energy of a closed thermodynamic system (in our case, the narrow sliding band) is equal to the amount of heat energy supplied and the mechanical work done on the system. In our case, there are no direct sources of heat supplied to the band and, therefore, the increase in internal energy of the sliding band is equal to the plastic straining work associated with sliding. The increase in internal energy manifests as an increase in temperature and, in view of the previous point, excess pore water pressures will be induced in the sliding band. The immediate consequence is a reduction of effective stresses and of the associated shear strength. This mechanism leads, under appropriate circumstances, to a complete loss of resisting strength available at the sliding surface. Then, the unbalanced driving forces lead to an accelerated slide motion.

5.10.3 Formulating coupled thermo-hydro-mechanical (THM) phenomena in the shear band

This chapter provides a step-by-step procedure to formulate THM problems of fairly general nature in a saturated porous material. The particular geometry of the shear band (narrow thickness and very large lateral extension) makes the problem one-dimensional in practice and facilitates its solution. Another implication is that the THM problem associated with the shear band may be formulated independently from the geometry of the slide and its specific kinematic mechanism. The solution given for the THM band problem may be applied to other slide configurations not covered in the chapter, whenever the effect of heat-induced pore pressure development is sought.

5.10.4 Dynamics of Vaiont

A seemingly convincing explanation for the accelerated motion of Vaiont relies on the development of excess pore pressures generated by the temperature increase of the sliding surface. This is a consequence of the slide motion itself. A key condition to explain the phenomenon is the existence of a basal sliding plane located in a layer of low permeability high plasticity clay. Under these conditions, the self-feeding mechanism of pore pressure generation in the sliding surface may eventually lead to very high sliding velocities (> 25 m/s), which are reached in a few seconds (~30 s) even if proper account is given to the self-stabilizing evolving geometry of the slide and even if progressive failure mechanisms potentially acting on internal shearing surfaces are not considered.

Although the two-wedge analysis described here provides a reasonable explanation for the final catastrophic motion of the slide, the previous history of landslide creep-like displacements (Figs. 2.3. 2.4 and 2.12) cannot possibly be reproduced with the model described in this chapter. Other phenomena such as viscous-strength components at the failure surface or the strength degradation of the rock mass could be invoked to approximate the measured velocities prior to failure. However, additional limitations can be identified both in the model and in the available information: the geometry was kept two-dimensional and as simple as possible; pore water pressures prevailing at the failure surface were never measured; the effect of previous rainfall regime is essentially unknown beyond the condensed information offered by Figure 2.13; the actual conditions (in particular, the continuity of the high plasticity clay layer) and a significant proportion of the sliding surface remain buried by the slide and are essentially unknown. Therefore, complexities and uncertainties around Vaiont are far from being solved. However, it remains as a fascinating case and a permanent source of inspiration in the field of landslide analysis.

5.10.5 Relevant parameters to understand the dynamics of the motion

Slide geometry and strength properties of the sliding surface(s) are not enough to understand the dynamics of Vaiont. Three parameters have been found important to explain the motion: the thickness of the sliding band, its permeability, and its (confined) stiffness. Permeability is the major player. The sensitivity analysis performed for a slide of medium-high dimensions (one tenth of Vaiont) has shown that, below a certain permeability threshold (established around 10^{-8} to 10^{-9} m/s for a "stiff" band ($m_v = 10^{-9}$ Pa^{-1}) and 10^{-9} to 10^{-10} m/s for a "soft" band ($m_v = 10^{-8}$ Pa^{-1})), the maximum pore pressure development in the shear-band, which is the value controlling the shear strength, is not much affected by the band thickness, within a reasonable range of values. Above this threshold permeability value, pore pressure dissipation is enough to de-activate the process of pore pressure build-up and, therefore, the slide does not accelerate. In other words, the threshold permeability identified marks the transition from a potentially risky slide to a safe one. Of course, this conclusion is valid for the slide geometry analyzed (a scaled Cross-section 5 of Vaiont) and it should not be extended to other sliding configurations without further analysis.

5.10.6 Extreme phenomena

In very large landslides (Vaiont case), when conditions for accelerated motion exist, there are critical combinations of band permeability and band thickness that result in a substantial and rapid increase in shear band temperature. This is a natural outcome of the formulation and it is a consequence of the availability of small – but not negligible – shear strength in the shear band and an increasing shear strain rate as sliding velocity increases. The permeability of the band in these cases is low enough to maintain a significant pore pressure in the band but high enough to maintain a non-negligible effective normal stress. The calculated temperatures (hundreds or even thousands of °C) are enough to induce water vapourization and rock melting. These phenomena are not covered by the formulation developed, which only explains water pressure increase as a result of thermal dilation effects. For this reason, no reliable conclusions should be derived from the high temperatures calculated for some critical values of the band permeability. However, the estimated sliding-band parameters, in the case of Vaiont, lead to a moderate increase in temperature (< 100°C), which is not able to vapourize the interstitial pore water of the clayey band.

When the size of the slide decreases, the temperature generated in the band also decreases because the work input into the band decreases. A reduction of Vaiont dimensions by a factor of 1/10 still leads to a very large slide (a few million cubic metres), which has been analyzed. Sliding band temperatures, in this case, are substantially lower. For an impervious band ($k < 10^{-10}$ m/s) or for a pervious band ($k > 10^{-8}$ m/s), maximum temperature increments are moderate (a few degrees). In extreme cases, for critical k values of the sliding band, it is unlikely for temperature to raise more than a few hundreds of °C. Rock melting is excluded in these cases and water vapourization, in clay rich naterials, in uncertain. Since most slides do not reach, in practice, such a volume (a few million cubic meters), it appears that water vapourization and rock melting are extreme phenomena that rarely occur in practice.

The fact that temperature increases will likely remain moderate or low in most slides does not prevent, however, the development of significant velocities. The reason is that the reduced increase in pore water pressure in those cases is also matched by a reduced normal effective stress on the sliding surface. Therefore, the condition of zero effective stress may also be reached during motion. However, the smaller the slide is, the shorter the sliding path necessary to substantially change its geometry, to evolve to another type of motion, or to be affected by another geometrical restriction to its motion. These considerations added to the reduced momentum of the slide tend to limit the danger associated with smaller slides.

5.11 Advanced Topics

Analytical and numerical procedures are available to investigate the run-out distances and velocities reached by landslides. In recent studies (Hungr, 1995; McDougall and Hungr, 2004; Quecedo *et al.*, 2004), the slide is idealised as a fluidized mass of soil and the Navier–Stokes equations are integrated in depth, adapted to the curved geometry of the surface, and solved for some rheological models adopted for the moving mass. In other approaches conceived for rockslides, discrete element approaches and hybrid continuum-discontinuum models have been developed (Eberhardt *et al.*, 2004).

However, Vaiont is a different case. Changes in the slide geometry during sliding did not imply a change in the fundamental sliding mechanism, which may be approximated by one or several rigid bodies bounded by "thin surfaces" subjected to intense shearing. This was the implicit approach of all the contributions mentioned in the chapter. Of particular relevance is the analysis presented by Vardoulakis (2002). He approximated the sliding surface by means of a circle. To ensure initial equilibrium conditions the basal friction angle in this case has to increase to 20°. Vardoulakis (2002) also introduced heat dissipation away from the sliding band and the volumetric contraction ("collapse" was the term used) of the clay on the band due to the increase in temperature. A closed-form solution for planar slides was given by Pinyol and Alonso (2009).

Appendix 5.1. Finite difference approximation of the system of Equations (5.29) to (5.33)

A forward finite difference procedure was developed to solve the system of Equations (5.29) together with the initial and boundary conditions given in Equations (5.30) – (5.33). Consider in Figure A5.1 the domain of integration. The one-dimensional spatial domain is subdivided into n small elements of thickness Δz. The z_i coordinate of any point is defined by an index i such that $z_i = i\Delta z$. The following indices define singular points: $i = n_0$ corresponds to $z = 0$; $i = n_e$ to $z = e$. The far boundary is located at a distance $z = L$, where $i = n_L$. The horizontal axis in Figure A5.1 corresponds to time. The system of equations will be solved for each time interval Δt.

Time derivatives at any time t, for $z = z_i$ can be approximated by (Forward Euler Method):

$$\left.\frac{\partial f}{\partial t}\right|_{z_i} \approx \frac{f(z_i, t+\Delta t) - f(z_i, t)}{\Delta t}, \tag{A5.1}$$

where $f(z,t)$ is a general function of position (z) and time (t).

The first and second derivative with respect to z will be approximated by a central difference

$$\left.\frac{\partial f}{\partial z}\right|_{t} \approx \frac{f(z_{i+1}, t) - f(z_{i-1}, t)}{\Delta z}, \tag{A5.2a}$$

$$\left.\frac{\partial^2 f}{\partial z^2}\right|_{t} \approx \frac{f(z_{i+1}, t) - 2f(z_i, t) + f(z_{i-1}, t)}{(\Delta z)^2}. \tag{A5.2b}$$

Once the numerical approximations of the derivatives (Eqs. (A5.1) and (A5.2)) are substituted into the system of Equations (5.29c,d,e), the following discrete set of equations is obtained:

$$\begin{aligned} &u_w(z_i, t+\Delta t) = u_w(z_i, t) + \\ &c_v \frac{\Delta t}{\Delta z^2}\left[u_w(z_{i+1}, t) - 2u_w(z_i, t) + u_w(z_{i-1}, t)\right] + \\ &+ c_H \Delta t\, H(t) \quad \text{for } i = n_1, n_e, \end{aligned} \tag{A5.3a}$$

$$u_w(z_i, t+\Delta t) = \\ u_w(z_i, t) + c_v^r \frac{\Delta t}{\Delta z^2}\left[u_w(z_{i+1}, t) - 2u_w(z_i, t) + u_w(z_{i-1}, t)\right] \text{ for } i = n_e, n_L \tag{A5.3b}$$

$$v_{\max}(t+\Delta t) = v_{\max}(t) + \frac{\Delta t}{M}\left[W \sin(\beta) - \tau_f(t)\right]. \tag{A5.3c}$$

In these equations,

$$c_v = \frac{k}{(m_v + n\alpha_w)\gamma_w} \tag{A5.4}$$

is the consolidation coefficient of the shear band material;

$$c_v^r = \frac{k_r}{(m_v^r + n\alpha_w)\gamma_w} \tag{A5.5}$$

is the consolidation coefficient of the material outside the shear band; and

$$c_H = \frac{(1-n)\beta_s + \beta_w n}{(m_v + n\alpha_w)\rho c_m} \tag{A5.6}$$

is a parameter that integrates the dilation coefficients of water and solid, the compressibility of soil skeleton and water and the mean specific heat of the soil.

Equations (A5.3a,b) are explicit mathematical expressions for the excess of pore pressure in a point z_i, at a given time ($t + \Delta t$), if the old values (at the previous time, t) in three points: point z_i and the points just above and below (z_{i-1} and z_{i+1}), are known. This calculating procedure is graphically illustrated in Figure A5.1. Equation (A5.3) provides the new value of the maximum velocity as a function of the old values (previous step) of maximum velocity and excess of pore pressure at $z = 0$. Heat (H) and effective frictional strengh (τ_f) are given by Equations (5.29a,b) at the previous time t. It appears, therefore, that a forward marching procedure has been devised to calculate the independent variables (excess pore water pressure and velocity). The procedure requires that initial and boundary values are defined.

Initial and boundary conditions must also be expressed in a numerical way. The symmetry condition at $z = 0$ (Eq. (5.31)), valid at any time, can be approximated by extending the domain with an additional interval from $z = 0$ to $z = z_{-1} = -\Delta z$. Then, if the excess pore pressure at $z = z_{-1}$, $u_w(z_{-1}, t)$ is forced to be

$$u_w(z_{-1}, t) = u_w(z_1, t) \tag{A5.7}$$

at any time, the condition

$$\left.\frac{\partial u_w}{\partial z}\right|_t = 0 \text{ in } z = 0$$

is automatically satisfied in view of Equation (A5.7).

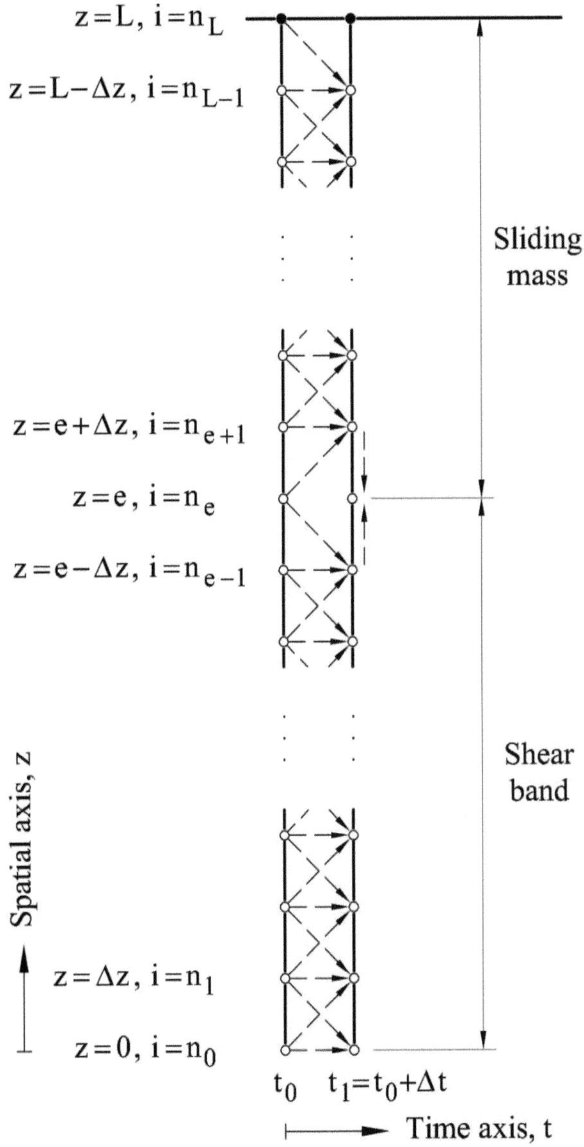

Figure A5.1 Dynamics of a planar infinite slope. Domain of integration.

The general expression (A5.3a) for $z = 0$ can now be written

$$u_w(z_0, t+\Delta t) = u_w(z_0, t) + c_v \frac{\Delta t}{\Delta z^2}\left[2u_w(z_1, t) - 2u_w(z_0, t)\right] + \Delta t\, c_H H(t). \tag{A5.8}$$

The numerical expression of the boundary condition at the edge of the shear band ($z = e$) (Eq. (5.32)), is obtained by means of a forward finite difference as follows:

$$k\frac{u_w\left(z_{n_e}, t\right) - u_w\left(z_{n_{e-1}}, t\right)}{\Delta z} = k_r \frac{u_w\left(z_{n_{e+1}}, t\right) - u_w\left(z_{n_e}, t\right)}{\Delta z}. \tag{A5.9}$$

This equality allows the calculation of excess pore water in $z = e$ at any time as a function of the values of excess pore water in the points just above and below (at the same time):

$$u_w\left(z_{n_e}, t\right) = \frac{k_r u_w\left(z_{n_e+1}, t\right) + k\, u_w\left(z_{n_e-1}, t\right)}{k_r + k}. \tag{A5.10}$$

The numerical expression of boundary condition at the upper limit of the discretization (Eq. (5.33)), where the excess of pore pressure must be zero, is simply

$$u_w\left(z_{n_L}, t\right) = 0. \tag{A5.11}$$

Regarding the initial conditions (Eqs. (5.30)), the numerical equivalents are given by

$$u_w(z_i, t_0) = 0 \text{ for } i \in [1, n_L], \tag{A5.12}$$

$$v(z_i, t_0) = 0 \text{ for } i \in [1, n_L], \tag{A5.13}$$

$$\theta(t_0) = \theta_{ref} \text{ for } i \in [1, n_L]. \tag{A5.14}$$

At the initial time (t_0), all values are known. The excess pore water pressure in the next time increment can be calculated by means of Equations (A5.3a,b) in $[n_0, n_e)$ and $(n_e, n_L]$, respectively. Note that n_e is not included in those intervals. However, the continuity condition, expressed in Equation (A5.10), provides the new value of excess pore pressure at $z = e$.

The value of the maximum velocity at the first time step is obtained by means of Equation (A5.3c) (with $v_{\max}(t_0) = 0$). Once velocity and excess of pore pressure are known at the new time step, the new value of heat and effective frictional strength can be calculated through Equations (A5.3a,b). Also, the temperature in the shear band can be obtained, at each time step, by means of Equation (5.23), writing it in a numerical form as follows:

$$\theta(t+\Delta t)=\theta(t)+\Delta t\frac{H(t)}{\rho c_m}. \tag{A5.15}$$

Summarizing, the numerical solution of the problem starts at the boundaries where the values of the variables are known at any time. The set of discretized equations provides, step by step, all the unknowns at any time and position of the domain of integration. Since initial conditions of excess pore pressure at any point, as well as the initial temperature are zero, the value of pore water pressure at the first time step ($t = \Delta t$) will be zero.

These approximations lead to a Forward Euler Method of integration which may be numerically unstable if Δt is larger than the stability limit, which is a function of material parameters and Δz. If the solution becomes unstable, the calculated values of the unknown function display an oscillatory behaviour in time that prevents convergence. Convergence of explicit integration schemes of standard parabolic equations (such as the consolidation equation) is achieved if the time and spatial increments satisfy the condition:

$$\frac{c_v \Delta t}{\Delta z^2} \le 0.5. \tag{A5.16}$$

This condition applies to the homogeneous part of the parabolic equation (Nakamura, 1991) and it may be thought that our field Equation (5.29c) leads to a similar relationship. Unfortunately, the "independent" term (proportional to heat input $H(t)$) in Equation (5.29c) is a function of pore pressure, through Equations (5.29a,b,e). Nevertheless, the preceding condition has been accepted as a reference in the calculations presented below. In general, care has been taken to check that the calculated pore pressures did not change for time steps below a certain value used in calculations.

The calculated velocity will increase as long as the sliding mass is unbalanced (driving forces exceed resisting ones). This is the case, even if the excess pore pressure at initial time is zero, because a positive increment of velocity will be calculated. Then, the positive value of generated heat, due to the velocity reached at the first time step, will result in an increment of temperature and excess pore pressure in the shear band. At the following time step, this positive excess pore pressure will reduce the effective frictional strength and will accelerate the slide mass. The slide will start to move in an accelerated motion.

The numerical procedure described above has been programmed in Fortran 90. The program is included in Appendix 5.2 to show all the details of the computational procedure and to allow the reader to perform its own calculations.

Appendix 5.2 Flowchart and Computer Program for the Dynamic Analysis of the Infinite Planar and Two-Wedge Slides

The numerical procedures described above for the dynamic analysis of the infinite slide and the two interacting wedges was programmed in Fortran 90. The complete code is included at the end of this appendix.

The program is subdivided in two main branches (Fig. A5.2) attending to the type of slide by means of two subroutines: *Infinite_planar_slide* and *Two_interacting_wedges*.

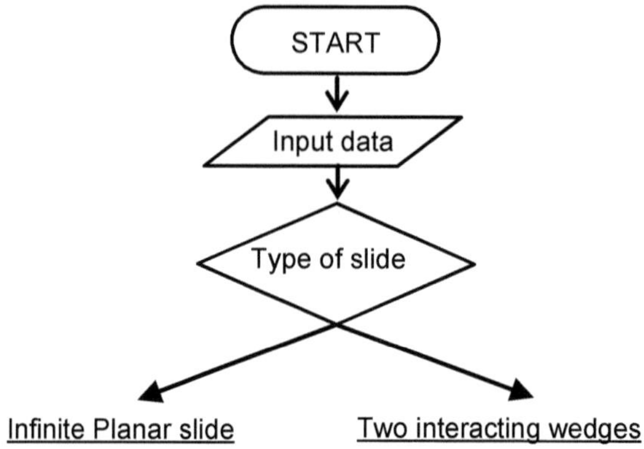

Figure A5.2 Flowchart of the main program.

The main program is structured in three parts. First, all the input parameters that can be modified by the user are defined: material parameters, reference temperature, type of the slide and its geometry, parameters for the discretization by finite differences, and control parameters. Regarding the discretization parameters, the user has to define the number of spatial steps in the shear band and in the sliding mass. As the thickness of the shear band is previously defined, the length of spatial increment (Δz) and the position of the coordinate of the upper boundary (L) is determined by default. The time interval (Δt) is defined, by default, by limiting the value of the stability coefficient ($c_v\,\Delta t/\Delta z^2$) to 0.3 in order to guarantee the stability.

Second, auxiliary parameters and constants are calculated by the program. In general, this part should not be modified by the user. Finally, in the third part of the main program, a subroutine is called depending on the type of slide.

The numerical procedure starts in the subroutines. Results are calculated and stored in external files, within the same subroutines. At intervals defined by the time frequency (*Dtw_time*), which is specified by the user in the main program, the values of velocity, excess pore pressure in the middle of the shear band, heat, displacement, shear strength, and temperature are written in data files. Excess pore pressure profiles for $z = 0$ to $z = L$ are also stored in external files following the time frequency (*Dtw_profile*) specified. The numerical procedure goes on until the maximum displacement (*displ_max*) is reached or until the slide velocity becomes zero (the landslide stops) for the case of two interacting wedges.

The implementation of the numerical procedures in the subroutines follows the flow chart diagrams included in Figures A5.3 and A5.4.

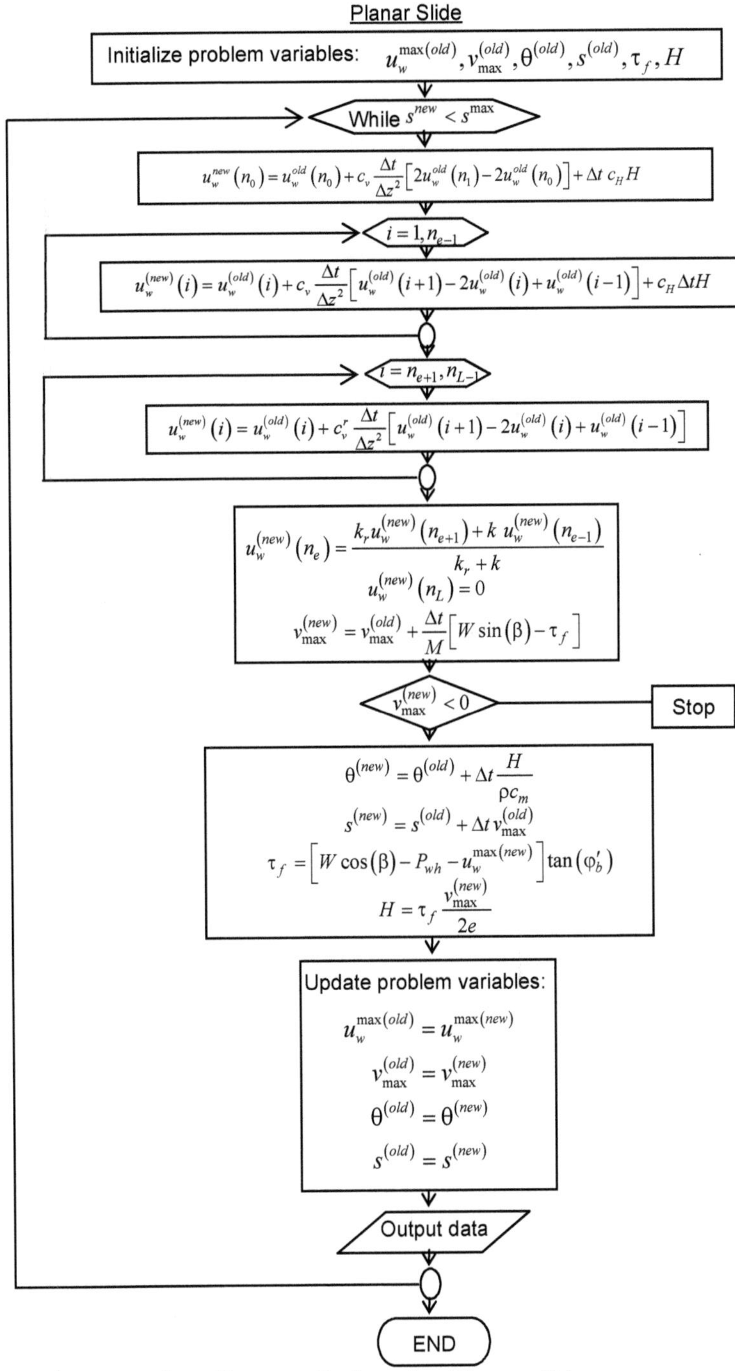

Figure A5.3 Flow chart diagram of subroutine *Planar_slide*.

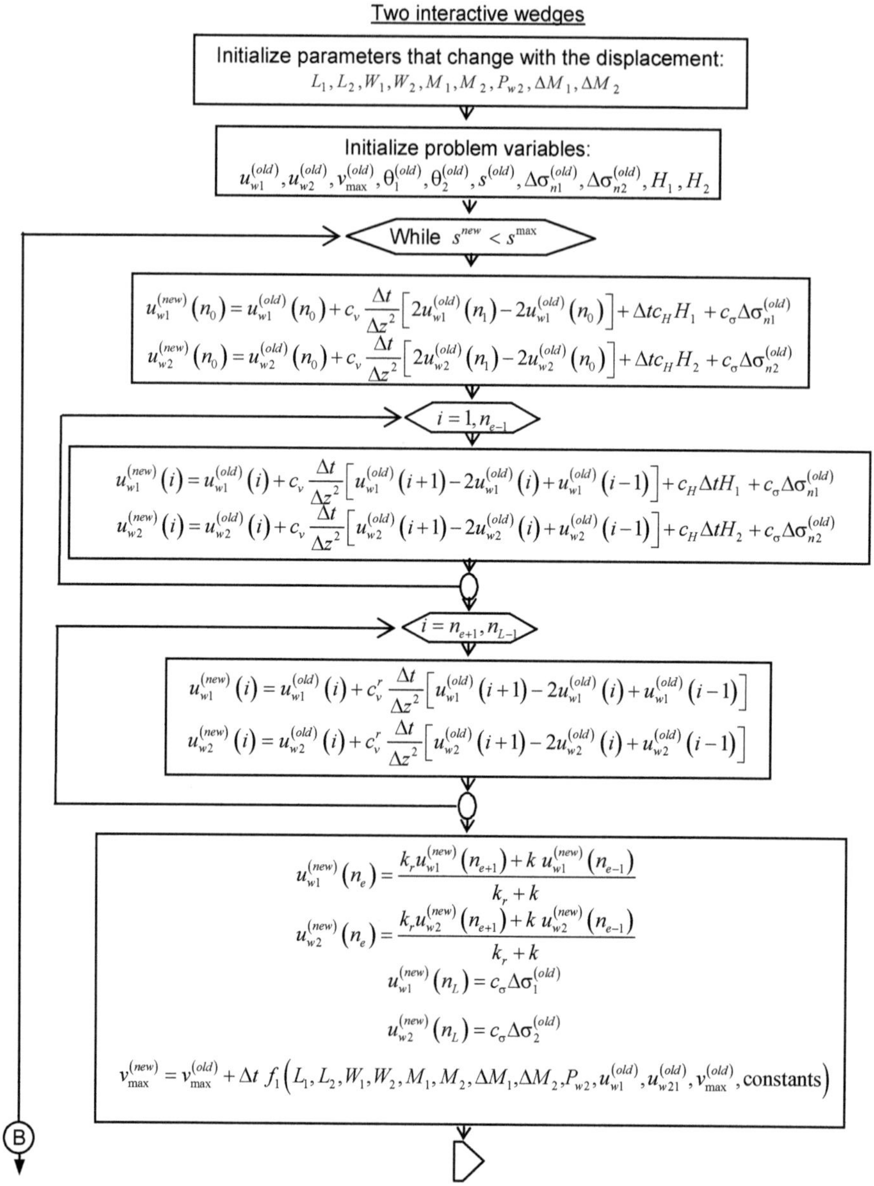

Figure A5.4 Flow chart diagram of subroutine *Two_interacting_wedges.*

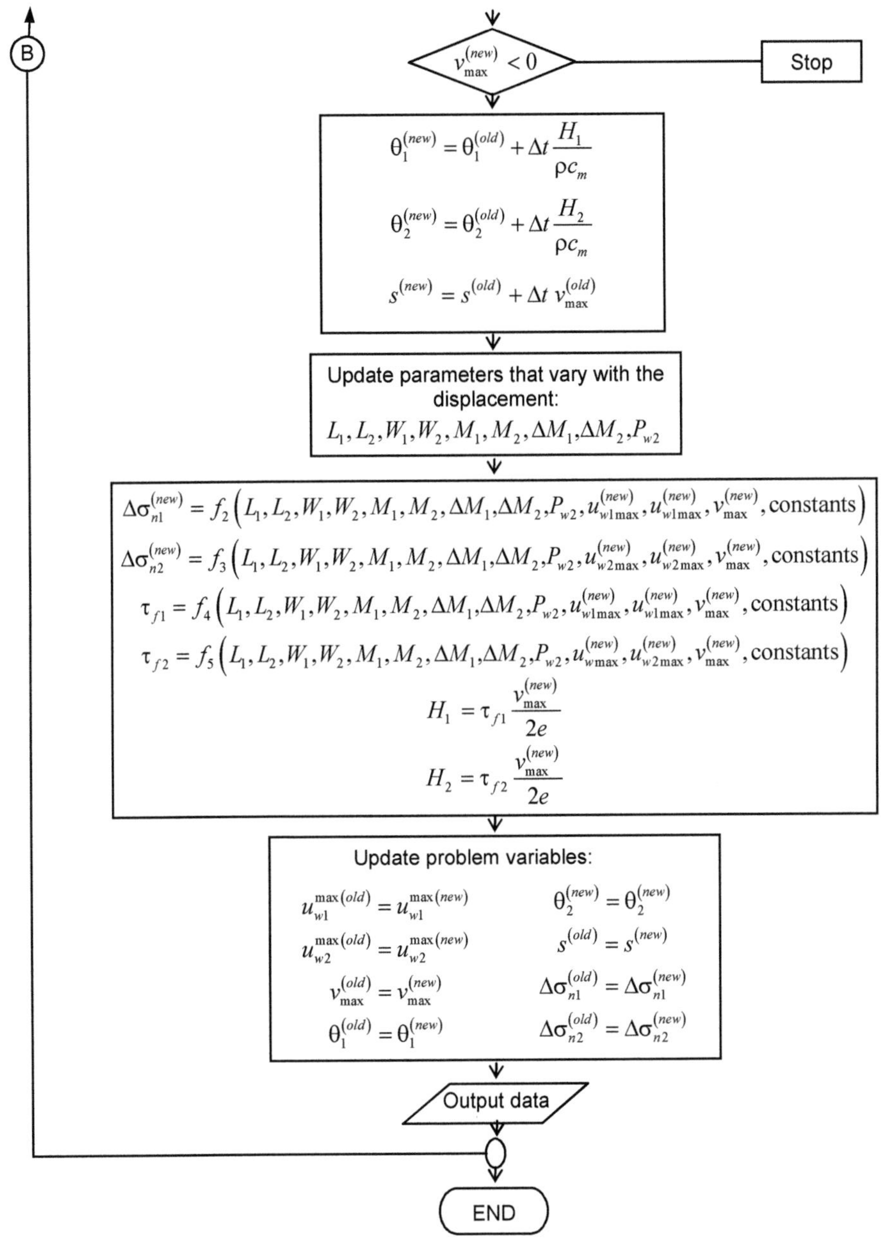

Figure A5.4 (continued) Flow chart diagram of subroutine *Two_interacting_wedges.*

```
!**********************************************************!
!                                                          !
!                        PROGRAM                           !
!                                                          !
!          VAIONT LANDSLIDE: DYNAMIC ANALYSIS              !
!                                                          !
!**********************************************************!

   implicit real*8 (a-h,o-z)

!**********************************************************!
! PARAMETERS TO BE DEFINED BY THE USER
!**********************************************************!
   pi =3.141592654

! MATERIAL PARAMETERS
!Water parameters
   delta_w = 1.e3         !kg/m3; density
   gamma_w = 9800.0       !N/m3; specific weight
   alpha_w = 5.e-10       !1/Pa; compressibility coefficient
   beta_w = 3.4e-4        !1/°C; thermal dilation coefficient
   c_w =4186.0            !J/(kg·°C); specific heat
!Solid parameters
   delta_s = 2.7e3        !kg/m3; density
   beta_s = 3.0e-5        !1/°C; thermal dilation coefficient
   c_s = 837.2            !J/(kg·°C); specific heat
!Shear band parameters
   zn_band = 0.2          !porosity
   zk_band = 1.e-11       !m/s; permeability
   zmv_band =1.5e-9       !1/Pa;1D compressibility coefficient
   fib=12.0*pi/180.       !rad; effective frictional angle in the band
!Sliding mass parameters
   zn_rock = 0.2 !porosity
   delta_rock = 2350.     !kg/m3; density
   gamma_rock =23500.     !N/m3: specific weight
   zk_rock = 1.e-11       !m/s; permeability
   zmv_rock = 1.5e-9      !1/Pa; 1D compressibility coefficient
   fir = 38.*pi/180.      !rad; effective frictional angle rock-rock
   coher =762.247e3       !Pa; cohesion rock-rock
   cc = 0.d0              !1/m; rock cohesión degradation rate with
   the displacement (if it is equal to zero, there is no degradation

! INITIAL CONDITIONS
!Initial excess pore pressure, velocity and displacement have been
!imposed equal to zero by default
   theta_ref =10.0        !°C; reference temperature

! TYPE OF SLIDE
   ntype_failure = 2      !1:infinite planar slide
                          !2:two interacting wedges

! GEOMETRIC PARAMETERS AND HYDROSTATIC FORCES
   e = 2.5e-3             !m; 2e = thickness of shear band
   if (ntype_failure.eq.1) then
    zHeight =240.         !m; thickness of planar landslide
    zHeight_w =120.       !m; height of phreatic level
    beta_slope = 9.490*pi/180. !rad; slope angle
   else if (ntype_failure.eq.2) then
```

```
     zH_wedge1 = 510.0    !m; height of upper triangular Wedge 1
     zH1_wedge2 = 260.0   !m; left height of lower Wedge 2
     zH2_wedge2 = 260.0       !m; right height of lower Wedge 2
     Base_wedge1 = 700.0      !m; horizontal length of Wedge 1
     zL2_0_p1 = 240.          !m; partial base length (1) of Wedge 2
     zL2_0_p2 = 320.          !m; partial base length (2) of Wedge 2
     alpha = 36.07*pi/180.        !rad; angle Wedge 1
     beta = atan(zH2_wedge2/zL2_0_p2)  !rad; angle Wedge 2
     zHeight_w = 120.1        !m; water level
    endif

! CONTROL VARIABLES
    displ_max = 400.0 !m; maximum displacement allowed in calculation

! DISCRETIZATION (FINITE DIFFERENCE)
    nze = 500          !num. of spatial intervals between z=0 and z=e
    zL = 0.02          !m; coordinate of upper boundary
    sfb = 0.3          !stability factor of the band for difference
                       !approximation

! CONTROL PARAMETERS OF OUTPUT DATA
    Dtw_profile=2.        !seconds; Time between successive writings
                          !of pore pressure profiles
    Dtw_time=1.           !seconds; Time between successive writings
                          !of problem variables varying in time

!*****************************************************************!
!            AUXILIAR PARAMETERS COMPUTED BY THE PROGRAM
!*****************************************************************!

!Auxiliary material parameter
    gamma_s = 9.8*delta_s        !N/m3; specific weight
    c_delta_band = (1.-zn_band)*delta_s*c_s+zn_band*c_w*delta_w
                                 !Pa/°C;specif heat*density
    cv_band = zk_band/(gamma_w*(zn_band*alpha_w + zmv_band))
                                 !1/s; consolidation coef. of the band
    c_delta_rock = (1.-zn_rock)*delta_s*c_s+zn_rock*c_w*delta_w
                                 !Pa/°C;specific heat*density
    cv_rock = zk_rock/(gamma_w*(zmv_rock+zn_rock*alpha_w))
                                 !1/s; consolidation coef. of the
                                 !sliding mass

!Auxiliary geometric parameters and hydrostatic forces
    if (ntype_failure.eq.1) then
     Pw = zHeight_w*gamma_w      !N; hydrostatic force on the base
     zmass=zHeight*delta_rock*cos(beta_slope)  !kg/m; mass per unit of
                                               !length
     weight = zHeight*gamma_rock*cos(beta_slope)  !N/m;total weight
                            !of the sliding mass per unit of length
    else if (ntype_failure.eq.2) then
     zL1_0 = sqrt(zH_wedge1*zH_wedge1+Base_wedge1*Base_wedge1)
                            !m; base of wedge 1
     zL2_0 =    zL2_0_p1+zL2_0_p2 !m; total base length of wedge 2
     delta = 0.5*(pi-alpha)
     shi = 0.5*alpha
     Pwint = 0.5*zHeight_w*zHeight_w*gamma_w/sin(delta)
                        !N; hydrostatic force on common shearing
                        !plane between wedges
```

```
      Pwf = 0.5*zHeight_w*zHeight_w*gamma_w/sin(beta)
                          !N; hydrostatic force on the right edge of
                          !lower wedge 2
      zl_shpl0 = zH1_wedge2/cos(shi)
                          !m; length of common shearing plane between
                          !wedges
      Area1_0 = 0.5*zL1_0*zl_shpl0*cos(shi)
                          !m2; initial area of wedge 1
      ddd = zH1_wedge2/tan(delta)
      Area2_0 =0.5*zH1_wedge2*ddd+
     . 0.5*(zH1_wedge2+zH2_wedge2)*(zL2_0_p1-ddd)+
     . 0.5*zH2_wedge2*zL2_0_p2   !m2: initial area edge 2
      Pw1_0 = 0.5*zHeight_w*zHeight_w*gamma_w/sin(alpha)
                          !N; hydrostatic force on base of wedge 1
      Pw2_0 = zHeight_w*zL2_0*gamma_w
                          !N; hydrostatic force on base of wedge 2
     endif

!Constants in the balance equations for mass and heat
     cv_band = zk_band/(gamma_w*(zn_band*alpha_w + zmv_band))
     c_heat = -((zn_band*beta_w+(1-zn_band)*beta_s))/
     . (c_delta_band*(zn_band*alpha_w + zmv_band))
     c_sigma_vertical = zmv_band/(zn_band*alpha_w + zmv_band)
     cv_rock = zk_rock/(gamma_w*(zn_rock*alpha_w + zmv_rock))

!Auxiliary parameter for the discretization by finite difference
     Dz = e/(nze-1)           !length of spatial intervals
     nz = int(zL/Dz)          !num. of spatial intervals
     Dt = sfb*Dz*Dz/cv_band   !seconds; time intervals

!Verification stability factor in sliding mass (rock)
     sfr = cv_rock*Dt/(Dz*Dz)
     write (6,*) 'Stability factor rock=',sfr  !Output data on screen
     write (6,*) 'Dz=', Dz,' Dt=',Dt !Output data on screen

!Auxiliary parameters of output data
     int_write_profile=Dtw_profile/Dt    !Number of time intervals
            !between successive writings of pore pressure profiles
     int_write_time=Dtw_time/Dt   !Number of time intervals between
            !successive writings of problem variables varying in time

!*********************************************************************!
!                        CALCULATION STARTS
!*********************************************************************!

     if (ntype_failure.eq.1) then
      call Infinite_planar_slide (Dz,Dt,nz,nze,e,
     .              theta_ref,
     .              cv_band,c_heat,cv_rock,
     .              zHeight,Pw,zmass,weight,
     .              sfb,sfr,zk_rock,zk_band,
     .              c_delta_band,c_delta_rock,
     .              fib,beta_slope,displ_max,
     .              int_write_profile,int_write_time)
     else if (ntype_failure.eq.2) then
      call Two_interacting_wedges (Dz,Dt,nz,nze,e,
      .             nt,tmax,
      .             v0,Ubt0,teta_ref,
```

```
     .              zL1_0,zL2_0,zh_cha0,
     .              Area1_0,Area2_0,
     .              zHeight,Em1,Em2,Pw1_0,Pw2_0,
     .              gamma_roca,delta_roca,
     .              alpha,beta,shi,fir,
     .              coher0,cc,
     .              cv_band,c_heat,c_sigma_vertical,cv_rock,
     .              sfb,sfr,zk_rock,zk_band,
     .              c_delta_band,gamma_w,
     .              fib_max,fib_min,v_max_fib,a_const,
     .              displ_max)
   endif

   end

!*******************************************************************!
!                 DYNAMIC ANALYSIS OF INFINITE SLIDE                !
!*******************************************************************!

   subroutine Infinite_planar_slide (Dz,Dt,nz,nze,e,
 .                     theta_ref,
 .                     cv_band,c_heat,cv_rock,
 .                     zHeight,Pw,zmass,weight,
 .                     sfb,sfr,zk_rock,zk_band,
 .                     c_delta_band,c_delta_rock,
 .                     fib,beta_slope,displ_max,
 .                     int_write_profile,int_write_time)

   implicit real*8 (a-h,o-z)
   dimension Utold(nz),Utnew(nz)

! OPEN FILES TO WRITE RESULTS
   open (unit=2, file='Velocity.dat', status='unknown')
   open (unit=3,
file='Excess_pore_pressure_z0.dat',status='unknown')
   open (unit=4, file='Temperature.dat', status='unknown')
   open (unit=7, file='Strength.dat', status='unknown')
   open (unit=8, file='Heat.dat', status='unknown')
   open (unit=9, file='Displacement.dat', status='unknown')

! WRITING TITLES IN RESULT'S FILES
   write (2,*) ' Time(s) ', 'Velocity (m/s) '
   write (3,*) ' Time(s) ', 'Excess pressure (MPa) '
   write (4,*) ' Time(s) ', 'Temperature (°C) '
   write (7,*) ' Time(s) ', 'Shear strength (MPa) '
   write (8,*) ' Time(s) ', 'Heat (MJ/s·m3) '
   write (9,*) ' Time(s) ', 'Displacement (m) '

   zero = 1.e-10

! INITIATE COUNTERS
   iw_profile = 0!Counter for writing pore pressure profiles
   iw_time = 0       !Counter for writing problem variables varying
in
                !time

! INITIALIZE PROBLEM VARIABLES
!Excess pore pressure
   do i=1,nz
```

```
     Utold(i) = 0.0
    enddo
!Velocity
    vtold = 0.0
!Temperature
    theta_told = theta_ref
!Displacement
    displ_told = 0.0
!Shear strength
    tau_failure = (weight*cos(beta_slope)-Pw)*tan(fib)
!Heat
    Heat=0.0

! CALCULATION PROCEDURE STARTS
    DO WHILE (displ_tnew.lt.displ_max)
    k=k+1
    t=k*Dt

!Excess pore pressure
    Utnew(1) = Utold(1)+sfb*(2.*Utold(2)-2.*Utold(1))-
  .   Dt*c_heat*Heat

    do i=2,nze-1
     Utnew(i) = Utold(i)+sfb*(Utold(i+1)-2.*Utold(i)+Utold(i-1))-
  . Dt*c_heat*Heat
    enddo

    do j=nze+1,nz-1
     Utnew(j) = Utold(j)+sfr*(Utold(j+1)-2.*Utold(j)+Utold(j-1))
    enddo

    Utnew(nz) = 0.0

    Utnew(nze) = (zk_rock*Utnew(nze+1)+zk_band*Utnew(nze-1))/
  . (zk_rock+zk_band)

!Velocity
    Vtnew = vtold+Dt*(weight*sin(beta_slope)-
  . tau_failure)/zmass

!Temperature
    theta_tnew = theta_told+Dt* Heat/c_delta_band

!Displacement
    displ_tnew = displ_told+Dt*vtold

!Shear strength
    tau_failure = (weight*cos(beta_slope)
  . -Pw-Utnew(1))*tan(fib)

    if (tau_failure.lt.zero)then
     write (6,*) 'tau_failure is negative'
     tau_failure = 0.0
    endif

!Heat
    Htnew = tau_failure*vtnew/(2.*e)
```

```
! UPDATE VARIABLES
!Excess pore pressure
    do m=1,nz
    Utold(m) = Utnew(m)
    enddo
!Velocity
    vtold=vtnew
!Temperature
    theta_told = theta_tnew
!Displacement
    displ_told = displ_tnew

! WRITING RESULTS AT SELECTED TIMES
    if ((k.eq.1).or.(mm.eq.k)) then
     iw_profile = iw_profile+1
     mm = int(int_write_profile)*iw_profile
     t = k*Dt
     write (6,*) k,' Excess pore pressure profile has been written'
     ifile = iw_profile+10
     write (ifile,*) 'Time(s) ',t
     do m = 1,nze
      z = (m-1)*Dz
      write (ifile,'(2(e15.5,1x))') z, Utnew(m)
      nmax = m
     enddo
     do n=1,nz-nze
      m = nmax+n
      z = e+n*Dz
      write (ifile,'(2(e15.5,1x))') z, Utnew(m)
     enddo
    endif

    if ((k.eq.1).or.(nn.eq.k)) then
     iw_time = iw_time+1
     nn = int(int_write_time)*iw_time
     t = k*Dt
     write (2,'(2(e15.5,1x))') t, vtnew
     write (3,'(2(e15.5,1x))') t, Utnew(1)*1.e-6
     write (4,'(2(e15.5,1x))') t, theta_tnew
     write (7,'(2(e15.5,1x))') t, tau_failure*1.e-6
     write (8,'(2(e15.5,1x))') t, Heat/1.e6
     write (9,'(2(e15.5,1x))') t, displ_tnew
    endif

   ENDDO

! CLOSE FILES OF RESULTS
   close (2)
   close (3)
   close (4)
   close (7)
   close (8)
   close (9)

   RETURN
   END
```

```
!*****************************************************************!
!        DYNAMIC ANALYSIS OF TWO INTERACTING WEDGES             !
!*****************************************************************!
    subroutine Two_interacting_wedges (Dz,Dt,nz,nze,e,
     .          nt,tmax,
     .          v0,Ubt0,teta_ref,
     .          zL1_0,zL2_0,zh_cha0,
     .          Area1_0,Area2_0,
     .          zHeight,Em1,Em2,Pw1_0,Pw2_0,
     .          gamma_roca,delta_roca,
     .          alpha,beta,shi,fir,
     .          coher0,cc,
     .          cv_band,c_heat,c_sigma_vertical,cv_rock,
     .          sfb,sfr,zk_rock,zk_band,
     .          c_delta_band,gamma_w,
     .          fib_max,fib_min,v_max_fib,a_const,
     .          displ_max)

    implicit real*8 (a-h,o-z)
    dimension U1told(nz),U1tnew(nz),U2told(nz),U2tnew(nz)

! OPEN FILES TO WRITE RESULTS
    open (unit=2, file='Velocity.dat', status='unknown')
    open (unit=3,
file='Excess_pore_pressure_z0.dat',status='unknown')
    open (unit=4, file='Heat.dat', status='unknown')
    open (unit=7, file='Displacement.dat', status='unknown')
    open (unit=8, file='Strength.dat', status='unknown')
    open (unit=9, file='Temperature.dat', status='unknown')

! WRITING TITLES IN RESULTS FILES
    write (2,*) ' Time(s) ', 'Velicity(m/s) '
    write (3,*) ' Time(s) ', 'U1 (MPa) ','U2 (MPa)'
    write (4,*) ' Time(s) ', 'Heat_Wedge1 (MJ/s·m3) ',
  . 'Heat_Wedge2 (MJ/s·m3) '
    write (7,*) ' Time(s) ', 'Displacement (m) '
    write (8,*) ' Time(s) ', 'Shear_strength_Wedge1 (MPa) ' ,
  . 'Shear_strength_Wedge2 (MPa)'
    write (9,*) ' Time(s) ', 'Temperature Wedge1(°) ' ,
  . 'Temperature Wedge2(°)'

    zero=1.e-10

! INITIALITE COUNTERS
    iw_profile=0  !Counter for writing pore pressure profiles
    iw_time=0     !Counter for writing problem variables varying in
                  !time

! INITIALIZE VALUES PARAMETERS THAT CHANGE WITH THE DISPLACEMENT
!Weights
    W1 = Area1_0*gamma_rock
    W2 = Area2_0*gamma_rock
!Mass
    zM1 = Area1_0*delta_rock
    zM2 = Area2_0*delta_rock
!Lengths
    zL1 = zL1_0
    zL2 = zL2_0
```

```
   zl_shpl = zl_shpl0
!Hydrostatic forces
   Pw1 = Pw1_0
   Pw2 = Pw2_0

! INITIALIZE PROBLEM VARIABLES
!Excess pore pressure
   do i=1,nz
    U1told(i) = 0.0
    U2told(i) = 0.0
   enddo
!Changes of notation
   U1 = U1told(1)
   U2 = U2told(1)
!Velocity
   vtold = 0.0
!Temperatures
   theta1_old = theta_ref
   theta2_old = theta_ref
!Displacements
   stold = 0.0
!Total vertical stress increments
   dSn1 = 0.0
   dSn2 = 0.0
!Mass increments
   dM1dt = 0.0
   dM2dt = 0.0
!Heat
   H1 = 0.0
   H2 = 0.0

! CALCULATION PROCEDURE STARTS
   DO WHILE (stnew.lt.displ_max)
   k=k+1

!Excess pore pressures
    U1tnew(1) = U1told(1)+sfb*2.*(U1told(2)-U1told(1))-
 .  c_heat*Dt*H1 + c_sigma_vertical*dSn1
    U2tnew(1) = U2told(1)+sfb*(2.*U2told(2)-2.*U2told(1))-
 .  c_heat*Dt*H2 + c_sigma_vertical*dSn2

   do i=2,nze-1
    U1tnew(i) = U1told(i)+
 .  sfb*(U1told(i+1)-2.*U1told(i)+U1told(i-1))-
 .  c_heat*Dt*H1 + c_sigma_vertical*dSn1
    U2tnew(i) = U2told(i)+
 .  sfb*(U2told(i+1)-2.*U2told(i)+U2told(i-1))-
 .  c_heat*Dt*H2 + c_sigma_vertical*dSn2
   enddo

   do j=nze+1,nz-1
    U1tnew(j) = U1told(j)+
 .  sfr*(U1told(j+1)-2.*U1told(j)+U1told(j-1)) +
 .  c_sigma_vertical*dSn1
    U2tnew(j) = U2told(j)+
 .  sfr*(U2told(j+1)-2.*U2told(j)+U2told(j-1)) +
 . c_sigma_vertical*dSn2
   enddo
```

```
    U1tnew(nz) = U1told(nz)+c_sigma_vertical*dSn1
    U2tnew(nz) = U2told(nz)+c_sigma_vertical*dSn2

    U1tnew(nze) = (zk_rock*U1tnew(nze+1)+zk_band*U1tnew(nze-1))/
 .  (zk_rock+zk_band)
    U2tnew(nze) = (zk_rock*U2tnew(nze+1)+zk_band*U2tnew(nze-1))/
 .                (zk_rock+zk_band)

!Changes of notation
    U1 = U1tnew(1)
    U2 = U2tnew(1)
    cg = zh_cha

!Velocity
   vtnew = vtold + Dt *
 .((W2 *tan(fib)** 2 * sin(shi) - 0.2D1 * cos(shi) *tan(fir)*
 #tan(fib)** 2 * Em1 *sin(shi) + cos(shi) *tan(fir)*tan(fib)** 2 * U
 #1 * zL1 + Pw2 *tan(fib)** 2 * cos(shi) *tan(fir)- W2 *tan(fib)** 2
 # * cos(shi) *tan(fir)+ U2 * zL2 *tan(fib)** 2 * cos(shi)*tan(fir)
 #+ cos(shi) ** 2 * Em1 -tan(fib)* cos(shi) ** 2 *tan(fir)* Em1 + 0.
 #2D1 * sin(shi) * cos(shi) * coher * cg + cos(shi) *tan(fib)* Em1 *
 # sin(shi) - cos(shi) *tan(fir)* Em1 * sin(shi) -tan(fib)* cos(shi)
 # *tan(fir)* dM1dt * vtold - dM2dt * vtold *tan(fib)* cos(shi) * ta
 #n(fir)+ 0.2D1* sin(shi) *tan(fib)** 2 * cos(shi) * coher * cg - Pw
 #2 *tan(fib)** 2 * sin(shi) - Em2 * cos(beta) *tan(fib)** 2 * cos(s
 #hi) *tan(fir)+ Em2 * cos(beta) *tan(fib)** 2 * sin(shi) +tan(fib)*
 # cos(shi) *tan(fir)* W1 * sin(alpha) + sin(shi) *tan(fir)*tan(fib)
 #* Pw1 - sin(shi) *tan(fir)* dM1dt * vtold + sin(shi) *tan(fir)* W1
 # * sin(alpha) +tan(fib)* sin(shi) * W1 * sin(alpha) + sin(shi) * t
 #an(fir)*tan(fib)*U1 * zL1 + sin(shi) *tan(fib)** 2 * U1 * zL1 - U2
 # * zL2 *tan(fib)** 2 * sin(shi) -tan(fib)* sin(shi) * dM1dt * vtol
 #d -tan(fib)* Pw2 * sin(shi) *tan(fir)+tan(fib)* W2 * sin(shi) * ta
 #n(fir)+dM2dt * vtold *tan(fib)* sin(shi) + dM2dt * vtold * sin(shi
 #) *tan(fir)+ dM2dt * vtold * cos(shi) + W2 *tan(fib)* cos(shi) - P
 #w2 *tan(fib)* cos(shi) + cos(shi) * dM1dt * vtold - cos(shi) * tan
 #(fib)*Pw1 -tan(fib)* U2 * zL2 * sin(shi) *tan(fir)+ sin(shi) * tan
 #(fib)** 2* Pw1 +tan(fib)* Em2 * cos(beta) * sin(shi) *tan(fir)+ Em
 #2 * cos(beta) *tan(fib)* cos(shi) - cos(shi) *tan(fib)* U1 * zL1 -
 # U2 * zL2 *tan(fib)* cos(shi) + cos(shi) *tan(fir)*tan(fib)** 2 *
 #Pw1 - cos(shi) * W1 * sin(alpha) - sin(shi) *tan(fib)** 2 * W1 * c
 #os(alpha) + cos(shi) *tan(fib)* W1 * cos(alpha) - cos(shi) *tan(fi
 #r)*tan(fib)** 2 * W1 * cos(alpha) - sin(shi) *tan(fir)*tan(fib)* W
 #1 * cos(alpha)) / (-zM1 * cos(shi) - zM2 * cos(shi) + zM1*tan(fib)
 # * sin(shi) + zM2 *tan(fib)* cos(shi) *tan(fir)+ zM1 * sin(shi) *
 #tan(fir)+zM1 *tan(fib)* cos(shi) *tan(fir)- zM2 *tan(fib)* sin(shi
 #) - zM2 * sin(shi) * tan(fir)))

    if (vtnew.lt.-zero) then
     write (6,*) 'The slide stops'
     stop
    endif

!Temperature
    theta1_tnew = theta1_old+Dt*H1/c_delta_band
    theta2_tnew = theta2_old+Dt*H2/c_delta_band

!Displacement
    stnew = stold+vtold*Dt
```

```
!Parameters that change with the displacement
   zL1 = zL1_0 - stnew
   zL2 = zL2_0 + stnew
   zl_shpl = zL1*zl_shpl0/zL1_0
   Area1= 0.5*zL1*zl_shpl*cos(shi)
   zInc_Area = Area1_0 - Area1
   Area2 = Area2_0 + zInc_Area
   W1 = Area1*gamma_rock
   W2 = Area2*gamma_rock
   zM1 = Area1*delta_rock
   zM2 = Area2*delta_rock
   Pw2 = Pw2_0 + z_Height_w*stnew*gamma_w
   coher = coher0*exp(cc*stnew)
   dM1dt = -delta_roca*zL1*zl_shpl*cos(shi)*vtnew/zL1
   dM2dt = -dM1dt

!Effective normal force on base of Wedge 1
  zN1 = -(Pw1*zM1* tan(fib) * cos(shi) * tan(fir)+ U1*zL1*zM1*
 #tan(fib) * cos(shi) * tan(fir)- 0.2D1*cos(shi)*tan(fir)*zM1 * tan
 #(fib)*Em1 * sin(shi) + 0.2D1 *sin(shi) *zM1 *tan(fib) * cos(shi)
 #* coher * cg + U1 * zL1 * zM1 * sin(shi) *tan(fir)+ Em1 * sin(shi)
 # * zM1 * cos(shi) - cos(shi) *tan(fir)* zM1 *tan(fib)* W2 + cos(sh
 #i) *tan(fir)* zM1 *tan(fib)* Pw2 - U1 * zL1 * zM1 * cos(shi) - zM2
 # * coher * cg + zM1 * coher * cg - W1 * cos(alpha)*zM1 *tan(fib)
 #* cos(shi) *tan(fir)+ sin(shi) * zM1 * dM2dt * vtold - sin(shi) *
 #zM2 * dM1dt * vtold + sin(shi) * zM1 * tan(fib)*W2 - sin(shi) * zM
 #1 * tan(fib) * Pw2 + Pw1 * zM1 *tan(fib)* sin(shi) + Pw1 * zM1*sin
 #(shi) *tan(fir)- Pw1 * zM2 * sin(shi) *tan(fir)+ sin(shi) * zM2 *
 #W1 * sin(alpha) - sin(shi) * zM1 *tan(fib)*U2 * zL2 + U1 * zL1 *
 #zM1 * tan(fib)*sin(shi) - U1 * zL1 * zM2 * sin(shi) *tan(fir)- Pw1
 # * zM1 * cos(shi) - Pw1 * zM2 * cos(shi) - W1 * cos(alpha) * zM1 *
 #tan(fib) * sin(shi) - cos(shi) *tan(fir)* zM2 * W1 *sin(alpha)+ s
 #in(shi) * zM1 *tan(fib)* Em2 * cos(beta) + cos(shi) *tan(fir)* zM2
 # * dM1dt * vtold - 0.2D1 * cos(shi) ** 2 * coher * cg * zM1 - cos(
 #shi) *tan(fir)* zM1 * dM2dt * vtold - W1 * cos(alpha) * zM1 * sin(
 #shi) *tan(fir)+ Em1 * zM2 *tan(fir)- cos(shi) *tan(fir)* zM1 * tan
 #(fib)*Em2 * cos(beta) + W1 * cos(alpha) * zM2 * sin(shi) *tan(fir)
 #- Em1 * zM1 *tan(fir)+ cos(shi) *tan(fir)* zM1 *tan(fib)* U2 * zL2
 # - U1 * zL1 * zM2 * cos(shi) + cos(shi) ** 2 * Em1 * zM1*tan(fir)
 #+ W1 * cos(alpha) * zM1 * cos(shi) + W1 * cos(alpha) * zM2 * cos(s
 #hi)) / (-zM1 * cos(shi) - zM2 * cos(shi) + zM1 *tan(fib)*sin(shi)
 # + zM2 *tan(fib) *cos(shi) *tan(fir)+ zM1 *sin(shi)*tan(fir)+ zM
 #1 *tan(fib) * cos(shi) *tan(fir)- zM2 *tan(fib) *sin(shi)-zM2 * s
 #in(shi) *tan(fir))

!Effective normal force on base of wedge 2
  zN2 =-(-Em2 *cos(beta) *zM2 *tan(fib)* cos(shi)*tan(fir)- Em2
 #* cos(beta) * zM1 * sin(shi) *tan(fir)+ Em2 * cos(beta) * zM1 * co
 #s(shi) - W2 * zM2 *tan(fib)* cos(shi) *tan(fir)+ U2 * zL2 * zM2 *
 #tan(fib)*cos(shi)*tan(fir)+ Em1 * sin(shi) * zM1 * cos(shi) - 0.2
 #D1 * cos(shi) *tan(fir)* zM2 *tan(fib)* Em1 * sin(shi) + 0.2D1 * s
 #in(shi) * zM2 *tan(fib)* cos(shi) * coher * cg - zM2 * coher * cg
 #+ zM1 * coher * cg + cos(shi) *tan(fir)* zM2 *tan(fib)* Pw1 - sin(
 #shi) * zM2 *tan(fib)* W1 * cos(alpha) + sin(shi) * zM2 *tan(fib)*
 #Pw1 + sin(shi) * zM1 * dM2dt * vtold - sin(shi) * zM2 * dM1dt * vt
 #old + Pw2 * zM1 * sin(shi) *tan(fir)- Pw2 * zM2 *tan(fib)* sin(shi
 #) - Pw2 * zM2 * sin(shi) *tan(fir)- W2 * zM1 * sin(shi) *tan(fir)+
 # W2 * zM2 *tan(fib)* sin(shi) + W2 * zM2 * sin(shi) *tan(fir)+ sin
```

```
 #(shi) * zM2 * W1 * sin(alpha) + sin(shi) * zM2 *tan(fib)* U1 * zL1
 # + U2 * zL2 * zM1 * sin(shi) *tan(fir)- U2 * zL2 * zM2 *tan(fib)*
 #sin(shi) - U2 * zL2 * zM2 * sin(shi) *tan(fir)- Pw2 * zM1 * cos(sh
 #i) + W2 * zM1 * cos(shi) - Pw2 * zM2 * cos(shi) + W2 * zM2 * cos(s
 #hi) + cos(shi) *tan(fir)* zM2 * W1 * sin(alpha) + Em2 * cos(beta)
 #* zM2 *tan(fib)* sin(shi) + Em2 * cos(beta) * zM2 * sin(shi) * tan
 #(fir)+ Em2* cos(beta) * zM2 * cos(shi) - cos(shi) *tan(fir)*zM2 *
 # dM1dt * vtold + cos(shi) *tan(fir)* zM2 *tan(fib)* U1 * zL1 + cos
 #(shi) *tan(fir)* zM1 * dM2dt * vtold - 0.2D1 * cos(shi) ** 2 * tan
 #(fir) * zM2 * Em1 + Em1 * zM2 *tan(fir)-cos(shi)*tan(fir)* zM2 * t
 #an(fib)*W1* cos(alpha) - Em1 * zM1 *tan(fir)+ 0.2D1 * cos(shi) **
 # 2 * coher * cg * zM2 - U2 * zL2 * zM1 * cos(shi) - U2 * zL2 * zM2
 # * cos(shi) + Pw2 * zM2 *tan(fib)* cos(shi) *tan(fir)+ cos(shi) **
 # 2 * Em1 * zM1 *tan(fir))/ (-zM1 * cos(shi) - zM2 * cos(shi) + zM1
 # *tan(fib)* sin(shi) + zM2 *tan(fib)* cos(shi) *tan(fir)+ zM1 * si
 #n(shi) *tan(fir)+ zM1 *tan(fib)* cos(shi) *tan(fir)- zM2 *tan(fib)
 #* sin(shi) - zM2 * sin(shi) *tan(fir))

    if (zN1_tnew.lt.zero) then
     zN1_tnew = 0.0
    endif
    if (zN2_tnew.lt.zero) then
     zN2_tnew = 0.0
    endif

!Total normal forces
    zN1tot = zN1+Pw1+U1*zL1
    zN2tot = zN2+Pw2+U2*zL2

!Total vertical stresses
    Sn1_tnew = -zN1tot/zL1
    Sn2_tnew = -zN2tot/zL2

!Total vertical stress increments
    if (k.eq.1) then
     dSn1_tnew = 0.0
     dSn2_tnew = 0.0
    else
    dSn1 = Sn1_tnew-Sn1_told
    dSn2 = Sn2_tnew-Sn2_told
    endif

!Shear strength
    Tau1 = (zN1/zL1) * tan(fib)
    Tau2 = (zN2/zL2) * tan(fib)

!Heat
     H1 = Tau1*vtnew/(2.*e)
     H2 = Tau2*vtnew/(2.*e)

! UPDATE PROBLEM VARIABLES
!Excess pore pressure
    do i=1,nze
     U1told(i) = U1tnew(i)
     U2told(i) = U2tnew(i)
    enddo
!Velocity
    vtold = vtnew
```

```
!Temperatures
   theta1_old = theta1_tnew
   theta2_old = theta2_new
!Displacement
   stold = stnew
!Total vertical stresses
   Sn1_told = Sn1_tnew
   Sn2_told = Sn2_tnew

! WRITING RESULTS AT SELECTED TIMES
    if ((k.eq.2).or.(mm.eq.k)) then
     write (6,*) k,' Profile results have been written'
     iw_profile = iw_profile+1
     mm=int(int_write_profile)*iw_profile
     t=k*Dt
     iarchivo=iw_profile+100
     write (iarchivo,*) 'Time(s) ',t
     do m=1,nze+10
      z=(m-1)*Dz
      write (iarchivo,'(2(e15.5,1x))') z, U1tnew(m)
     enddo
     do m =    nze+11,nz,10
      z = m*Dz
      write (iarchivo,'(2(e15.5,1x))') z, U1tnew(m)
     enddo
    endif
    if ((k.eq.1).or.(nn.eq.k)) then
     iw_time = iw_time+1
     nn = int(int_write_time)*iw_time
     t=k*Dt
     write (2,'(2(e15.5,1x))') t, vtnew
     write (3,'(3(e15.5,1x))') t, U1tnew(1)*1.e-6,U2tnew(1)*1.e-6
     write (4,'(3(e15.5,1x))') t, H1/1.e6,H2/1.e6
     write (7,'(2(e15.5,1x))') t, stold
     write (8,'(3(e15.5,1x))') t, Tau1 *1.e-6,Tau2*1.e-6
     write (9,'(3(e15.5,1x))') t, theta1_tnew,theta2_tnew
    endif

   ENDDO

   close (2)
   close (3)
   close (4)
   close (7)
   close (8)
   close (9)

   return
   end
```

Appendix 5.3 Parameters of the Balance Equations for the Dynamic Analysis of Two Interacting Wedges

The parameters that complete the dynamic equilibrium equation of the two wedges (Eq. (5.55)) are:

$$t_{W_1} = \left(-\tan\varphi_b' \tan\varphi_r' - \tan^2\varphi_b'\right)\sin\left(\frac{\alpha}{2}\right)\cos\alpha + \left(\tan\varphi_r' + \tan\varphi_b'\right)\sin\left(\frac{\alpha}{2}\right)\sin\alpha + \left(\tan\varphi_b' - \tan^2\varphi_b' \tan\varphi_r'\right)\cos\left(\frac{\alpha}{2}\right)\cos\alpha + \tan\varphi_b' \tan\varphi_r' \cos\left(\frac{\alpha}{2}\right)\sin\alpha - \cos\left(\frac{\alpha}{2}\right)\sin\alpha, \tag{A5.17}$$

$$t_{W_2} = \left(\tan\varphi_b' - \tan^2\varphi_b' \tan\varphi_r'\right)\cos\left(\frac{\alpha}{2}\right) + \left(\tan^2\varphi_b' + \tan\varphi_b' \tan\varphi_r'\right)\sin\left(\frac{\alpha}{2}\right) \tag{A5.18}$$

$$t_{P_{w\mathrm{int}}} = -2\tan\varphi_r' \sin\left(\frac{\alpha}{2}\right)\cos\left(\frac{\alpha}{2}\right)\left(1 + \tan^2\varphi_b'\right) \tag{A5.19}$$

$$t_{P_{wf}} = \left(\tan^2\varphi_b' + \tan\varphi_b' \tan\varphi_r'\right)\sin\left(\frac{\alpha}{2}\right)\cos\beta + \left(1 - \tan\varphi_b' \tan\varphi_r'\right)\cos\left(\frac{\alpha}{2}\right)\sin\beta + \left(\tan\varphi_b' - \tan^2\varphi_b' \tan\varphi_r'\right)\cos\left(\frac{\alpha}{2}\right)\cos\beta + \left(\tan\varphi_b' + \tan\varphi_r'\right)\sin\left(\frac{\alpha}{2}\right)\sin\beta \tag{A5.20}$$

$$t_{P_{w1}} = \tan^2\varphi_b \tan\varphi_r' \cos\left(\frac{\alpha}{2}\right) + \tan\varphi_b' \tan\varphi_r' \sin\left(\frac{\alpha}{2}\right) + \tan^2\varphi_b' \sin\left(\frac{\alpha}{2}\right) - \tan\varphi_b' \cos\left(\frac{\alpha}{2}\right) \tag{A5.21}$$

$$t_{P_{w2}} = \tan^2\varphi_b' \tan\varphi_r' \cos\left(\frac{\alpha}{2}\right) - \tan\varphi_b' \tan\varphi_r' \sin\left(\frac{\alpha}{2}\right) - \tan^2\varphi_b' \sin\left(\frac{\alpha}{2}\right) - \tan\varphi_b' \cos\left(\frac{\alpha}{2}\right) \tag{A5.22}$$

$$t_{u_{w1}} = \left(\tan^2 \varphi_b' \tan \varphi_r' - \tan \varphi_b'\right) \cos\left(\frac{\alpha}{2}\right) + \left(\tan^2 \varphi_b' + \tan \varphi_b \tan \varphi_r'\right) \sin\left(\frac{\alpha}{2}\right) \tag{A5.23}$$

$$t_{u_{w2}} = \left(\tan^2 \varphi_b' \tan \varphi_r' - \tan \varphi_b'\right) \cos\left(\frac{\alpha}{2}\right) - \left(\tan^2 \varphi_b' + \tan \varphi_b' \tan \varphi_r'\right) \sin\left(\frac{\alpha}{2}\right), \tag{A5.24}$$

$$t_{c_r} = 2 \cos\left(\frac{\alpha}{2}\right) \sin\left(\frac{\alpha}{2}\right) \left(1 + \tan^2 \varphi_b'\right), \tag{A5.25}$$

$$t_{dM_1dt} = \cos\left(\frac{\alpha}{2}\right) - \left(\tan \varphi_b' + \tan \varphi_r'\right) \sin\left(\frac{\alpha}{2}\right) - \tan \varphi_b' \tan \varphi_r' \cos\left(\frac{\alpha}{2}\right), \tag{A5.26}$$

$$t_{dM_2dt} = \cos\left(\frac{\alpha}{2}\right) + \left(\tan \varphi_b' + \tan \varphi_r'\right) \sin\left(\frac{\alpha}{2}\right) - \tan \varphi_b' \tan \varphi_r' \cos\left(\frac{\alpha}{2}\right), \tag{A5.27}$$

$$t_{M_1} = \left(\tan \varphi_r' + \tan \varphi_b'\right) \sin\left(\frac{\alpha}{2}\right) + \left(\tan \varphi_r' \tan \varphi_b' - 1\right) \cos\left(\frac{\alpha}{2}\right), \tag{A5.28}$$

$$t_{M_2} = -\left(\tan \varphi_r' + \tan \varphi_b'\right) \sin\left(\frac{\alpha}{2}\right) + \left(\tan \varphi_r' \tan \varphi_b' - 1\right) \cos\left(\frac{\alpha}{2}\right). \tag{A5.29}$$

Coefficients r and s in Equations (5.56) and (5.57) are given by

$$r_{W_1} = M_1 \tan \varphi_b' \cos \alpha \left(\cos\left(\frac{\alpha}{2}\right) - \tan \varphi_r' \sin\left(\frac{\alpha}{2}\right) - \tan \varphi_b' \sin\left(\frac{\alpha}{2}\right) - \tan \varphi_b' \tan \varphi_r' \cos\left(\frac{\alpha}{2}\right) \right) + M_2 \tan \varphi_b' \left(\tan \varphi_r' \cos \alpha \sin\left(\frac{\alpha}{2}\right) + \sin\left(\frac{\alpha}{2}\right) \sin \alpha - \tan \varphi_r' \cos\left(\frac{\alpha}{2}\right) \sin \alpha + \cos \alpha \cos\left(\frac{\alpha}{2}\right) \right), \tag{A5.30}$$

$$r_{W_2} = M_1 \tan^2 \varphi_b' \left(\sin\left(\frac{\alpha}{2}\right) - \tan \varphi_r' \cos\left(\frac{\alpha}{2}\right) \right), \tag{A5.31}$$

$$r_{P_{w\,\mathrm{int}}} = M_1 \tan \varphi_b' \sin\left(\frac{\alpha}{2}\right)\left(-\cos\left(\frac{\alpha}{2}\right) - s \tan \varphi_r' \sin\left(\frac{\alpha}{2}\right) - 2 \tan \varphi_r' \tan \varphi_b' \cos\left(\frac{\alpha}{2}\right) \right) + M_2 \tan \varphi_r' \tan \varphi_b', \tag{A5.32}$$

$$r_{P_{wf}} = M_1 \tan^2 \varphi_b' \cos\beta \left(\sin\left(\frac{\alpha}{2}\right) - \tan \varphi_r' \cos\left(\frac{\alpha}{2}\right) \right), \tag{A5.33}$$

$$r_{P_{w1}} = M_1 \tan \varphi_b' \left(-\cos\left(\frac{\alpha}{2}\right) + \tan \varphi_r' \sin\left(\frac{\alpha}{2}\right) + \tan \varphi_b' \sin\left(\frac{\alpha}{2}\right) + \tan \varphi_b' \tan \varphi_r' \cos\left(\frac{\alpha}{2}\right) \right) - M_2 \tan \varphi_b' \left(\cos\left(\frac{\alpha}{2}\right) + \tan \varphi_r' \sin\left(\frac{\alpha}{2}\right) \right), \tag{A5.34}$$

$$r_{P_{w2}} = M_1 \tan^2 \varphi_b' \left(\tan \varphi_r' \cos\left(\frac{\alpha}{2}\right) - \sin\left(\frac{\alpha}{2}\right) \right), \tag{A5.35}$$

$$r_{u_{w1}} = M_1 \tan \varphi_b' \left(\tan \varphi_r' \sin\left(\frac{\alpha}{2}\right) + \tan \varphi_b' \sin\left(\frac{\alpha}{2}\right) - \cos\left(\frac{\alpha}{2}\right) + \tan \varphi_b' \tan \varphi_r' \cos\left(\frac{\alpha}{2}\right) \right) - M_2 \tan \varphi_b' \left(\tan \varphi_r' \sin\left(\frac{\alpha}{2}\right) + \cos\left(\frac{\alpha}{2}\right) \right), \tag{A5.36}$$

$$r_{u_{w2}} = M_1 \tan^2 \varphi_b' \left(\tan\varphi_r' \cos\left(\frac{\alpha}{2}\right) - \sin\left(\frac{\alpha}{2}\right) \right), \tag{A5.37}$$

$$r_{c_r} = M_1 \tan \varphi_b' \left(\sin^2\left(\frac{\alpha}{2}\right) - \cos^2\left(\frac{\alpha}{2}\right) + 2 \tan \varphi_b' \sin\left(\frac{\alpha}{2}\right) \cos\left(\frac{\alpha}{2}\right) \right) - M_2 \tan \varphi_b' \cos^2\left(\frac{\alpha}{2}\right), \tag{A5.38}$$

$$r_{dM_1dt} = M_2 \tan\varphi_b' \left(\tan\varphi_r' \cos\left(\frac{\alpha}{2}\right) - \sin\left(\frac{\alpha}{2}\right)\right), \tag{A5.39}$$

$$r_{dM_2dt} = M_1 \tan\varphi_b' \left(-\tan\varphi_r' \cos\left(\frac{\alpha}{2}\right) + \sin\left(\frac{\alpha}{2}\right)\right), \tag{A5.40}$$

$$\begin{aligned} r_{M_1} &= \tan\varphi' \cos\left(\frac{\alpha}{2}\right) - \tan\varphi_r' \tan\varphi' \sin\left(\frac{\alpha}{2}\right) - \\ &\tan^2\varphi_b' \sin\left(\frac{\alpha}{2}\right) - \tan^2\varphi_b' \tan\varphi_r' \cos\left(\frac{\alpha}{2}\right), \end{aligned} \tag{A5.41}$$

$$\begin{aligned} r_{M_2} &= -\tan^2\varphi_b' \tan\varphi_r' \cos\left(\frac{\alpha}{2}\right) + \tan^2\varphi_b' \sin\left(\frac{\alpha}{2}\right) + \\ &\tan\varphi_b' \cos\left(\frac{\alpha}{2}\right) + \tan\varphi_b' \tan\varphi_r' \sin\left(\frac{\alpha}{2}\right), \end{aligned} \tag{A5.42}$$

and

$$\begin{aligned} s_{W_1} = M_2 \Bigg(&\tan\varphi_r' \tan\varphi_b' \cos\left(\frac{\alpha}{2}\right) \sin\alpha + \tan\varphi_b' \sin\left(\frac{\alpha}{2}\right) \sin\alpha - \\ &\tan^2\varphi_b' \sin\left(\frac{\alpha}{2}\right) \cos\alpha - \tan\varphi_r' \tan^2\varphi_b' \cos\left(\frac{\alpha}{2}\right) \cos\alpha \Bigg), \end{aligned} \tag{A5.43}$$

$$\begin{aligned} s_{W_2} &= M_1 \tan\varphi_b' \left(\cos\left(\frac{\alpha}{2}\right) - \tan\varphi_r' \sin\left(\frac{\alpha}{2}\right)\right) + \\ M_2 \tan\varphi_b' &\left(\tan\varphi_b' \sin\left(\frac{\alpha}{2}\right) - \tan\varphi_b' \tan\varphi_r' \cos\left(\frac{\alpha}{2}\right) + \tan\varphi_r' \sin\left(\frac{\alpha}{2}\right) + \cos\left(\frac{\alpha}{2}\right)\right), \end{aligned} \tag{A5.44}$$

$$\begin{aligned} s_{P_{w\,\mathrm{int}}} &= M_1 \tan\varphi_b' \left(\sin\left(\frac{\alpha}{2}\right) \cos\left(\frac{\alpha}{2}\right) - \tan\varphi_r' \sin^2\left(\frac{\alpha}{2}\right)\right) + \\ M_2 \tan\varphi_r' \tan\varphi_b' &\left(-\cos^2\left(\frac{\alpha}{2}\right) + \sin^2\left(\frac{\alpha}{2}\right) - 2\tan\varphi_b' \cos\left(\frac{\alpha}{2}\right) \sin\left(\frac{\alpha}{2}\right)\right), \end{aligned} \tag{A5.45}$$

$$\begin{aligned} s_{P_{w2}} &= M_1 \tan\varphi_b' \cos\beta \left(\cos\left(\frac{\alpha}{2}\right) - \tan\varphi_r' \sin\left(\frac{\alpha}{2}\right)\right) + \\ M_2 \tan\varphi_b' \cos\beta &\left(\cos\left(\frac{\alpha}{2}\right) - \tan\varphi_b' \tan\varphi_r' \cos\left(\frac{\alpha}{2}\right) + \tan\varphi_b' \sin\left(\frac{\alpha}{2}\right) + \tan\varphi_r' \sin\left(\frac{\alpha}{2}\right)\right), \end{aligned} \tag{A5.46}$$

$$s_{P_{w1}} = M_2 \tan^2 \varphi_b' \left(\sin\left(\frac{\alpha}{2}\right) + \cos\left(\frac{\alpha}{2}\right) \tan \varphi_r' \right), \tag{A5.47}$$

$$\begin{gathered} s_{P_{w2}} = M_1 \tan \varphi_b' \left(\tan \varphi_r' \sin\left(\frac{\alpha}{2}\right) - \cos\left(\frac{\alpha}{2}\right) \right) + \\ M_2 \tan \varphi_b' \left(\tan \varphi_b' \tan \varphi_r' \cos\left(\frac{\alpha}{2}\right) - \cos\left(\frac{\alpha}{2}\right) \tan \varphi_r' \sin\left(\frac{\alpha}{2}\right) - \tan \varphi_b' \sin\left(\frac{\alpha}{2}\right) \right), \end{gathered} \tag{A5.48}$$

$$s_{u_{w1}} = M_2 \tan^2 \varphi_b' \left(\sin\left(\frac{\alpha}{2}\right) + \tan \varphi_r' \cos\left(\frac{\alpha}{2}\right) \right), \tag{A5.49}$$

$$\begin{gathered} s_{u_{w2}} = M_1 \tan \varphi_b' \tan \varphi_r' \left(\sin\left(\frac{\alpha}{2}\right) - \cos\left(\frac{\alpha}{2}\right) \right) + \\ M_2 \tan \varphi_b' \left(\tan \varphi_b' \tan \varphi_r' \cos\left(\frac{\alpha}{2}\right) - \tan \varphi_b' \sin\left(\frac{\alpha}{2}\right) - \tan \varphi_r' \sin\left(\frac{\alpha}{2}\right) - \cos\left(\frac{\alpha}{2}\right) \right), \end{gathered} \tag{A5.50}$$

$$\begin{gathered} s_{c_r} = M_1 \tan \varphi_b' + \\ M_2 \tan \varphi_b' \left(-\sin^2\left(\frac{\alpha}{2}\right) + \cos^2\left(\frac{\alpha}{2}\right) + 2 \sin\left(\frac{\alpha}{2}\right) \tan \varphi_b' \cos\left(\frac{\alpha}{2}\right) \right), \end{gathered} \tag{A5.51}$$

$$s_{dM_1 dt} = -M_2 \tan \varphi_b' \left(\tan \varphi_r' \cos\left(\frac{\alpha}{2}\right) + \sin\left(\frac{\alpha}{2}\right) \right), \tag{A5.52}$$

$$s_{dM_2 dt} = M_1 \tan \varphi_b' \left(\tan \varphi_r' \cos\left(\frac{\alpha}{2}\right) - \sin\left(\frac{\alpha}{2}\right) \right), \tag{A5.53}$$

$$s_{M_1} = \tan \varphi_b' \left(\cos\left(\frac{\alpha}{2}\right) - \tan \varphi_r' \sin\left(\frac{\alpha}{2}\right) - \tan \varphi_b' \sin\left(\frac{\alpha}{2}\right) - \tan \varphi_b' \tan \varphi_r' \cos\left(\frac{\alpha}{2}\right) \right), \tag{A5.54}$$

$$\begin{gathered} s_{M_2} = \tan \varphi_b' \left(\cos\left(\frac{\alpha}{2}\right) - \tan \varphi_b' \tan \varphi_r' \cos\left(\frac{\alpha}{2}\right) + \right. \\ \left. \tan \varphi_b' \sin\left(\frac{\alpha}{2}\right) + \tan \varphi_r' \sin\left(\frac{\alpha}{2}\right) \right). \end{gathered} \tag{A5.55}$$

References

Alonso, E.E., Gens, A. and Lloret, A. (1992) The landslide of Cortes de Pallás Spain. A case study. *Géotechnique* 42 (4), 601–624.

Bock, K. (2001) Rock Mechanics Analyses and Synthesis: RA Experiment. *Mont Terri Technical Report 2000–02*. Q+S Consult.

Bossart, P., Meier, P.M., Moeri, A. Trick, T. and Mayor, J.C. (2002) Geological and hydraulic characterisation of the excavation disturbed zone in the Opalinus Clay of the Mont Terri Rock Laboratory. *Engineering Geology* 66 (1–2), 19–38.

Desrues, J. (1984) Sur l'application de la stéréophotogrammétrie à la mesure des grandes déformations. *Revue Française de Mécanique* 3, 55–63.

Eberhardt, E., Stead, D. and Coggan, J.S. (2004) Numerical analysis of initiation and progressive failure in natural rock slopes –the 1991 Randa rockslide. *International Journal of Rock Mechanics and Mining Sciences* 41 (1), 69–87.

EUR (2006) Heater Experiment: Rock and bentonite thermo-hydro-mechanical (THM) processes in the near-field of a thermal source for development of deep underground high-level radioactive waste repositories. *Project FIS5–2001–00024*. Project funded by the European Community under the 5th EURATOM Framework Programme (1998–2002). 106 pp.

Hendron, A.J. and Patton, F.D. (1985) The Vaiont slide, a geotechnical analysis basedon new geologic observations of the failure surface. *Technical Report GL–85–5*. Department of the US Army Corps of Engineers, Washington, DC.

Hungr, O. (1995) A model for the runout analysis of rapid flow slides, debris flows and avalanches. *Canadian Geotechnical Journal* 32, 610–623.

Leonards, G.A. (1987) Overview and personnal commentary. *Engineering Geology* 24, 577–612.

London Southbank University website, http://www.lsbu.ac.uk/water/phase.html

McDougall, S. and Hungr, O. (2004) A model for the analysis of rapid landslide runout motion across three-dimensional terrain. *Canadian Geotechnical Journal* 41, 1084–1097.

Morgenstern N.R. and Tchalenko, J.S. (1967) Microstructural observations of shear zones from slips in natural clays. *Proceedings Geotechnical Conference, Oslo*, 147–152.

Muñoz, J.J. (2007) Thermo-hydro-mechanical analysis of soft rock. Application to large scale heating and ventilation tests. *Doctoral Thesis*. Universitat Politècnica de Catalunya. Barcelona.

Nakamura, S. (1991) *Applied Numerical Methods with Software.* Prentice-Hall. New Jersey.

Nonveiller, E. (1987). The Vajont reservoir slope failure. *Engineering Geology* 24, 493–512.

Olivella, S., Gens, A. Carrera J. and Alonso, E.E. (1996) Numerical formulation for a simulator (CODE BRIGHT) for the coupled analysis of saline media. *Engineering Computations* 13 (7), 87–112.

Pinyol, N.M. and Alonso, E.E. (2009) Fast planar slides. A closed-form thermo-hydro-mechanical solution. International Journal for Numerical and Analytical

Methods in Geomechanics 34, 27-52.

Quecedo M., Pastor, M., Herreros, M.I. and Fernández Merodo, J.A. (2004) Numerical modelling of the propagation of fast landslides using the finite element method. *International Journal of Numerical Methods in Engineering* 59, 755–794.

Roscoe, K.H. (1970) The influence of strains in soil mechanics. *Géotechnique* 20, 129–170.

Scarpelli, G. and Wood, D.M. (1982) Experimental observations of shear band patterns in direct shear tests. *Proceedings of the IUTAM Conference on Deformation and Failure in Granular Media.* Delft, The Netherlands. Vermeer and Luger (eds.). Balkema, 473–484.

Thury, M. and Bossart, P. (1999) Mont Terri rock laboratory, a new international research project in a Mesozoic shale formation, in Switzerland. *Engineering Geology* 52 (3–4), 347–359.

Tika, T.E. and Hutchinson, J.N. (1999) Ring shear test on soil from the Vaiont landslide slip surface. *Géotechnique* 49 (1), 59–74.

Uriel Romero, S. and Molina, R. (1977) Kinematic aspects of Vaiont slide. *Proceedings of the 3rd International Conference of the ISRMR.* Denver, USA. National Academy of Sciences. 2B, 865–870.

Valdés Díaz-Caneja, J.M. (1964) Meditaciones sobre la catástrofe del Vaiont. *Boletín nº 20.* Servicio Geológico. Ministerio de Obras Públicas. Madrid

Vardoulakis, I. (1980) Shear band inclination and shear modulus of sand in biaxial tests. *International Journal for Numerical and Analytical Methods in Geomechanics* 4, 103–119.

Vardoulakis, I. (2002) Dynamic thermo-poro-mechanical analysis of catastrofic landslides. *Géotechnique* 52 (3), 157–171.

Voigt, B. and Faust, C. (1982) Frictional heat and strength loss in some rapid ladslides. *Géotechnique* 32 (1), 43–54.

Chapter 6
Dynamics of Dam Sliding: Aznalcóllar Dam, Spain

TABLE OF CONTENTS

E.E. Alonso et al., *Geomechanics of Failures. Advanced Topics*,
DOI 10.1007/978-90-481-3538-7_6, © Springer Science+Business Media B.V. 2010

Chapter 6

Dynamics of Dam Sliding: Aznalcóllar Dam, Spain

6.1 Introduction

Aznalcóllar dam, a 28 m high rockfill dyke, failed catastrophically on April 25, 1998 and triggered an uncontrolled flow of acid pyritic tailings whose volume was estimated in 5.5 Mm^3. Chapter 4 describes the failure, the geotechnical properties of stored tailings and foundation soils and the stability analysis performed to explain the reasons for the failure. The failure was described as a translational slide which carried with it the dyke itself. The central section of Aznalcóllar slide travelled forward 50 m until it came to rest (Figs. 4.5 and 4.6). The travelled distance decreased towards the North for a reason which will be explained later. The displacement at the breach zone was close to 20 m, always in the direction SSE (approximately perpendicular to the southern dyke direction). Figure 6.1 is a sketch of the relative motion between the northern and southern dykes at the initial breach section. The figure shows the measured displacement of the southern dyke at this point and the orientation of the opening, pointing towards the northeast direction. This geometry results in an estimation of the initial width of the breach close to 14 m. The two lips of the breach separated and, therefore, no shear resistance was involved in this motion.

The dam was literally torn apart by the relative motion between the immobile northern dyke and the southern slide. The 14 m wide initial breach allowed the immediate pouring of tailings and it increased in size as erosion proceeded. The intensity and destruction power of the ensuing flood is related to this width, which, to a large extent, controls the outgoing flow rate. Should the slide displacement have been longer, the width of the breach would have increased as well as the flood rate and its effect. Doñana National Park, located downstream, was barely hit by the actual flood. But a larger one would have certainly created a major catastrophe.

From our geotechnical perspective, the right question to be asked is: does a procedure exist to predict the slide runout? This question has no simple answer. Runout distances depend on the type of slide, the evolution of the slide material and the downstream topography. Aznalcóllar was simple in this regard: the dam and the upper sliding soil slab displaced as a rigid body with minor changes in geometry. This observation offers the possibility of performing a straightforward dynamic exercise directly based on Newton's second law.

In order to do so, however, acting and resisting forces on the slide should be established with some certainty. The critical point here is probably not to describe precisely all the details of the motion but to pinpoint the key aspects explaining the motion, even if they are described in an approximate manner. The static analysis described in Chapter 4 offers a good opportunity to isolate the main features of the moving slide.

The information now available is the travelled distance and attention is focused on the central section of the slide. Slide displacements in the central stretch, 300 m long, are rather constant and close to 50 m. Therefore, a two-dimensional analysis could be performed with limited error and the restraining effects of the slide borders will be disregarded.

The travelled distance is the fundamental information to validate the model. The problem may be formulated in these terms: if a (dynamic) model is built in such a way that the input data (soil strength parameters and pore water pressures in particular) is consistent with the data explaining the initiation of failure, an additional validation of the main hypothesis introduced could be achieved. In addition, a dynamic analysis also offers the possibility of investigating some unanswered questions about the motion: its acceleration, velocity, and time duration could be estimated.

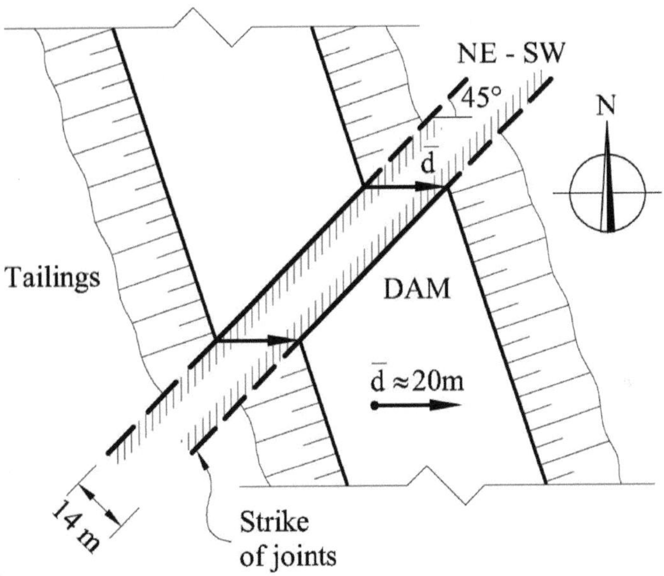

Figure 6.1 Geometry of the rupture breach (Alonso and Gens, 2006).

6.2 Conceptual Model

A few fundamental field observations, discussed in Chapter 4, are crucial to develop a conceptual model for the slide. They are:

a) position and geometry of the upstream limit of the slide, which is described as a vertical wall or "cliff" located at the toe of the upstream dam slope;
b) symptoms of liquefaction of tailings ;
c) reconstruction of the incipient failure mechanism based on the description of the geometry of the failed mass (Fig. 4.6).

This information has been integrated in Figure 6.2, which shows a free body diagram of the slide at a certain time *t* during the motion. The features of the model and its rationale are further explained by referring to Figure 6.2.

The geometry of the motion is conceived as a large rectangular wedge that transports the dam and a distal passive wedge that opposes the motion. This passive wedge increased in weight during the displacement in a process of accumulation of highly folded strata as the displacement took place.

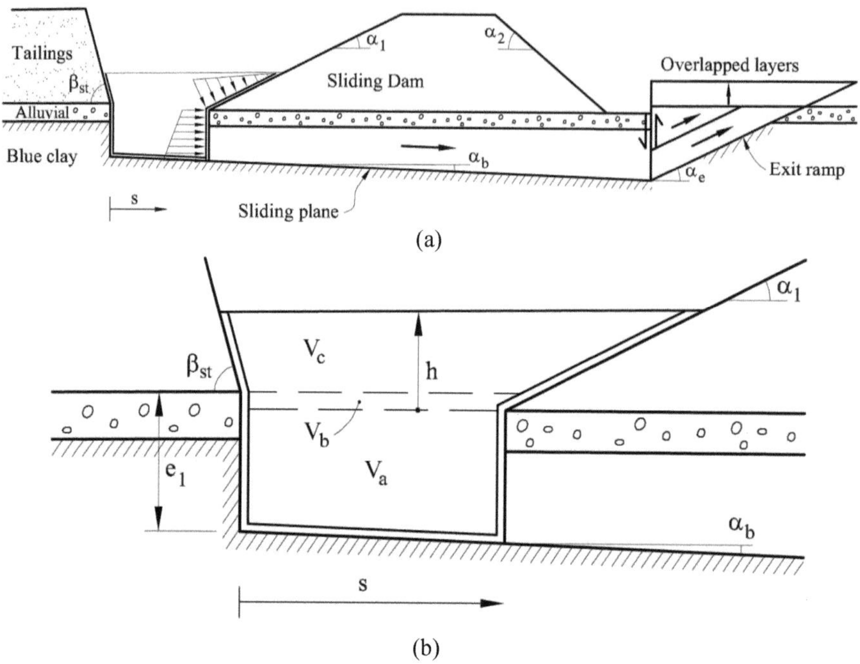

Figure 6.2 (a) Geometry of the slide in motion (Alonso and Gens, 2006); (b) Detail of the upper part.

The upstream sub-vertical limit plane remained stable at all times. The approximately basal failure surface follows a sedimentation plane that has a low dip angle (2°).Therefore, the depth of the failure plane increases in the direction of motion. The passive wedge brings the failure plane to the ground surface. According to the slide reconstruction in Figure 4.6, the passive wedge starts at a distance of roughly 55 m from the downstream toe of the dam.

As the slide moves forward, a large opening appears upstream since the vertical limiting wall has remained stable. The assumption made is that this opening of increasing size was filled by liquefied tailings initially located over the upstream slope. Evidence of tailing liquefaction was obtained immediately after the failure. Figure 6.3 shows the small volcanoes scattered in the exposed surface of the tailings immediately upstream of the dam. The picture was taken from the crest of the dam looking downwards into the pond a few hours after the failure. This observation suggests that, as the slide accelerated forward (because of the rapid loss of available strength on the sliding surface), the tailings wedge sitting over the upstream slope of the dam slid in the upstream direction and filled the opening gap. This process led to the tailings liquefaction at an early time. Beyond this time, the tailings occupying the opening gap upstream remained liquefied.

This scenario has two implications: one is that the acting forces against the slide are simply calculated as a hydrostatic pressure against the exposed surfaces "submerged" by the liquefied tailings and the second one is that the level of liquefied tailings should decrease because of the increasing size of the gap opened upstream.

Pore pressures on the sliding surface are initially known (see Chapter 4). However, when the slide displaces forward, an undrained loading (in the direction of motion, downstream of the dam) and a parallel unloading (in the upstream position of the dam) takes place. Pore pressures instantaneously generated will be assumed to be given by the increment of mean stress in points of the failure plane. This hypothesis has proven to be sufficiently accurate to reproduce actual observations in piezometers and has been described in Chapter 4. However, the calculation procedure will be simplified here if compared with the model developed in Chapter 4.

Figure 6.3 Mud volcanoes observed a few hours after the failure on the surface of the depressed basin, upstream of the slid dam (Courtesy of J. M. Rodriguez Ortiz).

The chosen variable selected to control the slide geometry is the horizontal displacement $s(t)$. The objective of the model developed is to calculate the slide displacement, velocity and acceleration. To do so, dynamic equilibrium equations will be written for the two wedges which define the motion following the conceptual model sketched in Figure 6.2a. Note that the mass of the two wedges changes during the motion (the main wedge loses the mass gained by the passive wedge). Before writing down the equations of motion let us discuss two important aspects which control the external forces on the slide: the tailings thrust and the pore pressure acting on the sliding surfaces.

6.3 Tailings Thrust against the Sliding Mass

Consider first the liquefaction of tailings. The volume of liquefied tailings should occupy the volume left upstream by the moving dam and accompanying foundation slab. The initial volume, at zero displacement, is limited by the upstream slope of the dam and the subvertical limiting wall of the slide (defined by an angle β_{st} in Fig. 6.2). Its volume will be given by

$$V_0 = 0.5h_0\left(h_0 \tan\left(90-\beta_{st}\right)+\frac{h_0}{\tan\alpha_1}\right), \tag{6.1}$$

where h_0 is the initial depth of tailings (27 m) and α_1 is the upstream dam slope (29°). The dam displacement results in an increasingly larger volume ready to be occupied by the liquefied tailings. This volume is defined by the three partial volumes (V_a, V_b and V_c) in Figure 6.2b. These volumes can easily be expressed in terms of h, the current height of tailings, the thickness, e_1, of the sliding foundation soil at the upstream limit of the slide (approximately 11.5 m following field observations; see also Fig. 4.16) and the dip angle of the sliding plane, α_b:

$$V_a = \left(e_1 - \frac{s\tan\alpha_b}{2}\right)s, \tag{6.2}$$

$$V_b = s^2\tan\alpha_b\left(1+\frac{\tan\alpha_b}{2\tan\alpha_1}\right), \tag{6.3}$$

$$V_c = \left(h\,m+f\right)\left(h-n\right), \tag{6.4a}$$

where

$$m = \frac{1}{2}\left(\frac{1}{\tan\alpha_1}+\tan\left(\frac{\pi}{2}-\beta_{st}\right)\right), \tag{6.4b}$$

$$f = \frac{s}{2}\left(2+\frac{\tan\alpha_b}{\tan\alpha_1}-\tan\alpha_b\tan\left(\frac{\pi}{2}-\beta_{st}\right)\right), \tag{6.4c}$$

$$n = s\tan\alpha_b. \tag{6.4d}$$

The condition of constant volume of liquefied tailings

$$V_0 = V_a + V_b + V_c \tag{6.5}$$

allows, in view of Equations (6.2) to (6.4), the calculation of the height of tailings,

h, as a function of the horizontal component (s) of the slide displacement. It will be also convenient to introduce the displacement s_b along the basal sliding surface because it offers some algebraic advantages to formulate the dynamic equation of motion. In addition, the relative displacement of the moving body and the blue clay underneath controls directly the development of residual conditions. The displacements s and s_b are related through

$$s = s_b \cos \alpha_b . \tag{6.6}$$

Since α_b is a very small angle (2°), s and s_b are almost identical but the distinction will be formally maintained in the remaining equations. The use of s and s_b is a simple a matter of algebraic convenience.

The condition of constant volume of liquefied tailings is not fulfilled in the sections close to the breach of the dam because the tailings that poured out of the pond contributed to reducing the height of the liquefied tailings acting against the moving slide. In fact, this situation may explain the smaller travelled distance in the sections close to the outlet (20 m), if compared with slide displacements of 50 m at some distance from the breach section. This distance probably marks the effect of the flow rate of tailings lost through the breach on the displacement reached by the slide.

The thrust of liquefied tailings against the upstream border of the slide is simple to calculate because pressures became hydrostatic. The horizontal component of the acting forces against the slide is

$$F_h = 0.5\gamma_t (h + e_1)^2 , \tag{6.7}$$

where γ_t is the specific weight of the liquefied tailings, which will be close to 31 kN/m^3 (see Chapter 4).

Let us now consider the initial force that induced the onset of instability. It was argued in Chapter 4 that this force could be approximated by an active Rankine state associated with a total height of (h+e_1). Then, Equation (4.10) could be taken as a first approximation for the initial value of the driving force. Sometime after the initiation of the motion, when the unstable wedge of tailings transforms into a liquefied mass, this initial force would increase substantially towards the value given in Equation (6.7) The displacement necessary to reach liquefaction is not known and it will be introduced as a parameter of the problem, ε. In this way, the evolution of the upstream horizontal force against the slide, F_h, is plotted in Figure 6.4.

The plot is defined by the initial force (F_{hi}) and peak (liquefaction conditions) force, by the displacement (ε) (a linear variation is assumed from s_b = 0 to s_b = ε), and by the force decay beyond the peak liquefaction force for increasing s values. The latter is given by Equation (6.7) in terms of h. Equation (6.5) provides a relationship between h and s.

The next step before formulating motion equations will be to calculate the pore water pressures acting against the sliding surface.

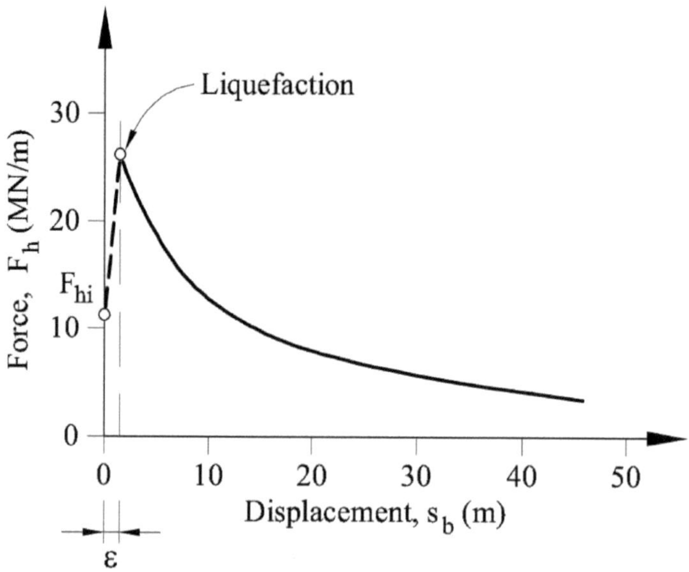

Figure 6.4 Evolution of tailings thrust against the moving dam and accompanying foundation soil, as the displacement s_b increases.

6.4 Pore Pressures during Motion

The distribution of pressures immediately before failure was calculated in Chapter 4. Figure 4.18 shows the estimated variation along the failure plane.

This variation was approximated by means of linear segments, as shown in Figure 6.5, for the initial position of the slide ($s = 0$). Water heights (in m) have their origin on the sliding surface. When the slide moves forward, instantaneous changes in pore pressure will be induced on the failure surface. They will be essentially given by the change in mean stress, which is, in turn, is given by the weight of the dam (upstream, however, the weight of tailings, decreasing with displacement, will define the new excess pore pressures). Since the sliding surface will be, soon after the initiation of the motion, under residual conditions no dilatancy effects are expected on the pore water pressure generation.

It will be assumed that the pore pressure at the position of the upstream toe of the slope, in excess of the hydrostatic value provided by the original phreatic level (located at the surface of the soil), will be proportional to the height of liquefied tailings. This pressure height will change from an initial value of e_1 when the height $h(s)$ of tailings evolves from $h = 27$ m to $h = 0$. Therefore, in connection with Figure 6.6

$$u_1 = e_1 + \frac{68 - e_1}{27} h(s) \text{ (m of water head).} \tag{6.8}$$

Under the dam crest (pressure u_2), the excess pore pressure will be essentially constant and equal to the initial value (~52 m of water column). It may be also expressed as the sum of two contributions: the hydrostatic component associated with the depth under the phreatic level at the center of the dam base (e_R) and the excess pore pressure

$$u_2 = u_3 = e_R + 38.6 \text{ (m of water head).} \tag{6.9}$$

At the position of the downstream toe of the dam the excess pore water pressure will also be maintained essentially equal to the original level during the motion and therefore

$$u_4 = e_2 + 10.5 \text{ (m of water head).} \tag{6.10}$$

where e_2 is the depth of the failure surface under the downstream toe of the dam.

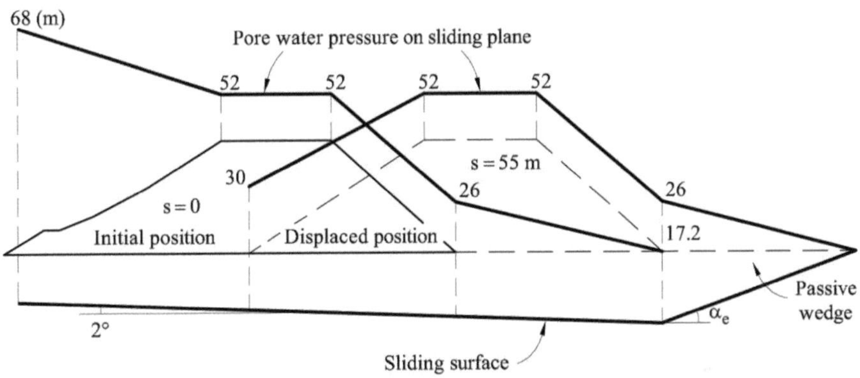

Figure 6.5 Assumed distribution of pore water pressure heads (in m) under the dam during sliding (based on Alonso and Gens, 2006).

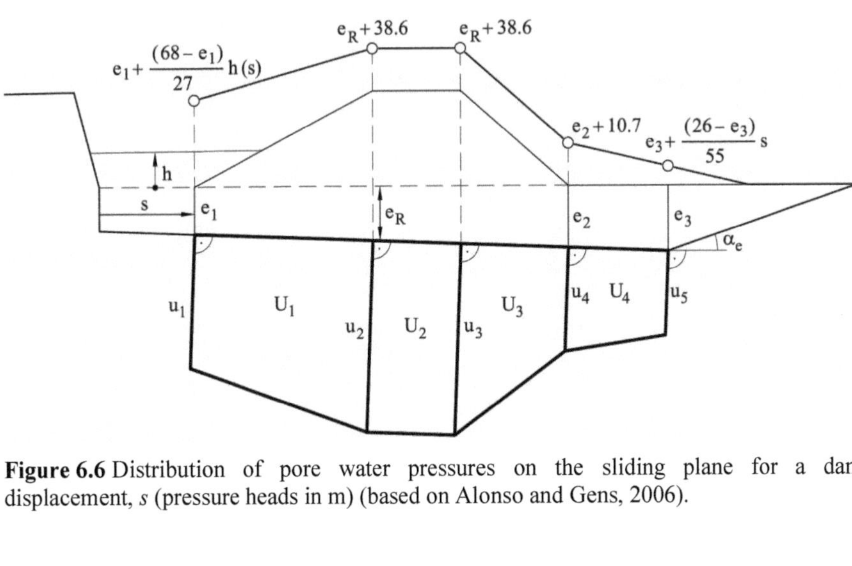

Figure 6.6 Distribution of pore water pressures on the sliding plane for a dam displacement, s (pressure heads in m) (based on Alonso and Gens, 2006).

A final control point to define the pore pressure variation is the position of the exit ramp of the passive wedge. This position, which is located 55 m downstream of the dam toe, "sees" the dam approaching. When the slide came to rest, the dam toe was on top of the origin of the exit ramp. In this position, the pore pressure at this point of the failure surface should be equal to the calculated pore pressure under the dam toe for $s = 0$. Again, for a linear relationship between the excess pore pressures and the displacement, s, one may calculate

$$u_5 = e_3 + \frac{(26 - e_3)}{55} s \approx e_3 + 0.16s \text{ (m of water head).} \tag{6.11}$$

Note that e_3 changes with displacement from its initial value (e_{3_0} = 17.21 m in Fig. 4.16) to a final value (15.29 m) when the dam displaces 55 m. See also next paragraph on the geometry of the motion and Figure 6.7. The current value of e_3 is given by

$$e_3(s) = e_{3_0} - \frac{e_{3_0} - e_2}{55} s \text{ (in m)} \tag{6.12}$$

The implicit assumption in the preceding proposal is that there was very limited dissipation of excess pore pressures at the position of the failure plane. This was indeed the case because of the very low clay permeability. Therefore, the existing distribution of pore pressures on the failure plane was taken as a "model" for the induced pore pressures during the rapid undrained loading associated with the slide motion.

The control points for the determination of water pressure provide an approximate distribution of pore pressures under the slide, at the position of the sliding plane, if a linear variation of pore pressures among them is assumed. Then, Figure 6.6 provides the instantaneous pore pressures for a displacement s.

A more sophisticated procedure to calculate the pore pressures under the moving dam could be devised. It was shown in Chapter 4 that the application of elasticity solutions was quite accurate. But in the context of the dynamic analysis developed here, this approach would lead to a cumbersome calculation procedure that is probably not warranted. It will be shown later that the solution is quite robust and not very sensitive to limited variations of the main parameters defining the problem.

The water pressure forces acting on the lower failure plane will now be computed by dividing the variation of pore pressure into four regions (U_1, U_2, U_3, and U_4) as illustrated in Figure 6.6. Water pressure forces are now simply calculated in terms of the "height" of the corresponding trapeziums:

$$U_1 = \frac{1}{2}\left[e_1 + \frac{(68 - e_1)h}{27} + 38.6 + e_R \right] \gamma_w 48.7, \tag{6.13a}$$

$$U_2 = (38.6 + e_R)\gamma_w 26, \tag{6.13b}$$

$$U_3 = \frac{1}{2}[38.6 + e_R + 10.5 + e_2]\gamma_w 33.3, \tag{6.13c}$$

$$U_4 = \frac{1}{2}\gamma_w [10.5 + e_2 + e_3 + 0.16s](55 - s), \tag{6.13d}$$

where e_2, e_R and e_3 are shown in Figure 6.6. The distances (in m) 48.7, 26, 33.3 and 55 define the length of pore pressure trapezoids along the sliding plane. If γ_w is given in kN/m^3, Equations 6.13 provide forces U_i in kN/m (e values in m).

6.5 The Motion Equation

6.5.1 Geometry of the motion

Figure 4.16 provides the initial state of the slide. The dam and a slab of soil underneath displace towards the passive wedge which slides on a plane inclined α_e. The kinematics of the motion have been simplified as shown in Figure 6.7. It will be assumed that the passive wedge (B) increases in size during the motion, i.e. it gains the mass "lost" by the main slide (A). The mass transfer takes place on the vertical plane which limits upstream the passive wedge.

Of course, other geometrical configurations are possible and they may lead to some differences in the calculations. However, it will be shown below that the entire formulations is quite insensitive to specific details. The idea here is to provide an acceptable mechanism, essentially consistent with field observations and analyze its implications.

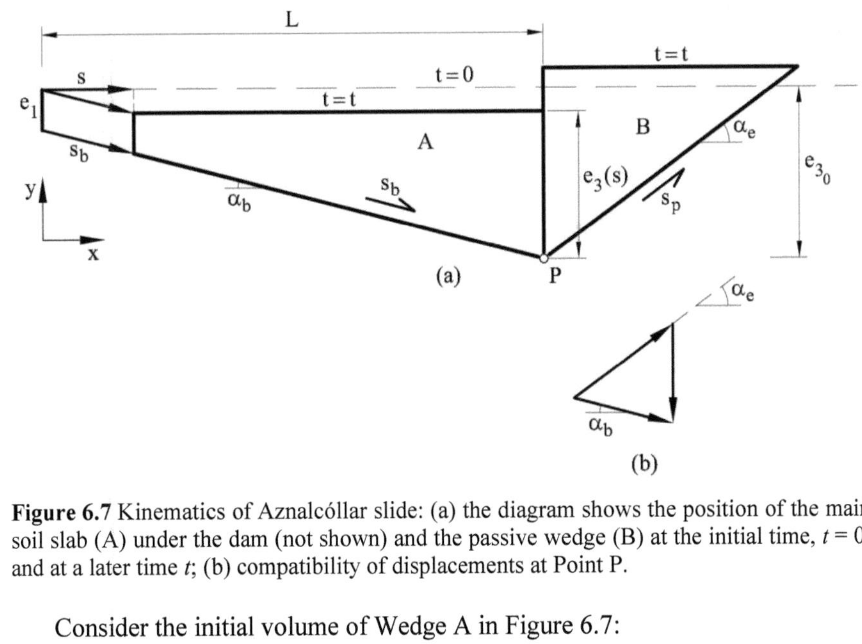

Figure 6.7 Kinematics of Aznalcóllar slide: (a) the diagram shows the position of the main soil slab (A) under the dam (not shown) and the passive wedge (B) at the initial time, $t = 0$, and at a later time t; (b) compatibility of displacements at Point P.

Consider the initial volume of Wedge A in Figure 6.7:

$$V_{A0} = \frac{e_1 + e_1 + L\tan\alpha_b}{2} L = L\left(e_1 + \frac{L\tan\alpha_b}{2}\right) \tag{6.14}$$

where L is the total horizontal length of the main slide ($L = 108 + 55 = 163$ m in Fig. 4.16).

In Figure 6.7 the displacement of the slide is characterized by the variable s (in the x-direction) or alternatively by s_b. Both are related through Equation 6.6.

Once the Wedge A has displaced s_b, its volume is given by

$$V_{A1} = \left(L - s_b\cos\alpha_b\right)\left(e_1 + \frac{L - s_b\cos\alpha_b}{2}\tan\alpha_b\right). \tag{6.15}$$

Therefore the variation of the volume of the wedge can be expressed as a function of the displacement by:

$$\Delta V_A = s_b\cos\alpha_b\left(e_1 + L - \frac{s_b\cos\alpha_b}{2}\tan\alpha_b\right) \tag{6.16}$$

which will be lost by the main wedge and added to the passive wedge.

6.5.2 Dynamic equilibrium of the passive wedge

Consider in Figure 6.8 the passive wedge at time t. The resultant of water pressures on the wedge were given in Equations (4.15) and (4.16) for the initial conditions. It will be assumed that the initial values will be maintained and therefore, with reference now to Figure 6.8,

$$U_{hp} = \frac{1}{2}\gamma_w e_{3_0}^2, \tag{6.17}$$

$$U_p = \frac{\gamma_w e_{3_0}^2}{2\sin\left(\alpha_e\right)}. \tag{6.18}$$

The current volume of the passive wedge at time t will be given by adding the volume gained by the wedge (Eq. (6.16)) to the initial volume. Therefore, the current weight of the passive wedge at time t, for a slide displacement s_b, will be given by

$$W_p = \gamma_{\text{soil}}\frac{e_{3_0}^2}{2\tan\alpha_e} + s_b\cos\alpha_b\left[e_1 + \tan\alpha_b\left(L - \frac{s_b\cos\alpha_b}{2}\right)\right] \tag{6.19}$$

where the first term provides the initial weight (Eq. (4.18)).

Since the mass of the passive wedge (M_p) is changing in time, the equation of dynamic equilibrium will have the general form:

$$\frac{\mathrm{d}}{\mathrm{d}t}\left(M_p\frac{\mathrm{d}s}{\mathrm{d}t}\right) = \frac{\mathrm{d}M_p}{\mathrm{d}t}\frac{\mathrm{d}s}{\mathrm{d}t} + M_p\frac{\mathrm{d}^2 s}{\mathrm{d}t^2} = \Sigma\mathbf{F}_s \tag{6.20}$$

where s is the generic direction of displacement and $\sum \mathbf{F}_s$ the resultant forces in the direction s.

Equation (6.20) will be now applied in the direction of the motion of the wedge (displacement s_p, see Fig. 6.8) and in the normal direction to s_p (no displacements in this direction).

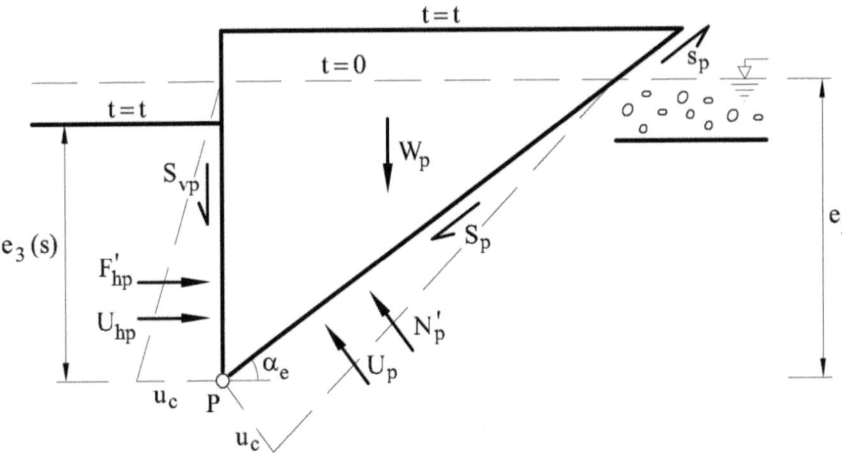

Figure 6.8 Geometry and forces acting on the passive wedge.

Equilibrium in direction parallel to s_p

$$F'_{hp}\left(\cos\alpha_e - \sin\alpha_e \tan\varphi'_b\right) + U_{hp}\cos\alpha_e - N'_p \tan\varphi'_b - W_p \sin\alpha_e = \frac{\mathrm{d}}{\mathrm{d}t}\left(M_p \frac{\mathrm{d}s_p}{\mathrm{d}t}\right). \tag{6.21}$$

Equilibrium in direction normal to s_p

$$F'_{hp}\sin\alpha_e + \cos\alpha_e \tan\varphi'_b + U_{hp}\sin\alpha_e + W_p\cos\alpha_e - U_p - N'_p = 0 \tag{6.22}$$

where shear forces have already been written in terms of effective normal forces according to the Mohr Coulomb's Law under residual conditions (cohesion equal to zero).

The relationship between the main slide displacement s_b and the wedge displacement s_p is derived by forcing the kinematic compatibility of the motion at Point P. In view of Figure 6.7b:

$$s_p = \frac{\cos\alpha_b}{\cos\alpha_e} s_b. \tag{6.23}$$

6.5.3 Dynamic equilibrium of the main wedge and dam

Reference is now made to Figure 6.9 which shows the forces on the main slide at a given time during the motion. The force F_h has already been discussed and it is given in Figure 6.4 and in Equations (4.10) and (6.7). Once the tailings are liquefied the vertical force F_v is controlled by the variable heights of tailings h (Fig. 6.2b). It is given by

$$F_v = \frac{h^2 \gamma_t}{2 \tan \alpha_1}. \tag{6.24}$$

At the start of the motion $h \approx H$ and the tailings are not liquefied yet but F_v will be controlled by the saturated weight of tailings and Equation (6.24) would be essentially valid also.

Dynamic equilibrium follows also Equation (6.20) which will be written in the direction of the motion (s_b) and in the normal to s_b.

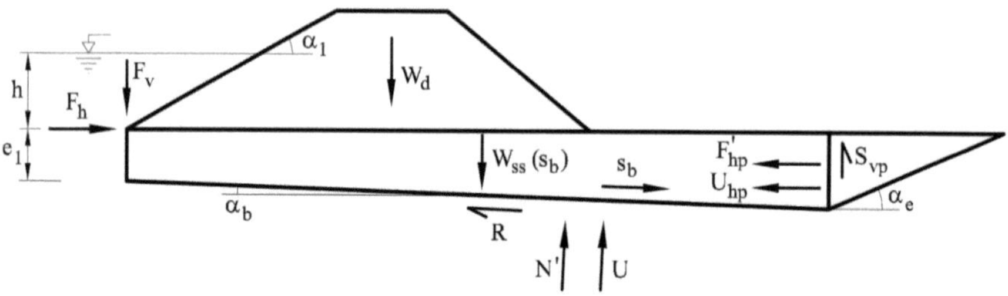

Figure 6.9 Geometry and forces acting on the main sliding wedge.

Equilibrium in direction parallel to s_b

$$\begin{aligned} &F_h \cos\alpha_b + F_v \sin\alpha_b + W_d \sin\alpha_b + W_{ss} \sin\alpha_b - N' \tan\varphi'_b - \\ &F'_{hp} \cos\alpha_b - U_{hp} \cos\alpha_b - F'_{hp} \tan\varphi'_b \sin\alpha_b = \frac{\mathrm{d}}{\mathrm{d}t}\left(M \frac{\mathrm{d}s_b}{\mathrm{d}t}\right) \end{aligned} \tag{6.25}$$

where the total mass M now includes the mass of the dam (M_d) and the variable mass of the soil slab (M_{ss}):

$$M = M_d + M_{ss} = M_d + \rho_s V_1 \tag{6.26}$$

where $V_1(s_b)$ is given in Equation (6.15) and the weight of the dam ($W_d = g\, M_d$) was given in Equation (4.12).

Equilibrium in direction normal to s_b

$$\begin{aligned} &-F_h \sin\alpha_b + F_v \cos\alpha_b + W_d \cos\alpha_b + W_{ss} \cos\alpha_b + F'_{hp} \sin\alpha_b + \\ &U_{hp} \sin\alpha_b - F'_{hp} \tan\varphi'_b \cos\alpha_b - N' - U = 0 \end{aligned} \tag{6.27}$$

where $U = U_1 + U_2 + U_3 + U_4$ (given in Eq. (6.13)).

6.5.4 Solution

The system of four Equations (6.21), (6.22), (6.25) and (6.27) must now be solved.

This is a system of nonlinear total differential equations and a numerical solution is unavoidable.

The strategy followed here to solve them is to write them in the form

$$a = \frac{\mathrm{d}v}{\mathrm{d}t} = f\left(s_b\left(t\right), v\left(t\right)\right) \tag{6.28}$$

where $v = \frac{\mathrm{d}s_b}{\mathrm{d}t}$ is the velocity of the main displacement in the direction of the sliding plane.

Equation (6.28) is simply discretized in a forward marching scheme in time as:

$$v_{new} = v_{old} + \Delta t\, f\left(s_{b\text{old}}, v_{\text{old}}\right) \tag{6.29}$$

which allows calculating the velocity of the slide if the acceleration is known at the beginning of the motion ($t = 0$). In fact, Equation (6.28) for $t = 0$, $s_b = 0$ and $v = 0$ provides the initial acceleration (a_i) essentially as a ratio between resultant forces and masses involved (see however the discussion below).

The system of Equations (6.21), (6.22), (6.25) and (6.27) was transformed after some elaboration into

$$\frac{\mathrm{d}}{\mathrm{d}t}\left[M\left(s_b\right)\frac{\mathrm{d}s_b}{\mathrm{d}t}\right] + m_4 \frac{\mathrm{d}}{\mathrm{d}t}\left[M_p\left(s_p\right)\frac{\mathrm{d}s_p}{\mathrm{d}t}\right] = F^* \tag{6.30a}$$

where

$$\begin{aligned} F^* &= F_h\left(s_b\right)m_1 + \left[F_v\left(s_b\right) + W_d + W_{ss}\left(s_b\right)\right]m_2 + \\ &U_{hp}\left(m_7 - m_1\right) + U_p m_5 - W_p\left(s_p\right)m_6 + U\left(s_b\right)\tan\varphi_b'. \end{aligned} \tag{6.30b}$$

The dependence of some variables on displacements, either s_b or s_p, has been indicated. The coefficients m_i are functions of geometry and friction angles as follows:

$m_1 = \cos\alpha_b + \sin\alpha_b t_4$; $m_2 = \sin\alpha_b - \cos\alpha_b t_4$; $m_3 = m_1 + m_2 t_4$;
$m_4 = m_3/t_1$; $m_5 = t_4 m_3/t_1$; $m_6 = t_3 m_3/t_1$; $m_7 = t_2 m_3/t_1$.

and

$$t_1 = \cos\alpha_e - \sin\alpha_e \tan\varphi_b' - \tan\varphi_b'\left(\cos\alpha_e \tan\varphi_b' + \sin\alpha_e\right);$$
$$t_2 = \cos\alpha_e - \sin\alpha_e \tan\varphi_b';$$
$$t_3 = \sin\alpha_e + \cos\alpha_e \tan\varphi_b';$$
$$t_4 = \tan\varphi_b'.$$

The term F^* in Equation 6.30 represents the unbalanced forces. Recalling Equation (6.23), Equation (6.30a) can be re-written as

$$\frac{dv_b}{dt} = \frac{F^* - v_b \left[\frac{dM}{dt} + m_4 \frac{\cos\alpha_b}{\cos\alpha_e} \frac{dM_p}{dt} \right]}{M + m_4 \frac{\cos\alpha_b}{\cos\alpha_e} M_p}. \tag{6.31}$$

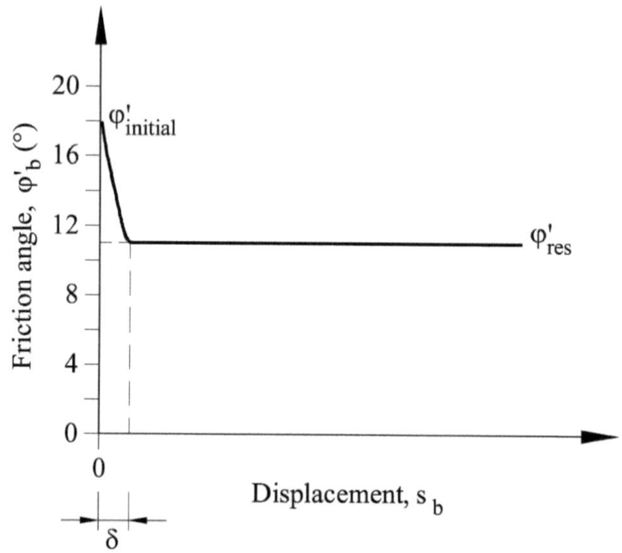

Figure 6.10 Reduction of average friction angle on basal plane as a function of sliding distance s_b.

Equation (6.31) has the form of Equation (6.28). It is interesting to analyze its structure. It simply tells that the acceleration of the main slide is the ratio of a generalized force and a generalized mass. Both include terms belonging to the main slide and the passive wedge. The mass in the denominator, in particular, is a weighted sum of the masses of the two wedges. The changing mass of both wedges introduces a term in the numerator which modifies the unbalanced forces F^*. The modification is proportional to the product of velocity and rate of mass changes. This term is a rate of change of momentum which only enters the second law of Newton when moving bodies change their mass. It was checked that in our case this term was quite small compared with the intensity of unbalanced forces

A final important point concerns the mobilized effective friction angle of the failure surface φ'_b. The boundary friction on the passive wedge was also defined by φ'_b. It was found in Chapter 4 that the average equilibrium value at the onset of the slide was close to 18° (and a zero effective cohesion). This friction value is intermediate between the peak friction angle of the blue clay (24.1°) and the residual angle (11°). The initiation of the motion implies further reductions in φ'_b. The rate of drop of φ'_b towards the residual value is unknown, but experimental evidence suggests that all that is required to reach residual conditions are a few centimetres, perhaps decimetres, of relative displacement. This uncertainty, however, is resolved by means of a displacement parameter, δ, defined in Figure 6.10. Parameter φ_b' is therefore defined in terms of the displacement s_b. From s_b = 0 to $s_b = \delta$, the friction, φ_b', decays linearly from the initial to the residual value. For $s_b \geq \delta$, $\varphi_b' = \varphi_{res}'$. δ is considered also as a model parameter.

6.6 Results

The motion Equation (6.28) was solved in an explicit manner (Eq. (6.29) and subsequent discussion). Starting at the equilibrium conditions given in Chapter 4, the slide was made unstable by increasing slightly the driving force F_{hi} (by reducing in 0.1° the friction angle of tailings).

An Excel sheet was written to perform the calculations. The accuracy of calculations was controlled by time increment Δt. Negligible errors were found for $\Delta t \leq 0.1$ s.

6.6.1 Set of model parameters

Model parameters were grouped into three groups: geometry, tailings, and blue clay, as follows:

Geometry:

β_{st}: Dip of the head scar of the slide within the tailings' deposit. Field observations indicate that it may vary between 70° and 90°.

e_R: Depth of the failure surface under the base of the dam at the centre of the dam base. It is close to 13.40 m.

α_b: Apparent dip of the failure surface. Based on field observations, it is estimated as $\alpha_b = 2°$.

α_e: Slope of the sliding plane under the passive wedge. Visual observations in trenches excavated at the foot of the slide indicate that $\alpha_e = 20°$.

Tailings

γ_t: Natural specific weight of liquefied tailings. In the southern pyrite lagoon the saturated unit weight of the deposited tailings is around 31 kN/m^3. This is the value adopted in calculations.

$F_{hi}(*)$: Initial horizontal thrust against the dam and the accompanying slice of moving soil. It was estimated as 11.3 MN/m for active conditions.

$\varepsilon(*)$: Necessary forward displacement of the dam to generate liquefaction conditions in the tailings' wedge. A value $\varepsilon = 1$ m was selected for the calculations reported below.

Blue clay

$\varphi'_{\text{initial}}(*)$: Average initial friction angle which ensures strict equilibrium before the failure. It was derived by limit equilibrium analysis ($\varphi'_{\text{initial}}$ = 18.09°).

φ'_{res}: Residual friction angle. It was derived from ring shear tests on remoulded blue clay and by direct shear tests on natural slickensided discontinuities. A value $\varphi'_{res} = 11°$ was adopted.

δ(*): Relative displacement along the failure plane needed to mobilize the residual friction angle. It is probably necessary to accumulate a few decimetres to achieve residual conditions. A value δ = 20 cm was selected for calculations.

Parameters with an asterisk (*) were found to have a very limited influence on the computed results. This was an interesting finding because there was some uncertainty about the correct value for the specific parameter involved. This is, in particular, the case of the two displacement parameters included in the model: δ and ε.

6.6.2 Motion of the dam

A representative "base case" was solved with a set of parameters, which are believed to represent the actual sliding conditions ($\varphi'_{initial} = 18.09°$; $\varphi'_{res} = 11°$; δ = 20 cm; $\gamma_t = 31$ kN/m^3; $F_{hi} = 11.3$ MN/m; $\varepsilon = 1$ m; $\beta_{st} = 70°$; $e_R = 13.40$ m; $\alpha_b = 2°$; $\alpha_e = 20°$).

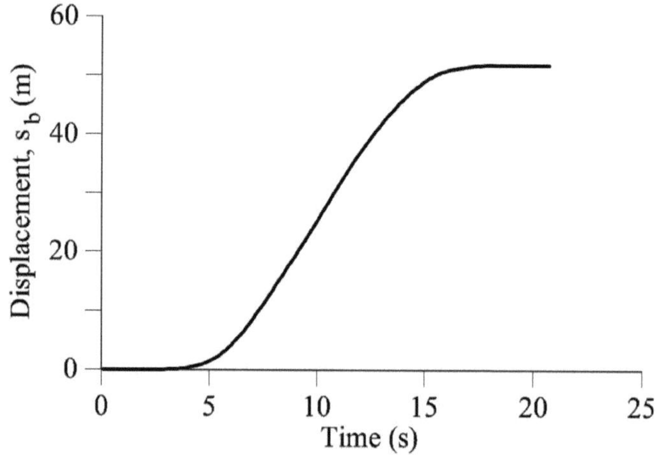

Figure 6.11 Computed evolution of dam displacement.

The distance travelled by the dam as a function of time is shown in Figure 6.11. According to the model, the slide stops 15 s after the initiation of the failure, when the dam has travelled 51.5 m. This latter figure is very similar to actual observations. The evolution of speed is shown in Figure 6.12. The dam reaches a maximum velocity of 6.2 m/s (22.3 km/h), 10 s after the start of the motion. The calculated acceleration is shown in Figure 6.13. The dam experiences a rapid acceleration during the first few seconds after the start. The evolution of forces, plotted in Figure 6.14, explains the acceleration. During the first instants of the motion, the liquefaction of tailings increases the driving force, which reaches values substantially higher than the total resistance offered by the base and the passive wedge (the evolution of both forces is also shown in the figure). The driving forces include also the component of slide weight in the direction of the motion, s_b. The positive net resultant force (driving minus resistant forces) accelerates the slide. The fall of the level of tailings at the upstream part of the slide (plotted in Fig. 6.15) progressively reduces the pushing force as well as the acceleration. Total resistance to sliding increases slightly with time. The main reason for the deceleration of the slide is therefore the continuous decrease in the driving force associated with the fall of the level of liquefied tailings and in a more limited manner by the restraint offered by the passive wedge.

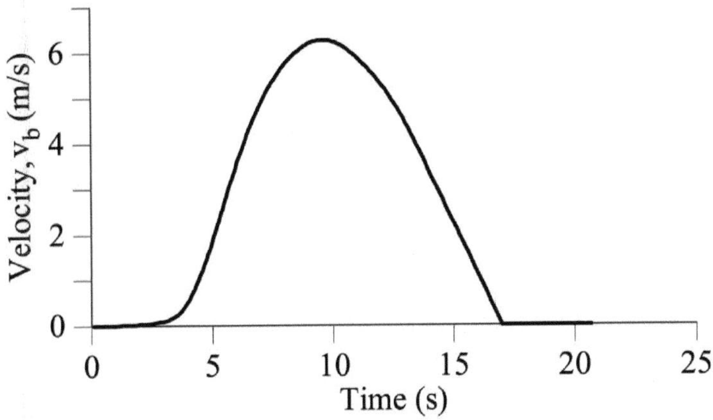

Figure 6.12 Computed slide velocity.

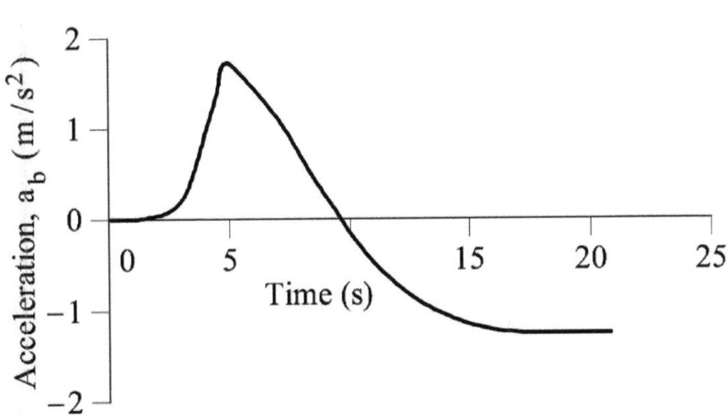

Figure 6.13 Computed slide acceleration.

The relevance of the stability of the tailings cliff left by the forward motion of the dam becomes now more apparent. A large volume of liquefied tailings would have implied a longer period of acceleration and higher velocities and dam displacements. Under these circumstances, the size of the breach opened in the dam would have increased dramatically, leading to a more violent flood.

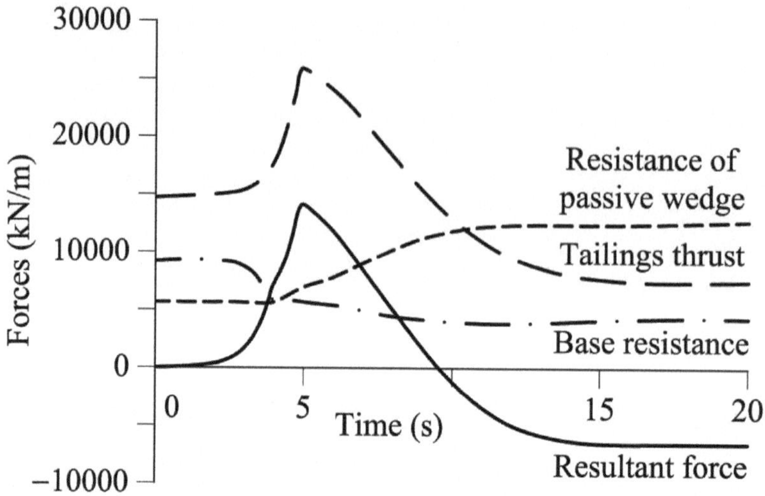

Figure 6.14 Computed variation of forces against the moving mass.

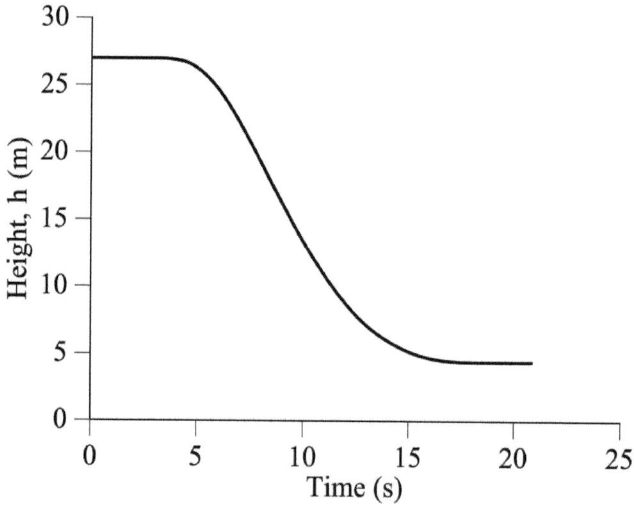

Figure 6.15 Evolution of height of liquefied tailings, acting against the upstream end of the slide.

The short-term stability of a quasi-vertical slope in the mass of tailings is explained by the cementation exhibited by them and its relatively low permeability. The rapid unloading caused by the forward slide displacement probably led to negative pore water pressures in the vicinity of the exposed cliff. In the medium and long term, tailings were able to maintain the initial subvertical slope although the strong erosion and, in some cases, local failures led to the partial collapse of the original slope. However, the volume added by these processes did not affect the motion of the slide, which was very rapid. In fact, at the end of the motion, see Figure 6.14, the available resistance was much higher than the driving force. Additional sources of eroded or destabilized tailings only marginally add to the force exerted against the upstream slide face.

The dam breach also involved the pouring of a massive flow of tailings mud outside the dam, thus effectively controlling the level of the tailings in the large, open basin upstream of the slipped dam.

This outpouring probably affected the motion of the slide in the vicinity of the breach. In fact, the rate of reduction of the height of liquefied tailings, a key aspect of the model developed, is controlled by the condition of volume conservation expressed in Equation (6.5). However, the flow of tailings out of the pond may be viewed as an additional "sink term" in this equation and, therefore, it contributes to reducing faster the level of liquefied pushing tailings. This, in turn, implies a smaller distance travelled by the slide. This fact may explain the progressive reduction of the dam displacement as the breach position is approached (Fig. 4.5).

The acceleration of the dam in the early stages helps to explain the fall of the wedge of tailings sitting on the mantle of red clay placed against the upstream rockfill slope. The presence of this clay, placed on top of a slope of 29°, facilitated the instability of the wedge of tailings. The calculated maximum acceleration (0.175 g) is similar to the acceleration induced by an earthquake of intensity 7–8 (MKS).

The calculated height of tailings in the upstream basin, when the slide stopped, is 4 m over the initial level of the foot of the upstream slope. The measured central profile of the slide, a few months after the failure, shows that the level of the tailings upstream of the failed dam was 7 m above the level of the foot of the dam. The difference, which is not considered to be very significant, is attributed to the accumulation of eroded tailings in the days following the rupture.

6.7 Sensitivity Analysis

It was mentioned that the results of the analysis were not particularly affected by some of the model parameters. A relatively large uncertainty could "a priori" be associated with a few of them (δ, ε). However, most of the model parameters were known with some confidence. It is therefore advisable to examine the sensitivity of the results presented. In the sensitivity analysis, which is summarized in Table 6.1, each one of the listed parameters was changed in turn, while keeping the other properties involved fixed to those values used in the base case.

The variable selected for comparison is the total distance travelled by the dam. This distance is a natural choice because it can be compared with the measured value in the field. An examination of Table 6.1 reveals that the analysis is robust in the sense that none of the parameters changed seems to have a decisive influence on the distance travelled by the dam. The values listed in Table 6.1 do not differ much from the actual travelled distance. This result adds confidence to the interpretation given for the observed displacement of the dam and the associated physical phenomena.

Table 6.1 Results of the sensitivity analysis in terms of the length travelled by the dam.

Parameter/Property	Value	Travelled distance (m)
Slope of sedimentation planes, α_b (°)	2 1	51.5 (Base case) 53.3
Dip of upstream scar, β_{st} (°)	80 90	48.7 46.0
Residual friction angle of clay, φ'_{res} (°)	10 12	54.9 48.2
Exit angle of passive wedge, α_e (°)	21 26	50.5 46.6
Depth of sliding surface under the base center of dam, e_R (m)	11 12 13	57.5 54.8 52.40
Specific weight of tailings, γ_t (kN/m^3)	29 30	48.78 50.15
Initial horizontal thrust of tailings, F_{hi} (MN/m)	16.9 22.6	51.8 52.0
Necessary forward displacement of the dam to generate liquefaction conditions, ε (m)	0.1 0.4	52.0 51.0
Initial friction angle, $\varphi'_{initial}$ (°)	17 15	52 52
Relative displacement needed to mobilize the residual friction angle, δ (m)	0.1 0.5	52 51.9

6.8 Lessons Learned

6.8.1 Factors contributing to the acceleration of the slide

Two phenomena contributed to the acceleration of the slide: the brittleness of the clay formation whose strength dropped to residual values during the first stages of the motion and the liquefaction of tailings, which significantly increased the thrust against the incipient slide.

6.8.2 Liquefaction of tailings

The static liquefaction of tailings could not be reproduced in undrained laboratory triaxial tests. However, the tailings slide on the upstream side of the dam was able to liquefy in a massive way a large volume of tailings, even if they exhibited some significant cementation.

6.8.3 Further validation of strength and water pressures

The motion of the slide and not only the strict condition of instability, may prove useful to establish material parameters, prevailing pore pressures and, in general, to better define the conditions before failure. In the case of Aznalcóllar, the back analysis of the motion provides an additional validation of clay strength and pore pressure conditions.

6.8.4 The risk of dam displacement

Landslide risk is often related to the motion after instability. In the case of Vaiont, the attained velocity was the key aspect. In the case of Aznalcóllar the risk was also associated with the displacement. The slide displacement was enough to open a breach in the rockfill dam and to trigger the flood. The flow rate pouring out of the pond was controlled by the breach width, which was directly related to the slide displacement. Longer running distances (a consequence, for instance, of the weaker nature of tailings, which could imply a higher volume of liquefied tailings acting against the upstream boundary of the slide) would have resulted in a wider breach and in a more catastrophic flood.

6.8.5 Unknown aspects of failures

The dynamic analysis provided answers to some unsolved questions (the slide, being at night, was not observed) such as its velocity, duration and acceleration.

6.9 Advanced Topics

The analysis of landslide motion after failure is seldom performed in engineering practice, although analytical and numerical procedures are available to investigate the runout distances and velocities reached by landslides. In some studies (Hungr, 1995; McDougall and Hungr, 2004; Quecedo *et al.*, 2004) the slide is idealized as a fluidized mass of soil and the Navier-Stokes equations are integrated in depth, adapted to the curved geometry of the surface, and solved for some rheological models adopted for the moving mass. In other approaches, conceived for rockslides, discrete element approaches and hybrid continuum-discontinuum models have been developed (Eberhardt *et al.*, 2004).

References

Alonso, E.E. and Gens, A. (2006) Aznalcóllar dam failure. Part 3: Dynamics of the motion. *Géotechnique* 56 (3), 203–210.

Eberhardt, E., Stead, D. and Coggan, J.S. (2004) Numerical analysis of initiation and progressive failure in natural rock slopes –the 1991 Randa rockslide. *International Journal of Rock Mechanics and Mining Sciences* 41 (1), 69–87.

Hungr, O. (1995) A model for the runout analysis of rapid flow slides, debris flows and avalanches. *Canadian Geotechnical Journal* 32, 610–623.

McDougall, S. and Hungr, O. (2004) A model for the analysis of rapid landslide runout motion across three-dimensional terrain. *Canadian Geotechnical Journal* 41, 1084–1097.

Quecedo, M, Pastor, M., Herreros, M.I. and Fernández Merodo, J.A. (2004) Numerical modelling of the propagation of fast landslides using the finite element method. *International Journal of Numerical Methods in Engineering* 59, 755–794.

Epilogue

Geotechnical designs should comply with two fundamental requirements:

- The solution has to be stable with an appropriate safety margin.
- Deformations and displacements should remain within acceptable limits.

The first requirement leads in a natural way to the determination of soil strength. The second is associated with soil stiffness. In both cases equilibrium conditions and strain compatibility should be satisfied. This is the classical approach, one that has evolved into an extremely rich set of available solutions and analysis procedures. Critical state theory (Schofield and Wroth, 1967) integrated strength and stiffness into a unified and powerful framework which resulted in a more advanced understanding of soil mechanics and eventually led to modern elastoplastic computation tools.

However, the backbone of Geotechnical Engineering does not concern the development of more or less sophisticated tools and theories. It remains in a previous step. When facing a real problem it is necessary to isolate its fundamental aspects and to achieve a correct conceptual representation of its nature. This phase requires abstract thinking, which is certainly assisted by a proper understanding of paradigms and theories of Soil and Rock Mechanics. The process of abstract thinking with the aim of identifying the key issues usually renounces to complexity and secondary details. To be successful, concepts should be simple and rooted on well established mechanical and physical knowledge. Only when the relevant mechanisms or phenomena which define the problem are well understood, additional sophistication may be added for a more accurate analysis or interpretation. This book remains in this first "simple" stage.

The correct identification of the essential traits of a geotechnical situation relies heavily also on accumulated experience and on educated intuition. But, how to educate intuition and how to transfer practical experience?

Failures have always been regarded with extraordinary attention by engineers. They have a powerful human and emotional side because of their implications but also because they are often associated with mistakes, errors and lack of understanding. These aspects are very close to human nature as well to the practice of engineering. Failures convey a clear message: they point towards the limitation of our knowledge and practices. In a sense, they are telling us the truth. Unlike theories and current methodologies they remain valid in time. In fact, they are permanent benchmarks which can be revisited to check our conceptual representations and models. They provide a permanent incentive to direct research and technological developments. In addition, they may become an effective educational tool. They serve to identify errors, if properly interpreted they provide strong clues of the fundamental nature of problems and they constitute a good validation check for theories and models.

These considerations were present when this book was conceived. Major failures have attracted continuous and sustained efforts to explain their fundamental aspects through current theories. They all serve the purpose of developing new concepts. Some of them have been regarded as critical events for the development of new disciplines (Rock Mechanic received a definitive impulse from failures such as Vaiont slide and Malpasset dam). But any failure, not necessarily a spectacular one, has also a great potential to become a permanent source of knowledge.

Geotechnical engineers often face the need to find acceptable and reliable solutions in a limited amount of time. Text book solutions or even more advanced theoretical or empirical solutions are often not available for specific cases. The solution has to be built form basic considerations and fundamental principles. Almost all the cases presented in this book share this characteristic.

The use of comprehensive finite element and finite difference programs for Geotechnical analysis is well established in current practice. Numerical methods provide comprehensive tools but they are not valid in all cases. The dynamic analysis of Vaiont can not be possibly approached by any commercially available finite element code for geotechnical analysis. The constrained displacement of St. Moritz landslide is difficult to model properly by finite element procedures. Progressive failure, a phenomenon which partially explains the failure of Aznalcóllar dam is currently a research topic.

It may be also argued that numerical analysis still requires a significant amount of time, often inacceptable in practice. It is expected and desirable that numerical methods would progressively become easier to use and faster. But the intention of this book is far from promoting a dispute between “simple” solutions and numerical methods. However:

- A thorough understanding of fundamental mechanics is essential to interpret results of complex numerical calculations
- A simple check through approximate methods helps to grasp the nature of the problem and provides confidence to numerical results
- In many occasions the precision offered by analytical and empirical methods is adequate. The real uncertainty does not remain on the calculation procedure but on material properties, their spatial variability and the unknown boundary and initial conditions

These are the general lessons which should be extracted from this book. It is felt also that there is still room in Geomechanics for simple and accurate theoretical analysis. Closed form solutions or those which require easily available computer tools (Excel, Maple) maintain their elegance and interest in practice. They are also good benchmark cases to validate numerical tools.

However, it is the learning experience provided by the process of analyzing a failure through relatively simple procedures the main objective of the book. The lessons learned in each of the cases discussed are useful irrespective of the method of analysis.

A final objective of the book has been to provide an enjoyable reading and a rewarding experience to the readers. Writing the book was also a first class and pleasant learning experience for the authors.

MIX
Papier aus verantwortungsvollen Quellen
Paper from responsible sources
FSC® C105338

If you have any concerns about our products,
you can contact us on
ProductSafety@springernature.com

In case Publisher is established outside the EU,
the EU authorized representative is:
Springer Nature Customer Service Center GmbH
Europaplatz 3, 69115 Heidelberg, Germany

Printed by Libri Plureos GmbH
in Hamburg, Germany